Environmental Soil-Landscape Modeling

Geographic Information Technologies and Pedometrics

edited by

Sabine Grunwald

CRC Press
Taylor & Francis Group
Boca Raton London New York

CRC Press is an imprint of the
Taylor & Francis Group, an **informa** business

A TAYLOR & FRANCIS BOOK

Handbook of Phytoalexin Metabolism and Action, edited by M. Daniel and R. P. Purkayastha

Soil–Water Interactions: Mechanisms and Applications, Second Edition, Revised and Expanded, Shingo Iwata, Toshio Tabuchi, and Benno P. Warkentin

Stored-Grain Ecosystems, edited by Digvir S. Jayas, Noel D. G. White, and William E. Muir

Agrochemicals from Natural Products, edited by C. R. A. Godfrey

Seed Development and Germination, edited by Jaime Kigel and Gad Galili

Nitrogen Fertilization in the Environment, edited by Peter Edward Bacon

Phytohormones in Soils: Microbial Production and Function, William T. Frankenberger, Jr., and Muhammad Arshad

Handbook of Weed Management Systems, edited by Albert E. Smith

Soil Sampling, Preparation, and Analysis, Kim H. Tan

Soil Erosion, Conservation, and Rehabilitation, edited by Menachem Agassi

Plant Roots: The Hidden Half, Second Edition, Revised and Expanded, edited by Yoav Waisel, Amram Eshel, and Uzi Kafkafi

Photoassimilate Distribution in Plants and Crops: Source–Sink Relationships, edited by Eli Zamski and Arthur A. Schaffer

Mass Spectrometry of Soils, edited by Thomas W. Boutton and Shinichi Yamasaki

Handbook of Photosynthesis, edited by Mohammad Pessarakli

Chemical and Isotopic Groundwater Hydrology: The Applied Approach, Second Edition, Revised and Expanded, Emanuel Mazor

Fauna in Soil Ecosystems: Recycling Processes, Nutrient Fluxes, and Agricultural Production, edited by Gero Benckiser

Soil and Plant Analysis in Sustainable Agriculture and Environment, edited by Teresa Hood and J. Benton Jones, Jr.

Seeds Handbook: Biology, Production, Processing, and Storage, B. B. Desai, P. M. Kotecha, and D. K. Salunkhe

Modern Soil Microbiology, edited by J. D. van Elsas, J. T. Trevors, and E. M. H. Wellington

Growth and Mineral Nutrition of Field Crops: Second Edition, N. K. Fageria, V. C. Baligar, and Charles Allan Jones

Fungal Pathogenesis in Plants and Crops: Molecular Biology and Host Defense Mechanisms, P. Vidhyasekaran

Plant Pathogen Detection and Disease Diagnosis, P. Narayanasamy

Any opinions, findings, conclusions, or recommendations expressed in this publication do not necessarily reflect the views of the editor or of the University of Florida.

CRC Press
Taylor & Francis Group
6000 Broken Sound Parkway NW, Suite 300
Boca Raton, FL 33487-2742

First issued in paperback 2019

© 2006 by Taylor & Francis Group, LLC
CRC Press is an imprint of Taylor & Francis Group, an Informa business

No claim to original U.S. Government works

ISBN-13: 978-0-8247-2389-7 (hbk)
ISBN-13: 978-0-367-39201-7 (pbk)

Library of Congress Card Number 2005050562

Library of Congress Cataloging-in-Publication Data

Environmental soil-landscape modeling : geographic information technologies and pedometrics / edited by Sabine Grunwald.
    p. cm. -- (Books in soils, plants, and the environment)
  Includes bibliographical references and index.
  ISBN 0-8247-2389-9 (alk. paper)
  1. Soils--Environmental aspects--Computer simulation. 2. Geographic information systems. I. Grunwald, Sabine. II. Books in soils, plants, and the environment (Taylor & Francis)

S596.E58 2005
631.4'01'13--dc22   2005050562

Visit the Taylor & Francis Web site at
http://www.taylorandfrancis.com

and the CRC Press Web site at
http://www.crcpress.com

Foreword

As with all natural sciences, the aim of soil science is to understand the functioning of the natural world, in this case the soil. Soil scientists want to be able to explain how soil has been formed, how it evolves, how it interacts with the other geospheres, and how and why it varies in space and time. To answer these fundamental questions, pedologists have developed a mental model of how soils are formed from their parent material in a particular climate over time. This mental model entails a system of relationships that describe how geology and climate interact with tectonics and geomorphology, by erosion, sedimentation, surface and subsurface flow, infiltration, weathering, soil–plant interaction, organic matter decomposition, pedoturbation, and so on. Under the influence of all these processes, a soil-landscape emerges and evolves.

The development of a model of soil-landscape formation is a major undertaking. The first efforts date back to more than a century ago, when the Russian pedologist V.V. Dokuchaev introduced the state factor equation. Since then, many extensions and refinements to the original formulation have been made, but none of these justify the claim that the mental model of soil-landscape formation has been made operational. The pedological literature presents us with quasi-mathematical equations in which the soil is represented as a function of the soil-forming factors, but the details of this function have only partially been revealed. We may know how the various processes work in principle; we may know whether feedbacks in the system are positive, negative, or absent; we may know the relative importance of the various processes in a given situation; but all of this knowledge is only available in a conceptual or descriptive form. So far, we have not succeeded in building a generic, quantitative, reproducible model that predicts the soil from its controlling factors in a satisfactory way.

Why have we not managed to do so? First and foremost, this is because the soil-landscape system is extremely complex. We do not yet sufficiently understand some of the mechanisms involved in soil-landscape formation and development. We also lack the means to observe the soil with sufficient resolution and accuracy. Finally, we have long lacked the tools and computational power to construct and apply detailed, high-resolution, quantitative soil-landscape models. However, recent developments in soil science and other disciplines have enabled us to overcome many of these impediments. For example, many innovative measurement techniques have been developed that allow us to observe the behavior of the soil in a way that was not previously possible. These techniques include spectromicroscopy, microarray technology, dielectric methods, and diffuse reflectance spectroscopy. Computational power continues to double every 18 months. Geographic information systems continue to extend their functionality with new capabilities, such as three-dimensional visualization and spatially distributed dynamic modeling. Also, many new mathematical and statistical analysis and modeling approaches have appeared from which the development of predictive soil-landscape models can benefit greatly. Examples are spatial stochastic simulation, wavelets, multiscale modeling, and space–time geostatistics. Many of these techniques are being developed in an active subdiscipline of soil science known as pedometrics, which may be characterized as the part of soil science that aims to develop and apply mathematical and statistical methods for the study of the distribution and genesis of soils.

These are exciting times for soil science. Many developments are taking place on numerous fronts. These developments are relevant to soil-landscape modeling and yield favorable conditions for true progress in soil-landscape modeling. But how do we keep track of the many developments and their potential for soil-landscape modeling? This book provides a solution.

This book presents the latest methodological developments in soil-landscape modeling. It provides a cross-section of many recently developed measurement tools, as well as computer-related and pedometric techniques that are instrumental for soil-landscape modeling. It also contains in-depth reviews of the history of soil-landscape modeling, thus presenting views from soil geography, soil genesis, and soil geomorphology. I consider this one of the strong points of this book. Quantitative soil-landscape modeling will only be successful if it is a joint venture between various disciplines within soil science, notably soil geography, soil genesis, and pedometrics. It is reassuring to see work from reputable soil scientists with such diverse backgrounds in soil science united in this book.

The editor is not only very experienced in environmental soil-landscape modeling, but she also has a clear overview of the entire field. Through her choice of authors and subjects, Sabine Grunwald has created a comprehensive and well-balanced book. I am confident that it will be a stimulus and source of inspiration for all soil scientists working on the development of an operational model of soil-landscape evolution.

Gerard B.M. Heuvelink
Wageningen

The Editor

Sabine Grunwald is an assistant professor in the Soil and Water Science Department, Institute of Food and Agricultural Sciences, University of Florida, Gainesville, U.S.A., where she teaches geographic information systems (GISs) and soil-landscape modeling. Dr. Grunwald received a Master of Science (1992) and Ph.D. (1996) in environmental science (Faculty of Agriculture) from the Justus von Liebig University, Giessen, Germany. She is the current vice chair of the Commission 1.5 Pedometrics of the International Union of Soil Sciences. Her research focuses on GIS and remote sensing applications to assess the impact of land use management on soil and water quality; the application of statistical and geostatistical methods for the purpose of analyzing soils distribution, variability, properties, and behavior (pedometrics); and the development of quantitative spatially explicit models to assess environmental quality (environmetrics).

Contributors

Denis Allard
Institut National de la Recherche
  Agronomique (INRA)
Unité de Biométrie
Domaine Saint Paul
Site Agroparc
Avignon, France

Richard W. Arnold (Retired)
U.S. Department of Agriculture
  (USDA)
Natural Resources Conservation
  Service (NRCS)
Washington, D.C., U.S.A.

Thomas F.A. Bishop
Biomathematics and Bioinformatics
  Division
Rothmasted Research
Harpenden, Hertfordshire, U.K.

David J. Brown
Montana State University
Bozeman, Montana, U.S.A.

Jean-Paul Chilès
École des Mines de Paris
Centre de Géostatistique
Fontainebleau, France

Sabine Grunwald
Soil and Water Science Department
University of Florida
Gainesville, Florida, U.S.A.

Gerard B.M. Heuvelink
Soil Science Centre
Wageningen University and Research
  Centre
Wageningen, The Netherlands

Sanjay Lamsal
Soil and Water Science Department
University of Florida
Gainesville, Florida, U.S.A.

R. Murray Lark
Biomathematics and Bioinformatics
  Division
Rothamsted Research
Harpenden, Hertfordshire, U.K.

Budiman Minasny
Faculty of Agriculture, Food and
  Natural Resources
The University of Sydney
New South Wales, Australia

Margaret A. Oliver
Department of Soil Science
The University of Reading
Reading, U.K.

Carolyn G. Olson
Soil Survey Division
USDA-NRCS
Washington, D.C., U.S.A.

Jeroen M. Schoorl
Laboratory of Soil Science and Geology
Department of Environmental Sciences
Wageningen University and Research
 Centre
Wageningen, The Netherlands

Scot E. Smith
School of Forest Resources and
 Conservation
University of Florida
Gainesville, Florida, U.S.A.

Judith J.J.C. Snepvangers
Netherlands Institute for Applied
 Geosciences
Utrecht, The Netherlands

Ingrid van Cleemput
Department of Soil Management and
 Soil Care
Ghent University
Ghent, Belgium

Marc van Meirvenne
Department of Soil Management and
 Soil Care
Ghent University
Ghent, Belgium

Antonie Veldkamp
Laboratory of Soil Science and Geology
Department of Environmental Sciences
Wageningen University and Research
 Centre
Wageningen, The Netherlands

Richard Webster
Rothamsted Research
Harpenden, Hertfordshire, U.K.

Richard P. Wolkowski
Department of Soil Science
University of Wisconsin–Madison
Madison, Wisconsin, U.S.A.

Ruifeng Xu
Department of Soil Science
University of Wisconsin–Madison
Madison, Wisconsin, U.S.A.

Wei Yue
Department of Soil Science
University of Wisconsin–Madison
Madison, Wisconsin, U.S.A.

A.-Xing Zhu
State Key Laboratory of Resources and
 Environmental Information Systems
Institute of Geographical Sciences and
 Natural Resources Research
Chinese Academy of Sciences
Beijing, China
and
Department of Geography
University of Wisconsin–Madison
Madison, Wisconsin, U.S.A.

Jun Zhu
Department of Soil Science
University of Wisconsin–Madison
Madison, Wisconsin, U.S.A.

Contents

Section I

History and Trends in
Soil-Landscape Modeling

1 What Do We Really Know about the Space–Time Continuum of Soil-Landscapes?

Sabine Grunwald

CONTENTS

ABSTRACT

Growing concerns about environmental quality, degradation of ecosystems, aforestation, and human-induced erosion and pollution lead to the necessity of a holistic perception of soil-landscapes. From this viewpoint, soils are viewed as part of an ecosystem, stressing their important functions as buffers interfacing land use activities and water resources. Numerous qualitative and quantitative methods have been proposed to study the distribution, behavior, and genesis of soils. Environmental soil-landscape modeling is a science devoted to understanding the spatial distribution of soils and coevolving landscapes as part of ecosystems that change dynamically through time. There are slow but persistent shifts from qualitative to quantitative soil-landscape modeling providing digital, accurate, precise, and nonbiased information about soils. The philosophical soil-landscape paradigms rooted in empiricism, which emerged in the early 20th century, are still valid, and in fact, they form the cornerstone to reconstructing soil-landscapes with pedometrics. In this chapter an overview of holistic soil-landscape modeling techniques is given that integrates numerous environmental factors.

1.1 INTRODUCTION

Soils vary in geographic space and through time. They are complex and difficult to observe. Yaloon[1] asked a provoking question: "Is soil just dirt, too commonplace to study?" This textbook proves the opposite is true. There has been a multitude of research resulting in a variety of conceptual, qualitative, and quantitative soil-landscape models. This textbook covers the history and concepts of soil-landscape modeling. Emerging geographic information technologies and their impact on soil-landscape modeling are given special attention. We introduce numerous pedometrical techniques to study the distribution, behavior, and genesis of soils. In this chapter relevant terminology is introduced and an overview of soil-landscape methodology provided.

Soils are interrelated with surface attributes (e.g., topographic attributes) and aboveground attributes (e.g., land cover and land use), and they interact dynamically with the lithosphere, atmosphere, hydrosphere, and biosphere, collectively forming the pedosphere. The functions and relevance of soils include (1) production of food and fiber, (2) absorbance, storage, and release of water, (3) geomembrance filtering and buffering solutes, (4) habitat for biota, (5) foundations for housing, transportation, and engineering structures, and (6) cultural and aesthetic pleasure. Soils can be viewed as functional units that are used for multiple purposes (e.g., precision agriculture, urban development, recreation, etc.) by multiple users. Understanding the spatial and temporal distribution of soils and soil characteristics is a prerequisite for optimizing economic profits while minimizing adverse impacts on soil and water quality. Environmental soil-landscape modeling is a science devoted to understanding the spatial distribution of soils and coevolving landscapes as part of ecosystems that change dynamically through time.

1.2 PHILOSOPHICAL PERSPECTIVES OF SOIL-LANDSCAPES

Perspectives of soil-landscapes are diverse and change over time. Passmore[2] discussed the contrasting attitudes of man toward nature and soils. Early Stoic (Greek)-Christian doctrine assumed that "everything is made for man's sake." Aristotle argues that "plants are created for the sake of animals and the animals for the sake of men" and that "nature is at its best when it fulfills men's needs. So to perfect nature is to humanize it — to make it more useful for men's purposes."[3] Kant points out that "man's relationship with nature is not subject to moral census."[4] Descartes' radical philosophy suggests that "men should render themselves the masters and possessors of nature."[2] There is a strong Judeo-Christian Western tradition that man is free to deal with nature as he pleases, since it exists only for his sake. Soils are viewed as substrate to produce food and are perceived as an object for mankind, purely a matter of utility. This metaphysical belief has lead to widespread degradation of soil-landscapes, some of which have subsequently been deemed inhabitable for many generations.

Philosophies contrasting such self-centered views of soil-landscapes emphasize stewardship and cooperation with nature and perceive humans as managers or stewards of soil-landscapes. Socrates and Plato pointed out that it is humanity's responsibility to care for the welfare of resources.[5] There are different traditions of stewardship. One perceives humans as part of soil-landscapes seeking harmonized coexistence, whereas the other tradition perceives man as steward "to develop land and to perfect it."[2] Eastern philosophies emphasize stewardship in which men should take all responsible steps not to destroy any living beings or nonliving objects. Here the interdependence between everything that exists is stressed. While some deemphasize the responsibility of humans by putting the future of nature and soils in God's hands, others emphasize a deep spiritual concern for wildlife, soils, and water resources.

Growing concerns about environmental quality, degradation of ecosystems, aforestation, and human-induced erosion and pollution lead to the necessity of a holistic perception of soil-landscapes. From this viewpoint, soils are viewed as part of an ecosystem, stressing their important functions as buffers interfacing land use activities and water resources. More radical conservationists emphasize the protection of natural ecosystems, excluding any kind of human interference. Others seek a balanced ecosystem where humans are an integral part of the system willing to make genuine sacrifices to protect natural soil-landscapes. The ability to preserve the diversity of vegetation, habitats, and wildlife species is dependent on the preservation of pedodiversity. Shifting human-centered to enviro-centered perspectives is vital for conservation management.

Soil scientists have focused on two contrasting concepts to study soil-landscapes, both of which are equally important.[6] The reductionist approach promotes ever more detailed studies of soil characteristics and pedological processes derived from field investigations in pedons, along transects and laboratory studies. The other approach develops and enunciates an integrative, unifying point of view encompassing and integrating previous observations and results. With the advent of geographic information systems (GIS), the latter approach has focused on integrating a variety of environmental factors that correlate with soil attributes. McKenzie and Austin[7] outlined this concept of environmental correlation predicting soil characteristics (e.g., clay content) with more readily observed environmental variables (e.g., terrain attributes) as predictors.

1.3 WHAT ARE SOILS, LANDSCAPES, AND SOIL-LANDSCAPES?

Soil is the material that forms at the interface of the atmosphere and the lithosphere that is capable of supporting plants. It is the unconsolidated mineral or organic material on the surface of the Earth that has been subjected to and shows effects of genetic and environmental factors of climate (including water and temperature effects), macro- and microorganisms, conditioned by relief, acting on parent material over a period of time.[8,9] A *landscape* is the fundamental trait of a specific geographic area, including its biological composition, physical environment, and anthropogenic or social patterns.[10] Ruhe[11] defines landscape from a geomorphologic perspective as "a collection of spatially related, natural landforms, usually the collective land surface that the eye can comprehend in a single view." *Land* is the entire complex of surface and near-surface attributes of the solid portion of the surface of the Earth, which are significant to human activities.[8] All land and water resource data have a spatial and temporal dimension with variable scale, extent, and resolution. Commonly, soil-landscape datasets are managed in soil information systems (SISs) and analyzed using GISs.

The pedon serves as the mapping unit for soil taxonomic classifications and site-specific pedological studies focusing on profile dynamics. In contrast, sustainable land resource management is focused on the landscape-scale balancing economic profitability, stewardship of natural resources, and social equity concerned with the quality of life.[10] Sustainable management is important to maintain the functions of a soil-landscape for economic, recreational, biological, cultural, aesthetic, and social values. An integrative, synergistic ecosystem approach is rooted in understanding system components and behavior, including soils. Deductive science generated extensive knowledge of how individual parts function (e.g., pedological processes). Yet understanding how the parts interact as a whole requires a holistic perspective. Soil-landscape modeling attempts to integrate soils, parent material, topography, land use and land cover, and human activities. The goal of soil-landscape modeling is to gain an understanding of the spatial distribution of soil attributes, characteristics of soils, and their behavior through time. Soil-landscapes can be defined in terms of (1) geomorphology and topography, the form and shape of a landscape; (2) land cover, the aboveground characteristics; (3) land use, the functions that a landscape performs; (4) soil attributes, the belowground characteristics; and (5) genesis, the formation of soil attributes due to pedological processes. According to Dijkerman,[12] there are various, sometimes overlapping purposes of soil maps and soil-landscape models, including:

- Observational — used to sample subsets of a population of soil bodies or to study their behavior
- Experimental — used to study the effect of controlled situations
- Descriptive — used to characterize the system under study
- Explanatory — used to understand relationships and behavior of the soil system
- Predictive — used to forecast relationships, attributes, and behavior of the soil system
- Scale — used to represent the system
- Idealized — used to simplify the system
- Analog — used to extrapolate and transfer knowledge

1.4 GEOGRAPHIC SPACE

Soils are natural bodies with attributes varying continuously in the space–time continuum across landscapes.[13] Soil attributes are variable in the horizontal and vertical dimension, forming three-dimensional soil bodies that change over time due to pedological processes.[14] Commonly, soil attributes are anisotropic vertically and laterally. Soils coevolve through the interaction of physical and chemical weathering, erosion and deposition, lateral and vertical transport processes, and biological processes. To capture changes of soil attributes, models have been developed that segregate the soil continuum.[15] Such models are formal abstractions of soil-landscape reality that formalize how geographic space is discretized into smaller spatial units for analysis and communication. Generally, it is necessary to divide geographic space into discrete spatial units, and the resulting tessellation is taken as a reasonable approximation of reality at the level of resolution under consideration. There are two types of spatial discretization methods used for soil-landscape modeling: (1) the crisp soil map unit model and (2) the continuous-field or pixel (square grid cell, raster) model.

1.4.1 CRISP SOIL MAP UNIT MODEL

This model has its roots in empiric observations combined with 19th-century biological taxonomy and practice in geological survey. Traditional soil-landscape models use crisp map units, which are defined by abrupt changes from one map unit to the other.[16,17] Each soil map unit is associated with a representative soil attribute set.[18] Horizons of these soil map units differ from adjacent and genetically related layers in physical, chemical, and biological attributes such as texture, structure, color, soil organic matter, or degree of acidity. As such, soil horizons and profiles of these attributes correspond to discrete, sharply delineated (crisp) units, which are assumed to be internally uniform. A soil surveyor uses his tacit knowledge, intuition, and understanding of soils to delineate these crisp map units. Factors such as topography, geology, geomorphology, vegetation, and historic information guide the identification of these so-called natural breaks, which become map unit boundaries. Variation within the classes is acknowledged, but is described qualitatively and usually in vague terms. The rationale for employing a crisp model is that if the variation within the classes is less than that in a soil region at large, then using the class mean of the soil attributes of interest as a predictor should be more precise than the regional mean. The model only has merit if the variance within the classes is less than the total variance.

Peuquet,[19] Goodchild et al.,[20] and Burrough and McDonnell[21] define crisp mapping units as *entities* or *objects* (Figure 1.1). These objects represent soil-landscape phenomena that are defined by the geographic location (where things are) and its attributes (what is present). On choropleth soil maps, the geographic locations of soils are defined in reference to other geographic features (e.g., other soils or landmarks) within geographic space. Crisp sets allow only binary membership functions (i.e., true or false); an individual is a member or is not a member of any given set as defined by exact limits. McBratney and de Gruijter[22] and Heuvelink and Webster[23] pointed out that crisp sets do not allow ambiguities, and they are too inflexible to take into account genuine uncertainty. Though the crisp data model is practical, it ignores spatial variation in both soil-forming processes and in the resulting soils.[24] The validity of the crisp soil model has been questioned and critically discussed repeatedly.[25–28] Butler[29] emphasized that successful classification of soils at the local level must represent what can be mapped at the chosen scale and is determined by the local landscape. Major drawbacks of the crisp soil model are that it ignores spatial autocorrelation within and across map units and assumes the same variation of attributes across space. Commonly, SIS such as the State Soil Geographic (STATSGO) Database, Soil Survey Geographic (SSURGO) Database, and Soil Data Mart (SDM) adopt the crisp soil map unit concept.[30] Thousands of soil maps exist that are based on this model.

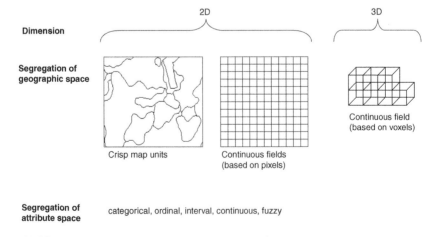

FIGURE 1.1 Models to segregate geographic and attribute space.

1.4.2 CONTINUOUS FIELD (PIXEL MODEL)

The continuous-field model displays the real world as a set of pixels or voxels (volume cells) (Figure 1.1). Burrough and McDonnell[21] argue that the continuous-field model is adequate for modeling natural phenomena that do not show obvious boundaries (e.g., soils). This spatial model has the potential to describe the gradual change of soil attributes formed by a variety of pedological processes within a domain. The spatial resolution or pixel size depends on the spatial variability of soil attributes. Geostatistical techniques have been applied in numerous studies to interpolate point observations and construct soil attribute pixel maps.[17,31,32] It is challenging to optimize the density and spatial distribution of observations across a domain to characterize soil-landscape reality without knowing the underlying spatial variability of soil attributes and operating pedological processes. If the spatial resolution of the continuous-field model is finer than the crisp soil map units, the former provides more detailed site-specific information. Comparisons between crisp and continuous pixel-based soil models have been presented by Heuvelink and Huisman[33] and Hengl.[34] In order to bridge the gap between the conventional crisp and the continuous-field model, several models have been proposed to deal with the situation in which there are discrete (abrupt) and continuous (gradual) spatial variations in the same area.[16,33,35–39]

1.4.3 TWO-DIMENSIONAL VS. THREE-DIMENSIONAL GEOGRAPHIC SPACE

Brady and Weil[40] point out that soils are three-dimensional bodies. However, soils are commonly represented in the form of two-dimensional maps.[41,42] In crisp soil maps, each map unit is associated with attribute tables that store the soil attribute datasets and metadata (reports) that describe the soil attributes. In pixel-based soil maps, each grid cell carries the code of a specific soil attribute value. Commonly, different grid themes are used to represent different soil attributes. Other soil-landscape representations use a 2 1/2-D design superimposing land use, soil maps, or other thematic maps over a digital elevation model (DEM) to produce a three-dimensional view.[43,44] Since this technique describes patterns on two-dimensional landscape surfaces rather than the spatial distribution of subsurface attributes (e.g., soil texture, soil horizons), it fails to address three-dimensional soil-landscape reality. McSweeney et al.[45] pointed out that two-dimensional soil models are lacking the ability to address three-dimensional soil-landscape reality. Only recently have three-dimensional reconstruction and scientific visualization techniques emerged to produce three-dimensional soil-landscape models. For example, the Cooperative Research Center for Landscape Evolution and Mineral Exploration (CRCLEME) constructed a three-dimensional regolith model of the Temora study area in Central New South Wales, Australia.[46] A three-dimensional soil horizon model in a Swiss floodplain was created by Mendonça Santos et al.[47] using a quadratic finite-element method. Sirakov and Muge[48] developed a prototype three-dimensional Subsurface Objects Reconstruction and

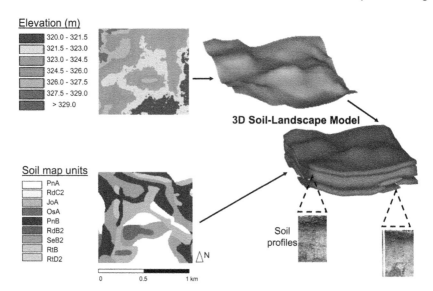

FIGURE 1.2 Elevation and soil data were used to create a three-dimensional soil-landscape model for a site in southern Wisconsin. (Reprinted from S. Grunwald and P. Barak, *Syst. Analy. Modeling Sim.,* 41, 755–776, 2001.) (See color version on the accompanying CD.)

Visualization System (3D SORS) in which two-dimensional planes are used to assemble three-dimensional subsurface objects. Grunwald et al.[14] presented three-dimensional soil-landscape models at different scales for sites in southern Wisconsin (Figure 1.2). Geographic reconstruction and scientific visualization techniques suitable for soil-landscape modeling were presented by Grunwald and Barak.[49] While complex, domain-specific prototype three-dimensional soil-landscape models have been developed, they have not been widely adopted yet for the following reasons: (1) input data requirements, (2) models are labor intensive to produce, requiring specialized software and programming skills, (3) lack of training and education, (4) preference of users for traditional crisp soil maps, and (5) lack of realistic abstraction of soil-landscapes.

1.5 ATTRIBUTE SPACE

Soil-landscape models employ different concepts to segregate attribute space. Some models focus on soil attributes while others aggregate soil attributes to form soil classes or taxa. The latter ones have been adopted to develop soil taxonomies in many different countries.[50] The rationale for adopting soil classes is that the soil cover can hardly be apprehended in its entirety. Hence, a subdivision is required in order to understand the different components of this continuum and to understand the relationships with the factors of their formation. There are problems with this approach, such as the establishment and ordering of classes and the ranking according to the importance of soil characteristics from high to low hierarchical levels, which vary widely among different soil taxonomies. The

U.S. Soil Taxonomy is a system developed using morphogenetic indicators (diagnostic horizons and attributes) as class criteria.[18] It is one of the most detailed soil taxonomies, which currently consists of 12 orders, 63 suborders, 319 great groups, 2484 subgroups, about 8000 families, and about 19,000 soil series in the U.S.[40,51]

Dudal[52] pointed out that existing soil classification systems enabled us to explain and characterize soil diversity in function of different sets of soil-forming factors. However, with the increasing demand for targeted soil information related to specific sites or specific purposes (e.g., assessment of phosphorus loads, forest management), technical soil surveys focusing on mapping of soil attributes rather than aggregated soil taxa are better suited. Dudal[52] stressed that due to the availability of GIS, there is no need to provide end users with aggregated soil taxa datasets. Burrough[53] points out that users can extract and aggregate soil attributes from GIS and soil datasets on demand targeting a specific application. In short, preference is given to raw field observations of soil attributes stored in a GIS rather than grouping of soil attributes to form soil classes.

Traditionally, soil mapping involves the collection of soil morphological attributes (e.g., soil texture, structure, consistency) derived from field observations and physical, chemical, and biological attributes derived from laboratory analyses. Often the sample support is small and samples are referred to as point observations. Such point observations made with augers or at excavated pits are labor intensive and costly. Emerging *in situ* techniques such as ground penetrating radar,[54] electromagnetic induction (EM),[55,56] profile cone penetrometers,[57,58] remote sensing,[59] and soil spectroscopy[60] are nonintrusive and allow for rapid data collection. While these new techniques are typically associated with greater uncertainty, they have the capability to complement traditional auger-based field observations.

Realistic soil-landscape reconstruction is highly dependent on soil and ancillary variable collection. The following criteria impact the soil-landscape modeling process:

1. **Attribute type**: Boolean (e.g., presence of redoximorphic features — yes or no, hydric or nonhydric soils), categorical (e.g., soil structure categories), ordinal (e.g., drainage classes ranging from very poor drainage to excessively well-drained soil), interval (e.g., soil texture), and continuous (e.g., bulk density in mg m^{-3}).
2. **Content of attributes**:
   a. **Soil attributes**: Describe soil characteristics.
      i. **Morphological, physical, chemical, and biological attributes**: Soil attributes are variable in geographic space. Soil attributes with a short range (small spatial autocorrelation) vary at close distances, whereas soil attributes with long range (large spatial autocorrelation) vary over long distances. The spatial autocorrelation of attributes is domain dependent. For example, the variability of a soil attribute (e.g., soil nitrogen) observed in a loess landscape is likely to differ from the same attribute observed in a fluvial landscape.

ii. **Soil classes**: Have a specific range in one or more particular attributes, such as texture, structure, drainage, acidity, or other. According to Wilding et al.,[61] attributes that are measured and closely calibrated to a standard (e.g., texture, color, pH, etc.) are commonly less variable than qualitatively assessed attributes (soil structure, consistency, porosity, root abundance, etc.).

b. **Topographic attributes**: Describe terrain characteristics.

i. **Primary topographic attributes**: Specific geometric attributes of the topographic surface calculated directly from a digital elevation model (DEM); these include slope, aspect, upslope drainage area, maximum-flow path length, profile curvature, plan curvature, and others.[62]

ii. **Secondary topographic attributes**: Secondary, or compound, attributes involve combinations of primary attributes and constitute physically based or empirically derived indices that can characterize the spatial variability of specific processes occurring in landscapes.[63] Examples of secondary attributes include the topographic wetness index, stream power index, and sediment transport index.

c. **Topographic classes**: A group of topographic attributes lumped into classes that form land surface classification systems. Ruhe[11] developed geomorphic slope units in a catenary way. Huggett[64] categorized terrain into four different basic slope shapes based on the convexity and concavity of the terrain. Conacher and Dalrymple[65] segregated three-dimensional units of a catena resulting in the nine land surface units based on soil morphology, mobilization, and transport of soil constituents, and redeposition of soil constituents by overland flow and throughflow or by gravity as mass movement. Pennock et al.[66] segregated terrain based on curvature, flow lines, and landscape position.

d. **Parent material**: The unconsolidated and more or less chemically weathered mineral or organic matter from which the solum of soils is formed by pedogenic processes.

e. **Land cover and land use**: Land cover describes the coverage of the land surface by vegetation, structures, pavement, or others. According to Turner and Meyer,[67] land cover is the biophysical state of the Earth's surface. In contrast, land use describes the function that land performs. Land use involves both the manner in which the biophysical attributes of the land are manipulated and the intent underlying that manipulation — the purpose for which the land is used.[67]

f. **Time**: Synoptic soil sampling provides a snapshot of the characteristics and distribution of soils within a soil mapping region. Repeated soil sampling (soil monitoring) provides information about soil attribute variability through time. Commonly, stable soil attributes (e.g., soil texture, particulate phosphorus) show less variation with time, whereas dynamic soil attributes (e.g., soil nitrate–nitrogen, soil temperature, water content, hydraulic conductivity, biological activity, exchangeable cations) are highly variable in time.[61]

3. **Sample support**: The volume of material or area of a sample. Observations that are made on small volumes of material (1 cm^3 to a few cubic meters), or areas of a few centimeters to a few meters, are considered point observations.[17]

4. **Geographic extent of observations**: Range from the molecular system, peds /aggregates, horizons, pedons, polypedons (multiple pedons), catena, to soil regions. The organization hierarchy of soil systems was outlined by Hoosbeek and Bryant.[68] McBratney et al.[69] provide an overview of small- and large-scale digital soil maps categorized by cartographic scale, pixel size, nominal spatial resolution, resolution *loi du quart*, and extent.

5. **Total number of observations**: Within the soil mapping region.

6. **Density of observations**: Varies from sparse to exhaustive (high-density, continuous) sampling. For example, profile cone penetrometers collect soil attributes (e.g., penetration resistance, soil moisture) continuously in small depth increments (e.g., 1 or 0.5 cm) along soil profiles. Satellite images provide exhaustive datasets covering large soil mapping regions with high spatial resolutions. For example, Landsat Enhanced Thematic Mapper (ETM) images have a grid resolution of 30 m. Hyperspectral images, such as NASA's Airborne Visible/Infrared Imaging Spectrometer (AVIRIS), with 224 contiguous spectral channels (bands), have a spatial resolution of 20 m, and IKONOS images have a resolution of 1 m in the panchromatic and 4 m in the multispectral resolutions. In contrast, auger-based point sampling is labor intensive and costly, and often fewer observations are collected to represent the soil-landscape.

7. **Sampling design**: Brus and de Gruijter[70] discuss model-based and design-based sampling. The design-based sampling approach is rooted in classical sampling theory where sample locations are selected by a predetermined random procedure. For example, simple random sampling and stratified random sampling are considered design-based sampling methods. Apart from measurement error, sampling is the only source of stochasticity considered in the design-based approach. This implies that the unknown value at any given location and time is considered fixed, not random. In the model-based approach, the soil-forming process that has led to the field of values of a particular attribute in the soil mapping region is modeled as a stochastic process. This implies that the sample does not necessarily have to be selected by a random procedure.

Sampling schemes comprise transect, random, stratified random, clustered, targeted, and grid sampling. Ideally, for mapping and subsequent analysis, observations should be located spatially distributed over the soil mapping area. Sampling schemes can take prior information into account derived from qualitative or quantitative soil maps, earlier point observations, or sampling barriers.[71]

1.6 PEDOMETRICS

Soil-landscapes are complex and diverse due to pedogeomorphological and hydrological processes acting over hundreds and thousands of years. These soil-forming and -destroying processes operate simultaneously in soils, and the resulting profile reflects the balance of these processes — present and past. The spatial distributions of subsurface attributes and processes in natural environments often vary at granularities ranging from pedons, hillslopes, to regions. Reconstructing soil-landscapes requires an interdisciplinary holistic approach. Pedometrics attempts to integrate knowledge from numerous disciplines, including soil science, statistics, and GIS (Figure 1.3). *Pedometrics* is a term coined by Alex McBratney — a neologism, derived from the Greek words πεδοζ (soil) and μετρον (measurement). Pedometrics is defined as the application of mathematical and statistical methods for the study of the distribution and genesis of soils (http://www.pedometrics.org).

Quantitative models that describe soil-landscapes are rooted in conceptual (mental) models. These conceptual soil-landscape models have currently evolved into complex quantitative models that utilize advanced mathematics and statistics, emerging soil mapping techniques, and computers that are capable of processing huge multidimensional datasets. Pivotal events that shaped soil-landscape modeling history are summarized in Figure 1.4.

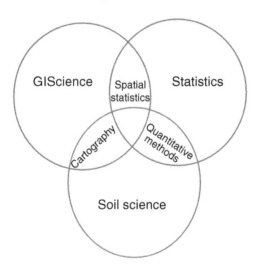

FIGURE 1.3 Pedometrics can be considered an interdisciplinary science integrating soil science, GIScience, and statistics (available from http://www.pedometrics.org).

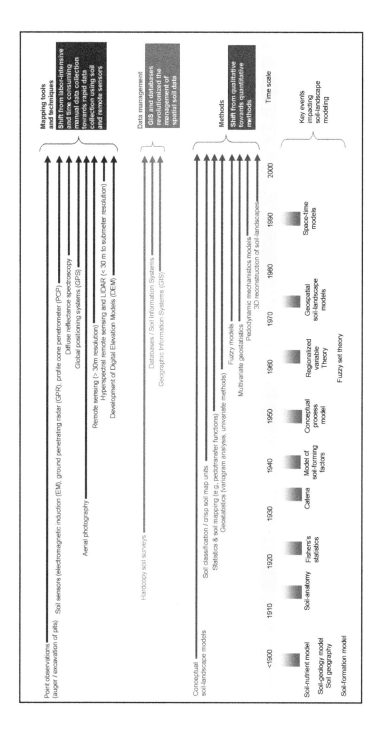

FIGURE 1.4 Pivotal events that shaped soil-landscape modeling history. Time periods and placement of events are approximate. (See color version on the accompanying CD.)

Since the 1980s, a dramatic change has taken place in our thinking about the utilization of natural resources. There has been an increased awareness of ecosystem health and environmental quality, and rate of resource consumption.[72] While soil survey in its traditional role is diminishing, the need for soil information is becoming more important in terms of sustainable land management and cycling of biogeochemicals. Pedometrics and environmetrics have grown closer. Environmetrics aims at fostering the development and use of statistical and other quantitative methods in the environmental sciences, environmental engineering, and environmental monitoring and protection. Spatially explicit, continuous, and quantitative soil-landscape data are required to assess environmental quality.

1.6.1 JENNY'S SOIL FACTORIAL MODEL

Soil mapping first relied merely on the intuition of soil surveyors to read a soil-landscape. Soil factor equations were introduced by Dokuchaev in Glinka[73] and affirmed and popularized by Jenny.[74,75] Jenny's model of soil-forming factors is rooted in previous work by Dokuchaev and describes soil as a function of climate, biological activities, topography, parent material, and time (Figure 1.5):

$$S = f (cl, o, r, p, t) \tag{1.1}$$

where
  S = soil attribute
  cl = climate factor
  o = organisms (biotic factor)
  r = relief (topographic factor)
  p = parent material
  t = time

Hudson[76] discussed the two reasons for the success of the soil factorial concept:

1. "A large number of adherents were excited by the idea that this apparently simple concept could be used as the basis for accurately locating soil boundaries and delineating bodies of soil anywhere in the world."
2. "There are no details. Nothing is stated, e.g., about the mechanics how soils vary, or which attributes vary in different climates. Lacking specifics it pointed the way to a wide variety of interesting problems for practitioners to solve."

Jenny[77] called the factors in Equation 1.1 independent variables. Independence inherently implies that factors change without altering other soil attributes. These independent variables define the soil system; i.e., for a given combination of cl, o, r, p, and t, the state of the soil system is fixed (only one type of soil exists under these conditions). In this interpretation of soil-forming factors, the

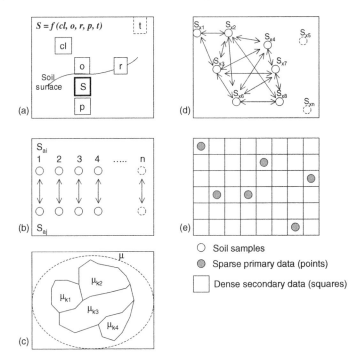

FIGURE 1.5 (a) Jenny's soil factorial model. (b) Pedotransfer functions without consideration of spatial autocorrelation. (c) Soil classification model. (d) Univariate geospatial model acknowledging spatial autocorrelation. (e) Multivariate geospatial model acknowledging spatial autocorrelation and covariation.

notions of "forming" or "acting" that connoted causal relationships have been replaced by the less ambiguous conceptions of "defining" or "describing" (*f* stands for "function of" or "dependent on"). Jenny[77] pointed out that in selecting *cl*, *o*, *r*, *p*, and *t* as the independent variables of the soil system, we do not assert that these factors never enter functional relationships among themselves. But he placed emphasis on the fact that the soil formers may vary independently and may be obtained in a great variety of constellations, either in nature or under experimental conditions.

Jenny's factorial model has been adopted in numerous soil survey programs worldwide that rely on tacit knowledge of soil surveyors. Jenny showed that the relationship between a certain soil property and a state factor can be investigated in a region in which one state factor is dominant compared to the combined contributions of the other state factors. These special cases are, depending on the dominant state factor, climo-, bio-, topo-, litho-, or chronofunctions. The conceptual factorial model Jenny developed provides the framework to develop quantitative descriptions of soil-landscapes.

Knowledge of the soil attributes produced by the interaction of environmental variables (geological, topographical, climatological, and biological variables) allows for the prediction of soil characteristics. The qualitative application of the factors of the soil formation model has been, and still is, the guiding paradigm of national soil survey programs around the world. Tacit knowledge of the soil surveyor is the basis for delineating soil boundaries. Soil surveyors use the qualitative model of soil-forming factors for mapping soils and rely on environmental factors as explanatory variables (e.g., landforms, local drainage, vegetation, parent material, etc.). Soil distribution is predicted on the basis of environmental correlation. These mental models can be complex and have considerable predictive power, but the results are not amenable to rigorous checking, repetition, and rational criticism by others unless the survey is repeated. It is difficult to distinguish between evidence and interpretation in survey reports. The quantification of soil-forming factors has been of major interest with the advent of digital terrain modeling and emerging mapping techniques such as remote and soil sensing. For example, Pennock et al.[78] related landform element complexes to physical (e.g., bulk density), chemical (e.g., inorganic carbon content), and biochemical (e.g., soil organic carbon) soil quality indicators. Hairston and Grigal[79] investigated relationships between topographic position and soil water, soil nitrogen, and tree growth on two regional landforms in Minnesota. Osher and Buol[41] related soil attributes to parent material and landscape position in eastern Madre de Dios, Peru, using 14 soil profiles. McKenzie and Austin[7] presented a quantitative approach to medium- and small-scale surveys based on soil stratigraphy and environmental correlation in the lower Macquarie Valley in Australia. The authors used a generalized linear model and a comprehensive set of soil attributes (e.g., soil morphological attributes, bulk density, texture, pH, electrical conductivity, cation exchange capacity) to characterize the soil region. Numerous studies showed that topography can be easily quantified and used successfully as a predictor of soil attributes. In reality, it is very difficult to assign quantitative values to some factors, particularly parent material and the biota. Processes of soil formation occur over timescales of tens to thousands of years. Hence, time is one of the factors that is very difficult to quantify. For example, Holzhey et al.[80] suggested that it took between 21,000 and 28,000 years to form the thick spodic horizons in North Carolina. Morrison and Frye[81] suggest that the formation of Churchill Geosol in Nevada can be envisaged as taking about 5000 years.

1.6.2 Catena Model

The catena[82] is a fundamental concept that explains the pattern of soils on hillslopes. Milne[82] coined the term to describe a repeating sequence of soils that occur from the top of a hillslope to the adjacent valley bottom. The catena concept includes both surficial stratigraphy and internal hillslope structure and lithology. Milne pointed out that soil formation along a hillslope is influenced by the parent material and pedological processes (e.g., erosion that redistributes soil material downslope). The catena is defined as a sequence of soils of about the same age, derived from similar parent material, and occurring under similar climatic conditions, but having different characteristics due to variation in relief and drainage.[15]

Bushnell[83] coined the term *toposequence*, which is often used synonymously with catena; however, the original meanings are not identical. A toposequence is a sequence of related soils that differ, one from the other, primarily because of topography as a soil formation factor. It is argued that changes in soil morphology can be mainly attributed to elevational position and local hydrology.

Huggett[64] further developed the catena concept, proposing that the basic three-dimensional unit of the soil-landscape is the first-order valley basin. The functional boundaries of this soil-landscape are defined as the atmosphere–soil interface, the weathering front at the base of the soil, and the drainage divides of the basin. Huggett[64] pointed out that the topographic boundaries of a drainage basin define the physical limits and directions of the basin, and thus control geomorphic processes such as erosion, deposition, surface runoff, and lateral subsurface flow. This conceptual model emphasizes soil hydrology and functional boundaries for soil mapping.

Sommer and Schlichting[84] presented a methodology of grouping catenas into three major categories: (1) transformation catenas, showing no gains or losses of the element or soil component under study; (2) leaching catenas, with losses in at least part of the catena and no accompanying elemental gains in other parts; and (3) accumulation catenas, showing gains in at least part of the catena but no losses elsewhere in the catena. Catenas that cover geomorphic units of distinctly different ages belong to the chronocatenas, a subcategory of either the transformation, leaching, or accumulation type. Though the catena concept acknowledges the continuum of soils along a hillslope, typically discrete entities (pedons) are identified and soils are segregated into crisp entities. Park et al.[85] presented a process-based terrain characterization model for a toposequence in southern Wisconsin. Its basic proposition is that soil distribution can be most efficiently identified by the separation of units where similar hydrological, geomorphological, and pedological processes occur.

1.6.3 PREDICTION OF A SOIL ATTRIBUTE FROM OTHER SOIL ATTRIBUTES

Soil-landscape modeling history was also shaped by the theory formalized by Fisher,[86] who emphasized the importance of random selection to ensure that estimates are unbiased. This resulted in the applications of many design-based soil sampling schemes. First, ordination techniques, multiple discriminant analysis, and other clustering and data fragmentation techniques were introduced, rooted in classical statistical theory.

Empirical observations recognized that selected soil attributes are interdependent. For example, drainage in soils is highly dependent on soil texture, soil structure, and bulk density. These relationships between collocated soil attributes can be identified using a training (calibration) dataset to develop pedotransfer functions (PTFs). Wösten et al.[87] described two types of PTF, namely, class and continuous PTFs predicting soil classes (e.g., textural classes, horizons) and continuous variables (e.g., hydraulic conductivity). McBratney et al.[88] considered the variable type of the predictor and predictor variables to distinguish different types of PTFs, ranging from hard and fuzzy classes, continuous or fuzzy variables, to mixed-class continuous variables.

Statistical methods such as multiple regressions, artificial neural networks, or classification and regression trees (CARTs) are commonly used to formalize the relationships to predict soil attributes or soil classes (Equations 1.2 and 1.3). The functions are commonly used to predict soil attributes at unsampled locations. Pedotransfer functions are attractive because they provide a way to predict soil attributes that are costly and labor intensive to collect from a new suite of soil attributes that are cheap and rapid to collect. Most PTFs consider observations as independent and ignore the spatial autocorrelation of soil attributes (Figure 1.5):

$$S_{aj} = f(S_{ai}) \tag{1.2}$$

$$S_{cj} = f(S_{ci}) \tag{1.3}$$

where
S_{aj} = soil attribute j
S_{ai} = soil attributes ($i = 1, 2, …, n$)
S_{cj} = soil class j
S_{cj} = soil classes ($i = 1, 2, …, n$)

Pedotransfer functions define relationships between different soil characteristics and attributes, such as the moisture retention characteristics and the pressure head–hydraulic conductivity.[89] Rawls et al.[90] estimated soil water properties and Grunwald et al.[58] soil physical properties using cone index data collected with a profile cone penetrometer.

1.6.4 SOIL CLASSIFICATION MODELS

Running parallel with, but without any formal connection, soil systematics was developing its own course. Its origins were intuitive and its proponents were trying to justify it, seeking to give it the same genetic respectability that biological and geological classifications seemed to have achieved. The lack of efforts to introduce statistical measures in soil survey programs is evident. For example, in current soil surveys in the U.S., no statistical techniques are used to optimize soil sampling designs and to document collected soil data (e.g., the variation of a soil attribute within a soil map unit). Nevertheless, soil classification systems still prosper around the world, supported by national soil surveying efforts to produce crisp soil maps.[50]

Adopting the classificatory approach, we assume that observations are taken from a statistically stationary population (i.e., the mean and variance of the data are independent of both the geographic location and the size of the support).[21] Other assumptions include that (1) the variations in the value of S

within the map units are random and not spatially contiguous, (2) all mapping units have the same within-class variance, which is uniform within the polygons, (3) all attributes are normally distributed, and (4) the most important changes in soil attribute values take place at boundaries, which are sharp, not gradual. Classification by homogeneous polygons assumes that within-unit variation is smaller than that between units. This conceptual model is commonly used in soil and landscape mapping to define homogeneous soil units, landscape units, ecotopes, etc. The model can be formalized using the following equation:

$$S(x_0) = \mu + \alpha_k + \varepsilon \tag{1.4}$$

where

$S(x_0)$ = value of the soil attribute at location x_0

μ = general mean of S over the domain of interest (soil mapping region)

α_k = deviation between μ and the mean μ_k of unit (class) k

ε = residual (pooled within-unit) error or noise

According to Webster,[91] there are three major types of classification models:

1. Intuitive classification models: The soil surveyor identifies class boundaries based on tacit knowledge.
2. Systematic classification (mathematical classification) models: Classification of multiple soil and environmental properties into classes, e.g., using multiple discriminant analysis, hierarchical clustering, or CART.
3. Spatial fragmentation classification models: Incorporation of the spatial position of observations into the grouping algorithm. Problems with the classificatory approach include a mismatch between multivariate and geographic space, where the geographic boundaries do not necessarily match with the attribute boundaries.

The value of the soil attribute classes in these models can be hard or fuzzy. Fuzzy sets were first introduced by Zadeh[92] and can be used to describe the uncertainty of data and boundaries between categories. Fuzzy methods allow the matching of individuals to be determined on a continuous scale instead of on a Boolean binary or an integer scale. In contrast to crisp sets, a fuzzy set is a class that admits the possibility of partial membership. Fuzzy c-means partition observations in multivariate space into relatively stable, naturally occurring groups.[93] Fuzzy set theory provides a rich framework to describe the vagueness of (spatial) data and information. McBratney and Odeh[94] provided an overview of applications of fuzzy sets in soil science that includes a numerical classification of soil and mapping, land evaluation, modeling and simulation of soil processes, fuzzy soil geostatistics, soil quality indices, and fuzzy measures of imprecisely defined soil phenomena.

1.6.5 GEOSPATIAL MODELS

Geospatial models predict soil attribute values at unsampled locations using soil attribute observations that are spatially distributed throughout the soil mapping domain. Global interpolation methods use all available observations to provide predictions for the whole area of interest, while local interpolators operate within a small zone around the point being interpolated to ensure that estimates are made only with data from locations in the immediate neighborhood.

Trend surfaces are the simplest geospatial model that requires fitting some form of polynomial equation through soil (or environmental) attribute values. These are least square methods that model the long-range spatial variation. Such models assume that the spatial coordinates are the independent variables and S (attribute of interest) is the dependent variable (Equation 1.5).[91] As trend surfaces are simplified representations of reality, it is difficult to ascribe any physical meaning to complex, higher-order polynomials. Therefore, the main use of trend surface analysis is not as an interpolator, but as a way of removing broad features of the data prior to using some complex local interpolator. The concept is based on partitioning the variance between trend and the residuals from the trend.

$$S(x_0) = \sum_{u=0}^{U} a_u f_u(x_0) + \varepsilon \qquad (1.5)$$

where
   a_u = unknown coefficients to be found by analysis
   f_u = known functions of spatial position (x_0)
   ε = uncorrelated error term

Local, deterministic interpolation methods focus on modeling short-range (local) variations. The interpolation involves (1) defining a search neighborhood around the point to be interpolated, (2) finding the observations within this neighborhood, (3) choosing a mathematical function to represent the variation over this limited number of points, and (4) evaluating it for the point on a regular grid. The procedure is repeated until all the points on a grid have been computed.[21] Local interpolation methods such as nearest neighbors (Thiessen or Vornonoi polygons), inverse distance weighting (IDW), and splines are described in Burrough and McDonnell.[21] Splines (local fitting functions) estimate values using a mathematical function that minimizes overall surface curvature, resulting in a smooth surface that passes exactly through the observation points, while at the same time ensuring that the joins between one part of the curve and another are continuous. Local interpolators in their simplest form are based on a linear interpolator, in which the weights are computed from a linear function of distance between sets of data points and the point to be predicted (Equation 1.6). The assumption of IDW is that the value of an attribute at an unsampled location is a distance-weighted average of data points occurring within a local neighborhood surrounding the unvisited point:[95]

$$\hat{S}(x_0) = \sum_{i=1}^{n} \lambda_i * S(x_i) \qquad \sum_{i=1}^{n} \lambda_i = 1 \qquad (1.6)$$

where

$\hat{S}(x_0)$ = predicted soil attribute value at unsampled location x_0

λ_i = weights

$S(x_i)$ = observed soil attribute value at locations x_i

Mercer and Hall in Webster[91] in 1911 at Rothamsted, UK, discovered that soil attributes show two distinct sources of variation: an autocorrelated component and a random one. The advent of computers and software starting in the 1960s and regionalized variable theory formalized by Matheron[96] resulted in the exponential growth in geostatistical applications applied to soil science. This spatial area evolved from variogram analysis, ordinary kriging, indicator kriging, and universal kriging to more advanced multivariate geostatistical applications, such as cokriging, regression kriging, and spatial stochastic simulations.[91,97] It was recognized that more complex spatial patterns could be modeled by treating soil variables as regionalized variables (Figure 1.5).

Since the environment and its component attributes, such as soil, result from many interactive physical, chemical, and biological processes that are nonlinear or chaotic, the outcome is so complex that the variation appears to be random.

If we adopt a stochastic view, then at each point in geographic space there is not just one value for an attribute, but a whole set of values. Thus, at a location x_0 a soil attribute, S, is treated as a random variable with a mean (μ), variance (σ^2), and cumulative distribution function (cdf). The set of random variables, $S(x_1)$, $S(x_2)$, … $S(x_n)$, constitutes a random function or a stochastic process.[17] Regionalized variable theory assumes that the spatial variation of any variable Z can be expressed as the sum of three major components (Equation 1.7)[21]: (1) a structural component, having a constant mean or trend that is spatially dependent; (2) a random, but spatially correlated component, known as the variation of the regionalized variable; and (3) a spatially uncorrelated random noise or residual term. The deterministic component m is dependent on some exogenous factors such as the cl, o, r, p, and t factors and can be described by a trend model, e.g., via universal kriging.

$$Z(x_0) = m(x_0) + \varepsilon'(x_0) + \varepsilon'' \qquad (1.7)$$

where

$Z(x_0)$ = value of a random variable at x_0

$m(x_0)$ = deterministic function describing the structural component of Z at x_0

$\varepsilon'(x_0)$ = stochastic, locally varying but spatially dependent residual from $m(x_0)$ — the regionalized variable

ε'' = a residual, spatially independent noise term having zero mean and variance

x_0 = geographic position in one, two, or three dimensions

Observations obtained close to each other are more likely to be similar than observations taken further apart from each other. This spatial autocorrelation of $\varepsilon'(x_0)$ is described by the semivariance γ. If γ is plotted as a function of the lag distance h, the semivariogram $\gamma(h)$ is obtained. In Equation 1.8, one implicit assumption is that the semivariance depends only on the separation distance h and not on the positions x_0 and $x_0 + h$ (stationarity assumption). $\gamma(h)$ is estimated as

$$\gamma(h) = \frac{1}{2N(h)} \sum_{i=1}^{N(h)} [Z(x_0) - Z(x_0 + h)]^2 \qquad (1.8)$$

where

γ = semivariance

h = distance (lag)

N = total number of datapairs

Regionalized variable theory is described in detail by Goovaerts,[31] Webster and Oliver,[17] and Chilès and Delfiner.[32] The semivariogram provides input for kriging, which is a weighted interpolation technique to create continuous (soil) prediction maps. Major limitations of the univariate geostatistical technique of kriging are due to the assumptions of stationarity, which are not often met by the field-sampled datasets and the large amount of data (>100 observations; better, >150 observations) to define the spatial autocorrelation.

1.6.6 MULTIVARIATE GEOSPATIAL MODELS

Jenny's factorial soil-landscape model inherently acknowledged relationships between soils and cl, o, r, p, and t factors. Univariate geostatistical models recognize spatial autocorrelation. We can unify both concepts to develop multivariate geospatial models considering spatial covariation and autocorrelation, forming complex models that describe soil-landscape reality (Figure 1.5).

McBratney et al.[69] expanded Jenny's model by explicitly introducing the spatial position n [x, y coordinates] and soil attributes (s) to predict a soil attribute (S_a) or soil class (S_c). They stressed the quantification of factors in Equations 1.9 and 1.10 using DEMs, remote sensing, soil sensing, and other emerging soil mapping techniques:

$$S_a[x, y, \sim t] = f(s[x, y, \sim t], cl[x, y, \sim t], o[x, y, \sim t], r[x, y, \sim t],$$
$$p[x, y], a[x, y], n) \tag{1.9}$$

$$S_c[x, y, \sim t] = f(s[x, y, \sim t], cl[x, y, \sim t], o[x, y, \sim t], r[x, y, \sim t],$$
$$p[x, y, \sim t], a[x, y], n) \tag{1.10}$$

where

S_a = soil attribute
S_c = soil class
s = soils, other attributes of the soil at a point
a = age, the time factor
n = space, spatial position defined by the x coordinate (easting) and y
coordinate (northing)

McBratney's model includes soil as a factor because soil can be predicted from its attributes, or soil attributes from its class or other attributes. The factor s can be derived from a prior map, remote sensing, proximal sensing, or expert knowledge. Implicit in the SCORPAN model are the spatial coordinates x, y and an approximate or vague time coordinate $\sim t$. This time coordinate can be expressed as "at about some time t." McBratney et al.[69] stress that Equations 1.9 and 1.10 are empirical in nature; i.e., there has been evidence of relationships between factors. They caution that the direction of causality is not defined. For example, vegetation is dependent on soil while soil is also dependent on vegetation. Another issue of using an empirical rather than a deterministic model is the nonuniqueness of S_c and S_a. For example, the soil attribute S_a at a specific geographic location and time might be explained by different factor combinations of s, cl, o, r, p, and a.

Below the SCORPAN model is extended to represent soil classes and attributes in three-dimensional geographic space considering the depth coordinate (z) explicitly:

$$S_c[x, y, z, \sim t] = f(s[x, y, z, \sim t], cl[x, y, z, \sim t], o[x, y, z, \sim t],$$
$$r[x, y, z, \sim t], p[x, y, z, \sim t], a[x, y, z], [x, y, z]) \tag{1.11}$$

$$S_a[x, y, z, \sim t] = f(s[x, y, z, \sim t], cl[x, y, z, \sim t], o[x, y, z, \sim t],$$
$$r[x, y, z, \sim t], p[x, y, z, \sim t], a[x, y, z], [x, y, z]) \tag{1.12}$$

This three-dimensional SCORPAN model was adopted to reconstruct soil-landscape models in three dimensions in southern Wisconsin[14] (Figure 1.2) and for a space–time simulation of water table dynamics in a flatwood landscape in northeast Florida.[98] We can extend the three-dimensional SCORPAN model even further by including scaling behavior of factors. Multiscale soil-landscape models do exist, but they are difficult to reconstruct with the present soil mapping technology and knowledge.

Hybrid modeling applications combine environmental correlation and geostatistics.[93] Hybrid techniques assume that soil variation is composed of deterministic and stochastic (empirical) components. With the advent of high-resolution and high-quality proximal sensing techniques of soil, land use/land cover, and topographic attributes such as electromagnetic induction (EM), soil video imaging systems, time domain reflectometry, airborne gamma-radiometry, and multispectral remote sensing, we are capable of rapidly collecting and integrating multiple exogenous variables to predict soils. For example, Odeh et al.[99] used a DEM to predict soil attributes (depth to solum, depth to bedrock, topsoil gravel content, subsoil clay content) comparing multilinear regression, ordinary kriging, global universal kriging using restricted maximum likelihood, cokriging, and regression kriging. The latter methods are multivariate geostatistical hybrid techniques.

Cokriging is a multivariate extension of kriging that often combines a sparsely measured primary variable (or target variable) with a denser set of a secondary variable that is spatially cross-correlated (Figure 1.5).[100] Typically, the more expensive-to-measure target soil variable is predicted using the cheaper-to-measure soil, vegetation, topographic, or other variables. For example, soil texture (target variable) is predicted using slope or other topographic attributes derived from a high-resolution DEM.

Regression kriging is a hybrid method that uses multiple regressions, regression trees, generalized linear models, or others to describe the deterministic component in Equation 1.7. The residuals are modeled as the spatially varying but dependent component. Regression kriging was employed by Knotters et al.[101] to predict soil layer depths and by Carré and Girard[102] to predict soil types. According to Odeh et al.,[103] regression kriging performed better than multilinear regression, ordinary kriging, universal kriging, and isotopic and heterotopic cokriging to predict depth of solum, depth to bedrock, topsoil gravel, and subsoil clay.

Kriging with an external drift uses an ancillary variable to represent the trend $m(x_0)$ in Equation 1.7, which is modeled as a linear function of a smoothly varying secondary (external) variable instead of a function of the spatial coordinates. This method requires that the relation between primary trend and secondary variable is linear and makes physical sense.[97] Other multivariate geostatistical techniques are described in Goovaerts,[31] Chilès and Delfiner,[32] and Wackernagel.[100]

1.6.7 SPACE–TIME MODELS

Other soil-landscape models stress the time component. Simonson's process model explained soil formation by the interaction of four processes: additions, removals, translocations, and transformations.[104] This model stresses pedological processes rather than the resulting soil attributes. For example, the eluviation of aluminum and iron from top O, A, and E horizons and the immobilization of these metals in short-range-order complexes with organic matter in the underlying B horizon is a process known as podzolization. Other conceptual pedogenic process models are based on a mass balance approach considering the accumulation of gains and losses of elements or tracers within a soil-landscape with defined boundaries.[105,106] Both models provide an important conceptual framework for understanding soil formation. However, neither model establishes functional boundaries for segregating the soil continuum into natural landscape units.

Since the 1990s, models that explicitly include time have been developed, including time series analysis (Equation 1.13), state–space modeling, space–time geostatistics, and spatial state–space approach, the last one based on physical laws and observations.[23] Time series analysis is based on an empirical statistical approach. In the simplest case, the measured series is treated as the realization of a stationary random process $Z(t)$:

$$Z(t) = m + \varepsilon(t) \tag{1.13}$$

where
  $Z(t)$ = realization of a stationary random process
  m = global mean
  t = temporally autocorrelated random residual with a mean of zero and variance characterized by its autocovariance function $C(s) = Cov[(t), (t + s)]$, where s denotes the lag and t is time

Several methods have been proposed to integrate both spatial and temporal variability into one model (Equation 1.14). Hoosbeek et al.[107] treated time as a third dimension added to two-dimensional space. They plotted a spatiotemporal semivariogram of leaching simulated with a mechanistic model (decision support model for agrotechnology transfer [DSSAT]). Journel[108] discussed possible problems associated with integrating spatial and temporal dimensions into one semivariogram model and pointed out that although some directional dependence (anisotropy) may exist, spatial phenomena in general show no ordering, whereas a notion of past, present, and future exists for temporal phenomena. Measured data may represent a unique realization for the past and present, but in the case of the future, a truer stochastic dependence exists. Because of this inequality, the additive space–time model cannot be used for interpolation. However, it is useful for describing and depicting the combined spatial and temporal variability of soil data:

$$Z(x_0, t) = m(x_0, t) + \varepsilon(x_0, t) \tag{1.14}$$

where
  $Z(x_0, t)$ = random variable dependent on location x_0 and t
  $m(x_0, t)$ = deterministic trend (in the simplest case a constant equal to a global mean)

The linkage between hydrology and soil formation has been acknowledged in numerous studies.[109, 110] Though the conceptual soil-landscape models recognize dynamic hydrologic flow as a major soil-forming factor, the attempt to segregate soils rather than to address their continuity has resulted in numerous soil classification systems. Since almost all hydrologic models use stochastic or mechanistic methods to describe water flow, the development of pedology and hydrology has split into different directions.

Pedodynamic-deterministic modeling of soil processes is focused on predicting future conditions of the soil. Heuvelink and Webster[23] pointed out that calibration and validation of these models are delicate and scale dependent. Equation 1.15 formalizes the concept of process-based deterministic models:

$$Z(t) = f(t) \qquad\qquad (1.15)$$

where
$f(t)$ = physical-deterministic part of the model

Quantitative soil process models, also referred to as pedodynamic models, are defined by "the quantitative integrated simulation of physical, chemical, and biological soil processes acting over short time increments in response to environmental factors."[111] Quantitative simulation models are based on the assumption that the state of each system at any moment can be quantified, and that changes in the state can be described by rate or differential equations. State variables are variables such as soil organic matter and have dimensions of length, number, volume, weight, energy, or heat content. Driving variables quantify the effect of external factors on the system and are not influenced by processes within the system. Their values must be monitored continuously (e.g., meteorological variables). Rate variables indicate the rate at which state variables change. Their values are determined by the state and driving variables according to equations that are based on process knowledge.[107] Hoosbeek and Bryant[111] developed the ORTHOD model, which simulates hydrology and pedologic processes (e.g., solute movement, microbial decomposition or organic matter, mineral dissolution, adsorption, ion exchange, and the precipitation of amorphous aluminum silicates), which lead to the formation of Typic Haplorthods. This model is specifically focused on one soil type out of thousands of others. A rudimentary mechanistic model for soil production and landscape development was proposed by Minasny and McBratney.[112] Their quantitative soil process model was constrained to the processes weathering and erosion.

Quantitative soil process models are computationally complex. The deterministic modeling approach requires a complete understanding and mathematical formulation of pedogenesis. In addition, the interaction between processes occurring over long periods has to be understood before such models can be developed. More research is necessary to unravel the mysteries of soil phenomena through time.

The time dimension is different from the space dimension because time has a direction, it moves forward only, processes take place in a sequence, and phenomena are almost never isotropic in the space–time domain. In the real world, the assumption of temporal stationarity is never a trivial one, and in many instances, it cannot be simply modeled by direct applications of kriging (inter-polation) in the space–time domain. In contrast, mechanistic models often fail because we lack (1) a complete understanding of the physical, chemical, and biological ecosystem processes; (2) an appreciation for proper scaling issues; and (3) adequate input data to run our models. No matter how complex the mechanistic model structure, typically there are deviations between model predictions and independent observations. Alternative models are mixed pedodynamic-determin-istic/stochastic models (Equation 1.16) where partial process knowledge is inte-grated with a stochastic component:[23]

$$Z(t) = f(t) + \varepsilon(t) \qquad (1.16)$$

Even more complex models acknowledge that the state at time t depends on state at time $t - 1$, which is considered in the Kalman filtering[113] state–space approach. The strength of the state–space approach lies in its ability to merge the formulation of pedodynamic-deterministic models with observations.[23] Applications of soil-landscape models that consider the variation of soil attributes in space and through time are still rare due to data input requirements.

1.7 CRITICAL REMARKS

In many countries around the world soil survey programs are still actively employ-ing the factorial soil-landscape paradigm using tacit knowledge of soil surveyors. At the same time, many issues related to environmental quality, site-specific farm management, and land resource management have emerged that cannot be addressed successfully with available two-dimensional crisp soil taxa maps. The demand for quantitative soil-landscape models that focus on the prediction of soil properties rather than soil classes is increasing. There are slow but persistent shifts from qualitative to quantitative soil-landscape modeling providing digital, accurate, precise, and nonbiased information about soils. The philosophical soil-landscape paradigms rooted in empiricism, which emerged in the early 20th century, are still valid, and in fact, they form the cornerstone to reconstructing soil-landscapes with pedometrics.

In this chapter numerous quantitative soil-landscape modeling techniques have been presented that use a holistic approach to integrate numerous environ-mental factors. Hence, the term *environmental soil-landscape modeling* was pro-posed to emphasize that soil predictions are interdependent on environmental factors such as *cl*, *o*, *r*, *p*, and *t*. In addition, we can incorporate knowledge of how soil attributes behave in space and through time to improve reconstructing soil-landscapes.

While soils and soil attributes generally vary continuously across soil-land-scapes and change through time, we measure the soil at only a finite number of geographic locations and at times with only small supports. Predictions are made from observation datasets to predict soil attributes at unsampled geographic locations. Because no measurement and model is perfect, uncertainties exist that need to be quantified. An inherent goal is to reduce prediction errors and uncertainty of quantitative soil-landscape models.

Recently, soil proximal and remote soil mapping techniques have improved greatly. Soil information systems are slowly migrating from entity to raster-based implementations. Hardware and software no longer impose limitations on developing quantitative soil-landscape models. Advanced statistical and mathematical techniques are available to test and compare conceptual soil-landscape models using stochastic simulations.[33] Virtual soil-landscapes and space–time applications can be implemented lending techniques developed in computer science.[14]

Phillips[114] argues that aggregate or probabilistic predictions are possible, but not deterministic predictions of individual soils. This implies that we currently cannot predict individual soil attributes accurately; instead, we can only predict broad-scale general behavior. In short, despite the fact that no two pedons are exactly alike, similarities do exist and commonalities can be identified. Phillips showed that intrinsic variability within homogeneous landscape units is more important in determining the pedodiversity of the study area than is the extrinsic variability associated with measurable differences in topography, parent material, and vegetation/land use. Despite the fact that our soil mapping techniques have improved tremendously, a completely deterministic approach to reconstruct soil-landscapes seems to be currently out of reach.

Although generic relationships between soil attributes and environmental factors have been identified, they are domain dependent. The relationship between soil and slope might be strong in one landscape setting but weak in another. Similarly, many soil attributes are nonstationary; their spatial variation and covariation with exogenous environmental attributes are domain dependent and potentially change through time. Therefore, no universal equation exists that fits all soil-landscapes. Rather, we have to strengthen our efforts to develop quantitative soil-landscape models that are customized based on our observations. Comparative studies that test different statistical and mathematical modeling techniques are desirable to improve our understanding of soil-landscapes.

Scaling behavior of soil and environmental factors confounds quantitative relationships between factors. Scale-dependent characteristics include (1) attributes and processes emerge at different scales, (2) nonlinear behavior of processes, (3) threshold dependency to trigger a process, (4) nonsimilarity of soil attributes and processes at different scales, (5) varying dominant processes at different scales, and (6) response to external disturbances interrelated to intrinsic processes. It is too simple to assume that we can describe soil regions by aggregating pedon descriptions.

The predicament entailed by the complexity of soil-landscapes requires a synergistic approach, integrating knowledge not only from pedology, but also geography, mathematics, computer science, ecology, hydrology, and others. Interdisciplinary collaboration will be key to reconcile deductive and inductive science to take soil-landscape modeling to the next level.

ACKNOWLEDGMENTS

This research was supported by the Florida Agricultural Experiment Station and approved for publication as Journal Series R-10601.

REFERENCES

1. DH Yaloon. Down to earth. *Nature* 407: 301, 2000.
2. J Passmore. *Man's Responsibility for Nature*. Duckworth Publ., London, 1980.
3. WD Ross. *Aristotle*, 5th ed. Methuen, London, 1949.
4. RC Solomon, KM Higgins. *A Short History of Philosophy*. Oxford University Press, New York, 1996.
5. WKC Guthrie. *Greek Philosophy*. Methuen, London, 1950.
6. DH Yaalon, RW Arnold. Attitudes towards soils and their societal relevance: then and now. *Soil Sci* 165: 5–12, 2000.
7. NJ McKenzie, MP Austin. A quantitative Australian approach to medium and small scale surveys based on soil stratigraphy and environmental correlation. *Geoderma* 57: 329–355, 1993.
8. SSSA (Soil Science Society of America). *Glossary of Soil Science Terms*. SSSA, Madison, WI, 1997. Available at http://www.soils.org.
9. SW Buol, FD Hole, RJ McCracken, RJ Southard. *Soil Genesis and Classification*. Iowa State University Press, Ames, 1997.
10. A Young. *Land Resources: Now and for the Future*. Cambridge University Press, London, 1998. Available at http://www.land-resources.com.
11. RV Ruhe. *Quarternary Landscapes in Iowa*. Iowa State University Press, Ames, 1969.
12. JC Dijkerman. Pedology as a science: the role of data, models and theories in the study of natural soil systems. *Geoderma* 11: 73–93, 1974.
13. DJ Mulla, AB McBratney. Soil spatial variability. In *Handbook of Soil Science*, ME Sumner, Ed. CRC Press, New York, 2000.
14. S Grunwald, P Barak, K McSweeney, B Lowery. Soil-landscape models at different scales portrayed in Virtual Reality Modeling Language. *Soil Science*, 165: 598–614, 2000.
15. DA Wysocki, PJ Schoeneberger, HE LaGarry. Geomorphology of soil-landscapes. In *Handbook of Soil Science*, ME Sumner, Ed. CRC Press, New York, 2000, pp. E5–E72.
16. M Voltz, R Webster. A comparison of kriging, cubic splines and classification for predicting soil attributes from sample information. *J Soil Sci* 31: 505–524, 1990.
17. R Webster, MA Oliver. *Geostatistics for Environmental Scientists*. John Wiley & Sons, New York, 2001.
18. Soil Survey Staff. *Keys to Soil Taxonomy*, 8th ed. U.S. Government Printing Office, Washington, DC, 1998.
19. D Peuquet. Presentations of geographic space: towards a conceptual synthesis. *Ann Assoc Am Geogr* 78: 375–394, 1988.

20. M Goodchild, G Sun, S Yang. Development and test of an error model for categorical data. *Int J Geogr Inf Syst* 6: 87–104, 1992.
21. PA Burrough, RA McDonnell. *Principles of Geographical Information Systems: Spatial Information Systems and Geostatistics.* Oxford University Press, New York, 1998.
22. AB McBratney, JJ de Gruijter. A continuum approach to soil classification by modified fuzzy k-means with extragrades. *J Soil Sci* 43: 159–175, 1992.
23. GBM Heuvelink, R Webster. Modeling soil variation: past, present, and future. *Geoderma* 100: 269–301, 2001.
24. PA Burrough, PFM van Gaans, R Hootsmans. Continuous classification in soil survey: spatial correlation, confusion and boundaries. *Geoderma* 77: 115–135, 1997.
25. R Webster, HE De La Cuanalo. Soil transect correlograms of North Oxfordshire and their interpretation. *J Soil Sci* 26: 176–194, 1975.
26. S Nortcliff. Soil variability and reconnaissance soil mapping: a statistical study in Norfolk. *J Soil Sci* 29: 403–418, 1978.
27. DJ Campbell, DG Kinninburgh, PHT Beckett. The soil solution chemistry of some Oxfordshire soils: temporal and spatial variability. *J Soil Sci* 40: 321–339, 1989.
28. WD Nettleton, BR Brasher, G Borst. The tax adjunct problem. *Soil Sci Soc Am J* 55: 421–427, 1991.
29. BE Butler. *Soil Classification for Soil Survey.* Oxford University Press, Oxford, 1980.
30. Soil Survey Staff. Soil taxonomy: a basic system of soil classification for making and interpreting soil surveys. In *USDA-SCS Agric. Handbook*, Vol. 436. U.S. Government Printing Office, Washington, DC, 1999.
31. P Goovaerts. *Geostatistics for Natural Resources Evaluation.* Oxford University Press, New York, 1997.
32. JP Chilès, P Delfiner P. *Geostatistics: Modeling Spatial Uncertainty.* Wiley-Interscience Publ., New York, 1999.
33. GBM Heuvelink, JA Huisman. Choosing between abrupt and gradual spatial variation? In *Quantifying Spatial Uncertainty in Natural Resources: Theory and Applications for GIS and Remote Sensing*, HT Mowrer, RG Congalton, Eds. Ann Arbor Press, Chelsea, MI, 2000, pp. 111–118.
34. T Hengl. Pedometric Mapping: Bridging the Gap between Conventional and Pedometric Approaches. Ph.D. thesis, Wageningen University, Netherlands, 2003.
35. A Stein, M Hoogerwerf, J Bouma. Use of soil-map delineations to improve (co)-kriging of point data on moisture deficits. *Geoderma* 43: 163–177, 1988.
36. BGM Heuvelink, MFP Bierkens. Combining soil maps with interpolations from point observations to predict quantitative soil attributes. *Geoderma* 55: 1–15, 1992.
37. AS Rogowski, JK Wolf. Incorporating variability into soil map unit delineations. *Soil Sci Soc Am J* 58: 403–418, 1994.
38. P Goovaerts, AG Journel. Integrating soil map information in modeling the spatial variation in continuous soil attributes. *Eur J Soil Sci* 46: 397–414, 1995.
39. GBM Heuvelink. Identification of field attribute error under different models of spatial variation. *Int J GIS* 10: 921–935, 1996.
40. NC Brady, RR Weil. *The Nature and Attributes of Soils*, 13th ed. Prentice Hall, Upper Saddle River, NJ, 2002.
41. LJ Osher, SW Buol. Relationship of soil attributes to parent material and landscape position in eastern Madre de Dios, Peru. *Geoderma* 83: 143–166, 1998.

42. EN Bui. Soil survey as a knowledge system. *Geoderma* 120: 17–26, 2004.
43. A Su, SC Hu, R Furuta. 3D Topographic Maps for Texas, 1996. Available at http://www.csdl.tamu.edu/~su/topomaps/.
44. M Hogan, K Laurent. Virtual Earth Science at USGS (U.S. Geological Survey), 1999. Available at http://virtual.er.usgs.gov/.
45. K McSweeney, PE Gessler, BK Slater, RD Hammer, GW Peterson, JC Bell. Towards a new framework for modeling the soil-landscape continuum. In *Factors in Soil Formation*, SSSA Special Publication 33. SSSA, Madison, WI, 1994.
46. CRCLEME (Cooperative Research Center for Landscape Evolution and Mineral Exploration). Annual Report, 1999. Available at http://leme.anu.edu.au.
47. ML Mendonça Santos, C Guenat, M Bouzelboudjen, F Golay. Three-dimensional GIS cartography applied to the study of the spatial variation of soil horizons in a Swiss floodplain. *Geoderma* 97: 351–366, 2000.
48. NM Sirakov, FH Muge. A system for reconstructing and visualizing three-dimensional objects. *Comput Geosci* 27: 59–69, 2001.
49. S Grunwald, P Barak. The use of VRML for virtual soil-landscape modeling. *Syst Anal Model Simulation* 41: 755–776, 2001.
50. H Eswaran, T Rice, R Ahrens, BA Stewart. *Soil Classification: A Global Desk Reference*. CRC Press, New York, 2003.
51. USDA-NRCS (United States Department of Agriculture–Natural Resources Conservation Service). *2003 Keys to Soil Taxonomy*, 9th ed. USDA-NRCS, Washington D.C., 2003.
52. R Dudal. How good is our soil classification? In *Soil Classification: A Global Desk Reference*, H Eswaran, T Rice, R Ahrens, BA Stewart, Eds. CRC Press, New York, 2003, pp. 11–18.
53. PA Burrough. The technologic paradox in soil survey: new methods and techniques of data capture and handling. *Enschede ITC J* 3: 15–22, 1993.
54. TJ Gish, WP Dulaney, KJ Kung, CST Daughtry, JA Doolittle, PT Miller. Evaluating use of ground-penetrating radar for identifying subsurface flow pathways. *Soil Sci Soc Am J* 66: 1620–1629, 2002.
55. JA Doolittle, ME Collins. A comparison of EM induction and GPR methods in areas of karst. *Geoderma* 85: 83–102, 1998.
56. DJ Inman, RS Freeland, JT Ammons, RE Yoder. Soil investigations using electronmagnetic induction and ground-penetrating radar in southeast Tennessee. *Soil Sci Soc Am J* 66: 206–211, 2002.
57. B Lowery, RT Schuler. Duration and effects of compaction on soil and plant growth in Wisconsin. *Soil Tillage Res* 29: 205–210, 1994.
58. S Grunwald, DJ Rooney, K McSweeney, B Lowery. Development of pedotransfer functions for a profile cone penetrometer. *Geoderma* 100: 25–47, 2001.
59. JR Jensen. *Remote Sensing of the Environment: An Earth Resource Perspective*. Pearson Education, New York, 2000.
60. HW Siesler, Y Ozaki, S Kawata, HM Heise. *Near-Infrared Spectroscopy*. Wiley VCH, Weinheim, Germany, 2002.
61. LP Wilding, J Bouma, DW Goss. Impact of spatial variability on interpretive modeling. In *Quantitative Modeling of Soil Forming Processes*, SSSA Special Publication 39, RB Bryant, RW Arnold, Eds. SSSA, Madison, WI, 1994.
62. JP Wilson, JC Gallant, Eds. *Terrain Analysis: Principles and Applications*. John Wiley & Sons, New York, 2000.

63. ID Moore, RB Grayson, AR Ladson. Digital terrain modeling: a review of hydrological, geomorphological, and biological applications. *Hydrological Processes* 5: 3–30, 1991.

64. RJ Huggett. Soil landscape systems: a model of soil genesis. *Geoderma* 13: 1–22, 1975.

65. AJ Conacher, JB Dalrymple. The nine-unit land surface model: an approach to pedogeomorphic research. *Geoderma* 18: 1–154, 1977.

66. DJ Pennock, BJ Zebarth, E deJong. Landform classification and soil distribution in hummocky terrain, Saskatchewan, Canada. *Geoderma* 40: 297–315, 1987.

67. BL Turner, WB Meyer. Global land use and land cover change: an overview. In *Changes in Land Use and Land Cover: A Global Perspective*, WB Meyer, BL Turner II, Eds. Cambridge University Press, Cambridge, MA, 1994, pp. 3–10.

68. MR Hoosbeek, RB Bryant. Towards the quantitative modeling of pedogenesis: a review. *Geoderma* 55: 183–210, 1992.

69. AB McBratney, ML Mendonça Santos, B Minasny. On digital soil mapping. *Geoderma* 117: 3–52, 2003.

70. DJ Brus, JJ de Gruijter. Random sampling or geostatistical modeling? Choosing between design-based and model-based sampling strategies for soil. *Geoderma* 80: 1–44, 1997.

71. JW van Groeningen, W Siderius, A Stein. Constrained optimization of soil sampling for minimization of the kriging variance. *Geoderma* 87: 239–259, 1999.

72. AR Mermut, H Eswaran. Some major development in soil science since the mid-1960s. *Geoderma* 100: 403–426, 2001.

73. KD Glinka. *The Great Soil Groups of the World and Their Development*. Edwards Bros., Ann Arbor, MI, 1927 (translated from German by C.F. Marbut).

74. H Jenny. *Factors of Soil Formation*. McGraw-Hill, New York, 1941.

75. H Jenny. *EW Hillgard and the Birth of Modern Soil Science*. Farallo Publ., Berkeley, CA, 1961.

76. BD Hudson. The soil survey as paradigm-based science. *Soil Sci Soc Am J* 56: 836–841, 1992.

77. H Jenny. *Factors of Soil Formation: A System of Quantitative Pedology*. Dover Publ., New York, 1992.

78. DJ Pennock, DW Anderson, E de Jong. Landscape-scale changes in indicators of soil quality due to cultivation in Saskatchewan, Canada. *Geoderma* 64: 1–19, 1994.

79. AB Hairston, DF Grigal. Topographic variation in soil water and nitrogen for two forested landforms in Minnesota, USA. *Geoderma* 64: 125–138, 1994.

80. CS Holzhey, RB Daniels, EE Gamble. Thick Bh horizons in the North Carolina coastal plain. II. Physical and chemical attributes and rates of organic additions from surface sources. *Soil Sci Soc Am Proc* 39: 1182–1187, 1975.

81. RB Morrison, JC Frye. Correlation of the Middle and Late Quarternary Successions of the Lake Lahontan, Lake Bonneville, Rocky Mountain, Southern Grean Plains, and Eastern Midwest Areas, Nevada Bureau of Mines Report 9. Nevada Bureau of Mines, 1965, 45 p.

82. G Milne. Normal erosion as a factor in soil profile development. *Nature* 138: 541–548, 1936.

83. TM Bushnell. Some aspects of the soil catena concept. *Soil Sci Soc Am Proc* 7: 466–476, 1942.

84. M Sommer, E Schlichting. Archetypes of catenas in respect to matter: a concept for structuring and grouping catenas. *Geoderma* 76: 1–33, 1997.

85. SJ Park, K McSweeney, B Lowery. Identification of the spatial distribution of soils using a process-based terrain characterization. *Geoderma* 103: 249–272, 2001.
86. RA Fisher. *Statistical Methods for Research Workers.* Oliver and Boyd, Edinburgh, Scotland, 1925.
87. JHM Wösten, PA Finke, MJW Jansen. Comparison of class and continuous pedotransfer functions to generate soil hydraulic characteristics. *Geoderma* 66: 227–237, 1995.
88. AB McBratney, B Minasny, SR Cattle, RW Vervoort. From pedotransfer functions to soil inference systems. *Geoderma* 109: 41–73, 2002.
89. JHM Wösten, CHJE Schuren, J Bouma, A Stein. Functional sensitivity analysis of four methods to generate soil hydraulic functions. *Soil Sci Soc Am J* 54: 832–836, 1990.
90. WJ Rawls, DL Brakensiek, KE Saxton. Estimation of soil water properties. *Trans ASAE* 1316–1320, 1982.
91. R Webster. The development of pedometrics. *Geoderma* 62: 1–15, 1994.
92. LA Zadeh. Fuzzy sets. *Inf Control* 8: 338–353, 1965.
93. AB McBratney, IOA Odeh, TFA Bishop, MS Dunbar, TM Shatar. An overview of pedometric techniques for use in soil survey. *Geoderma* 97: 293–327, 2000.
94. AB McBratney, IOA Odeh. Application of fuzzy sets in soil science: fuzzy logic, fuzzy measurements and fuzzy decisions. *Geoderma* 77: 85–113, 1997.
95. E Isaaks, RM Srivastava. *An Introduction to Applied Geostatistics.* Oxford University Press, New York, 1989.
96. G Matheron. *Les variables régionalisées et leur estimation. Une application de la théorie de functions aléatoires aux scienes de la nature.* Masson, Paris, 1965.
97. P Goovaerts. Geostatistics in soil science: state-of-the-art and perspectives. *Geoderma* 89: 1–45, 1999.
98. V Ramasundaram, S Grunwald, A Mangeot, NB Comerford, CM Bliss. Development of an environmental virtual field laboratory. *J Comput Educ* 45: 21–34.
99. IOA Odeh, AB McBratney, DJ Chittleborough. Spatial prediction of soil attributes from landform derived from a digital elevation model. *Geoderma* 63: 197–214, 1994.
100. H Wackernagel. *Multivariate Geostatistics.* Springer, Berlin, 2003.
101. M Knotters, DJ Brus, JH Oude Voshaar. A comparison of kriging, co-kriging and kriging combined with regression for spatial interpolation of horizon depth with censored observations. *Geoderma* 67: 227–246, 1995.
102. F Carré, MC Girard. Quantitative mapping of soil types based on regression kriging of taxonomic distances with landform and land cover attributes. *Geoderma* 110: 241–263, 2002.
103. IOA Odeh, AB McBratney, DJ Chittleborough. Further results on prediction of soil properties from terrain attributes: heterotopic cokriging and regression-kriging. *Geoderma* 67: 215–226, 1995.
104. RW Simonson. Outline of a generalized theory of soil genesis. *Soil Sci Soc Am Proc* 23: 152–156, 1959.
105. GH Brimhall, WE Dietrich. Constitutive mass balance relations between chemical composition, volume, density, porosity, and strain in metsomatic hydrochemical systems: results on weathering and pedogenesis. *Geochim Cosmochim Acta* 5: 567–587, 1987.
106. OA Chadwick, RC Graham. Pedogenic processes. In *Handbook of Soil Science*, ME Sumner, Ed. CRC Press, New York, 1999, pp. E41–E72.

107. MR Hoosbeek, RG Amundson, RB Bryant. Pedological modeling. In *Handbook of Soil Science*, ME Sumner, Ed. CRC Press, New York, 2000, pp. E-77–E-111.

108. AG Journel. Geostatistics, models and tools for the earth sciences. *Math Geol* 18: 119–140, 1986.

109. AL Steinwand, TE Fenton. Landscape evolution and shallow groundwater hydrology of a till landscape in central Iowa. *Soil Sci Soc Am J* 59: 1370–1377, 1996.

110. ME Sumner, Ed. *Handbook of Soil Science*. CRC Press, New York, 2000.

111. MR Hoosbeek, RB Bryant. Developing and adapting soil process submodels for use in the pedodynamic orthod model. In *Quantitative Modeling of Soil Forming Processes*, SSSA Special Publication 39, RB Bryant, RW Arnold, Eds. SSSA, Madison, WI, 1994.

112. B Minasny, AB McBratney. A rudimentary mechanistic model for soil production and landscape development. *Geoderma* 90: 3–21, 1999.

113. RE Kalman. A new approach to linear filtering and prediction problems. *Trans Am Soc Mech Eng J Basic Eng* 82D: 35–45, 1960.

114. JD Phillips. Contingency and generalization in pedology as exemplified by texture-contrast soils. *Geoderma* 102: 347–370, 2001.

2 Soil Survey and Soil Classification

Richard W. Arnold (Retired)

CONTENTS

ABSTRACT

The basic functional relationships of factors, processes, and soil properties enable us to recognize and delineate unique combinations as soil-landscapes. In the U.S., soil survey models of landscapes have commonly been unwritten qualitative explanations of field observations linked to abstract concepts of classification through soil series. The catena concept has been a useful tool to predict occurrence of soils in many landscapes. A concern with soil maps is a lack of quantitative information about map unit composition and details of soil property distributions. The fact that soil maps are prepared at different scales reinforces the lack of a satisfactory continuum model linking spatial scales of soil-landscape observations. Classification schemes have embraced quantitative diagnostic soil horizons and features, but variations among systems have hindered global correlation of defined taxa. There is a solid foundation of qualitative soil-landscape models based on tacit knowledge and supported by quantitative taxa based on well-described pedons. New pathways and techniques to understand functionally related soil-landscapes build on this foundation and give rise to fresh insights into the nature and behavior of the pedosphere.

2.1 INTRODUCTION

When pedologists talk about reading a landscape, they refer to how our senses interpret the scenery around us. What we learn about soil-landscapes depends largely upon the kinds of questions we ask. Hole and Campbell[1] provided a comprehensive review of concepts and studies of soil-landscape analysis in the U.S.

Many soil-landscape explanations that evolved in the 1950s and 1960s were based on concepts of geomorphologists like W.M. Davis, W. Penck, L.C. King, and W.D. Thornbury and soil scientists like B.E. Butler and R.V. Ruhe. It was not a true paradigm shift for soil science; however, it was a major revolution of thinking, looking, seeing, and comprehending soil-landscapes for a new generation of field soil scientists.

2.1.1 WHAT IS SOIL?

Our ancestors thought of soil as something plants grew in, and that there were good soils and poor soils for specific plants. Soils could also be dug to create canals or make embankments or even burial mounds. These were utilitarian concepts associated with use. During investigations of soils for taxation purposes in Russia in the late 1870s, V.V. Dokuchaev concluded that soils should be thought of as independent natural entities whose features resulted from the interaction of climate, biota, relief, and parent rock over time.[2] In 1936, C.E. Kellogg[3] proposed a functional relationship for soils in which he used the term *age*, instead of *time*, but it was not until H. Jenny's 1941 treatise[4] that American scientists rapidly accepted the concept of soils as functionally related to climate, organisms, relief, parent materials, and time. Time, for many scientists, is necessary for processes and so is not a factor like the others. However, if one considers age as it relates to time, then it seems appropriate to let it remain. The simple format $S = f(cl, o, r, p, t)$ is, in retrospect, the basic premise of soil science as we know it today.

2.1.2 IMPLICATIONS OF A BASIC PREMISE

The relationship of factors → processes → properties has often been called the paradigm of soil science because many relationships, models, hypotheses, and concepts are derived from the factorial equation and its connection of factors and properties.

The main implication for landscape modeling is that each soil-forming factor has geographic expression through time. Because a soil serves as an integrator and recorder of the events and processes of its evolution, the complex history and explanations of changes in a soil are to be obtained from the soil itself. Landscapes are therefore the keepers of functional relationships that give credibility to the soil survey.

Patterns of the overlap of soil-forming factors are imprinted on landscapes. Where the patterns remain sufficiently long, changes in soil morphology occur and are recorded, and repeating sets of properties are recognized as kinds of soils. It is a matter of learning to read the landscape.

2.1.3 MAJOR DILEMMAS

The basic premise of soil science is a simply stated relationship. It is apparent that much of the complexity of soils is derived from the ambiguity of concepts and terms used to specify the components involved. For example, how should the soil-forming factors be subdivided? What do we mean by climate? Is it external and part of the atmosphere? Is it internal in a soil, with soil air, soil temperature, and soil moisture as necessary criteria?

What is parent material of a soil? In going from regions to local areas there is a tendency to have more options for definitions of parent materials. At each larger scale of observation more variability is recognized to separate features in the landscape of interest.

Within a soil profile there often is evidence of variations of materials that influence, or have influenced, the development of the profile. Lithologic discontinuities usually are associated with environmental changes, but biomantles such as tree throw and termite mounds may also produce changes equivalent to those of sediment differences.

As soils are observed in landscapes, differences in soil morphology are associated with landscape positions. At what scale do these changes occur? If observations are generalized, what features become inclusions? Competent field soil scientists throughout the world recognize very similar sets of variations at map scales from 1:1000 to 1:15,000. At smaller scales all delineations are combinations of several kinds of soils. The implication of scale dependency is that soil-landscapes are also scale dependent, yet very little attention has been given to this perspective as far as modeling is concerned.

Depending on the property being recognized, changes may occur over a few decimeters, meters, or even tens of meters, and mentally varying widths of boundaries can be visualized. Sharp boundaries are usually represented fairly well on maps, but variable-width boundaries have no standards for representation. Fridland[5] summarized concepts related to size and shape of map unit delineations, including indices of properties of delineation boundaries.

Each definition carves a piece out of a universe and creates two classes: those entities that are included in the definition and those that are excluded. When we define soils we create soils and not-soils. When we define Chernozems, we create Chernozems and not-Chernozems. A class has two important features: a central concept of the class and the boundaries with other classes. The central concept is usually a cluster of narrowly ranging soil properties that define and, in part, describe the entity of interest. All classes share this concept. The criteria used to recognize an entity become the definition of that class and establish a range suitable for the purpose of the classes. In mutually exclusive schemes, the boundaries among classes are rigidly fixed; thus, membership is either yes or no. When ranges of multiple properties that overlap are acceptable, boundaries become flexible and joint membership can be considered. Such membership is often expressed as percent of degree of belonging in each of the overlapping classes. This concept is commonly referred to as fuzzy logic-based classification.

Landscapes are variable in space and in time. Less attention has been given to defining soil-landscapes than soil profiles; consequently, standards for central concepts and associated boundaries are less well established. Definitions are therefore major sources of dilemma as soil-landscapes are deciphered. The training and experience of soil-landscape observers become important considerations in understanding patterns.

Spatial variability exists at all scales, and its representation on maps indicates that processes of generalization have been employed. Some features are emphasized and even exaggerated, whereas others are minimized or ignored, depending on the purpose of preparing and presenting information in the form of maps.

A classic example of generalization[6] considers the products of grouping the delineations or the legend, or both, to obtain generalized soil maps. We learn that if a detailed map legend is generalized to reduce the number of units, it may or may not be possible to generalize a map. It depends on the geographic association of the detailed components. On the other hand, if a detailed map is generalized to create larger areas on the map, it may or may not be possible to generalize the legend. Again, it depends on the kinds of components that are geographically associated. To produce a generalized map with a generalized legend, a tremendous amount of information must be discarded or kept only as supporting data.

By design, the detailed soil surveys in the U.S. generally show landscapes that emphasize soil types as specific subdivisions of soil series. There are numerous variations of the kind and detail of soil map units used throughout the U.S., including units of nonsoil components used to present a more comprehensive view of the geosphere, and not only the pedosphere.

Modeling soil-landscapes is a complex process, and specifying the purposes and definitions of the classes being used is critical to evaluating the adequacy and utility of such landscape models. We are sometimes unsure of not only what to measure, but also which tools to use for measuring. Soil-landscapes are scale dependent; consequently, some tools and techniques may be more appropriate for one range of scales than another. The lack of standard protocols, terms, and techniques tends to reduce the consistency of interpretations among studies of landscapes. Whether soil-landscapes are conceptual or defined geographic bodies, quantified diagnostics and methods of measurement are needed for consistency and eventual correlation of soil-landscapes.

With all measurements there are concerns about accuracy and precision. Precision implies that a set of observations of the same phenomena will cluster closely around a given value (whether accurate or not), and accuracy implies that the cluster of observations includes to some degree an acceptable value that represents the real entity or central concept of the entity of interest.

2.2 SOIL SURVEY

2.2.1 LANDSCAPE DISTRIBUTION

Soil surveys portray a perspective of the geography of soil morphology of certain kinds of soil profiles and provide information about landscape features at various spatial scales.

In 1927 Prof. Bushnell[7] commented that the odds were greatly against the soil surveyor to know the facts. Soil mapping, he noted, was possible only because men could examine a profile at one point and successfully predict its occurrence where surface indications were similar. He observed that the surveyors could seldom map profiles; rather, they separated map units as slight depressions, low flats, convex rises, slopes, light and dark surfaces of soils, and other features of the landscape. Thus, the majority of soil-landscapes are delineated using external features of topography and location, and the naming depends on the relationship of soil morphology to landscape features.

To a field soil scientist the first hole or observation of a soil is a reference — a starting point relating observed soil features of a profile to a point in space. Using the genetic model, it is assumed that similar soil profiles will radiate from this point until some combination of factors and the processes they influence are strong enough to create a change that can be recognized as different soils. Do map unit delineations represent soil-landscapes? Assume for the moment that they do.

What is the distribution of soil-landscapes in a survey area? Delineations have neighbors with degrees of contrast, in addition to angles and distances between delineation centroids. If some frequencies of spatial distribution are random, how does one develop a meaningful model of pedogenesis? Traditionally, these aspects of modeling have not been employed in soil science. However, in this textbook a variety of emerging techniques to map soil patterns are addressed.

2.2.2 MAPPING LEGENDS

In the U.S., soil series were traditionally the basic units of soil delineations. Within a county, or similar area of interest, the kinds of parent materials are limited by the local geologic events and landforms, as are the types of vegetation communities. Superimposed on combinations of vegetation and materials are different degrees of wetness, and when combined together with profiles, the basis for a mapping legend of soil series becomes apparent. Field mapping legends were very powerful and useful models of soil-landscapes. In fact, they were at the heart of pedogenesis as understood at the time of their development and use.

In most surveys components of the whole landscape (geosphere) are delineated and named. Many nonsoil features are mapped, such as mine spoil and salt slicks, rock talus, urban areas, and special point features, including schools, churches, and so forth.

2.2.3 DESCRIPTIONS OF MAJOR SOILS

Standardized descriptions of soil profiles are a fundamental contribution to soil science and soil-landscape analysis.[8] They can be read, understood, and interpreted by competent soil scientists everywhere. As technology improved, Munsell colors replaced qualitative terms, field and lab pH values were added, and standardized terminology of texture, structure, consistency, and the presence of special features such as roots, concretions, coarse fragments, clay coating, carbonates, and so forth, were systematized. The National Cooperative Soil Survey (NCSS) standards for soil profile descriptions are provided in the *Soil Survey Manual* and the *Field Book for Descriptions* and are available online.[9] Official soil series descriptions[9] represent the central concepts of the soil series in the U.S.; however, the soil series described in a survey area report represent the central concepts of the series in that survey area.

Combinations of soil series with or without nonsoil components are also described. Very few measurements of proportions of actual components are made. However, guidelines suggested limiting inclusions that would behave differently to less than 15%. In many reports statements were given about the assumed kind and amount of inclusions in specific map units in a survey area. As such, they now serve as starting points for refinement regarding accuracy and precision of soil variability.

2.2.4 MAP UNIT INTERPRETATIONS

Soil survey interpretations, as information technology, evolved over the years and became the centerpiece of each report. The accumulated knowledge about the common behavior of soils of an area for major kinds of uses influenced the design of mapping legends and the format of the interpretations of the map units.

Because the earlier published surveys used soil series as landscape units, the interpretations were given for soil series. The commonly adapted crops were duly noted as well as the expected average yields when several levels of management were employed. In the 1960s, engineering practices related to drainage, irrigation, and road and building construction were considered to be important enough to provide estimates of soil suitability for a number of localized practices, including septic tank use. In the 1950s, a land capability classification was developed[10] and became one of the most popular and widely recognized interpretations of soil-landscapes. With the emphasis on soil erosion in the U.S., the series map units were modified to more appropriately reflect field conditions of susceptibility to erosion, as well as to show actual conditions, such as areas with slight, moderate, and severe erosion. The kinds of interpretations for map units of a modern survey are provided in the *National Soil Survey Handbook* of NCSS, maintained by USDA and available online.[9]

2.2.5 DESCRIPTIVE SOIL-LANDSCAPE MODELS

Physiography can be used to stratify landscapes and to organize soil observations by river and stream valleys, slopes, undulating uplands, and depressions. Additional subdivisions are provided by landforms such as floodplains, terraces and alluvial fans in the valleys, and shale-, sandstone-, and limestone-derived parent materials in the uplands.

Soil survey legends for mapping were locally derived and implemented. Details of kinds of parent materials, vegetation types and their transitions, marked climate variations, and degrees of expression of horizons and other features differed enough from place to place so that legends for field use varied from area to area. Field legends were designed to help surveyors easily recognize the main features of soils in their landscapes and consistently provide the same name or number to each relationship.

A catena is a chain-like topographic continuum linking uplands to lowlands. The concept was adapted in the U.S. to refer to drainage members in the same parent material.[11] The concept was accepted by the National Soil Survey, and where appropriate, it was readily used. The catena concept recognized the continuity of wetness in landscapes, but techniques to handle the transitional nature of this continuum were either not available or inadequate for consistent application by field personnel.[12] Catenary soil-landscape relationships have been one of the more powerful tools of soil-landscape modeling, especially in the humid and subhumid areas, where use and management of land are commonly dependent on moisture conditions.

2.2.6 PHASES

In the U.S., soil survey was initiated in the Department of Agriculture to identify soils suitable for major crops and for specialty crops like tobacco and grapes. Soon it was apparent that landscape features were important to the management of land, mainly water management. Water and wind erosion became the driving forces for new national conservation policies in the U.S. during the 1930s, and a different type of soil survey was undertaken by the newly created Soil Conservation Service.[13]

By convention, phases are attributes of soils not usually considered as criteria within a classification system. For example, stoniness and rock outcrops are not part of soil by definition, yet these features are important to the management and use of soils. Slopes as they influence water movement and erosion are important considerations, but seldom are they part of formal classification. Phases are vital to provide additional information of interest to users. Classes of conditions that may be considered for phases are described in the *Soil Survey Manual* of the NCSS.[9]

2.2.7 SOIL SURVEY MAPS

Traditionally, soil maps have been made to show the distribution of soils to facilitate the transfer of technology, mainly agricultural, and to portray patterns of order in nature. Because soils are multicomponent systems having numerous properties, delineations of soil bodies also represent distributions of included soil properties to varying degrees. Profile textures, nature of parent materials, drainage or wetness conditions, and kinds and arrangements of horizons can often be related to specific soil map units.

Land use objectives for which soil surveys are used differ widely in both kinds and levels of generalization. Some objectives are to predict soil performance for specific plants, like corn or cotton. Others require prediction of soil abilities for general land use classes, such as cropping, range, or forestry, and some require evaluation of alternative uses in land use planning. Each soil use objective has a set of limiting soil attributes that are critical.

There are practical limits to the number and minimum size of delineations on maps relative to the legibility of identification symbols and colors, importance of features and soils to be separated, and levels of spatial accuracy that can be achieved. In the U.S., a national map accuracy standard for maps of 1:20,000 or smaller is that not more than 10% of well-defined location points should be in error by more than 0.5 mm.[14] A line width of 0.3 mm (a common pen or pencil line on a 1:24,000 scale map) is about 7 m on the ground, and a location error of 0.5 mm represents 12 m on the ground. Errors of location and internal composition increase markedly as delineation size decreases or complexity (amount of dissection) of a delineation increases.

Thus, decisions about map scales depend on purpose of the survey, degree of precision of boundaries for locations, maximum amount of permissible delineation error, minimum size with an acceptable error of boundary placement, and the degree to which the minimum area of interest is acceptable to the user. Modeling soil-landscapes for pedogenesis and modeling those for land use or management practices generally require different map scales.

It is useful to consider two kinds of map unit variability, whether spatial or temporal. One kind is systematic and its pattern of occurrence is recognizable and mappable; the other appears to be random, unpredictable and unmappable. This is, in part, due to scale where similarities are included yet not all contrasts can be delineated because either the scale of the base map is too small or the exact location on the map is uncertain.

Random variability, once recognized, can be described as occurring in a soil-landscape unit. However, it is not possible to accurately predict its occurrence. An interesting question is: Does the complexity or dissection of a delineation of a map unit convey useful information about the composition of variability within the delineation?

Numerous maps showing the distribution patterns of interpretations derived from basic soil maps are available. Perhaps the most famous are the land capability maps that were included in farm and ranch plans. The green-yellow-red traffic light colors were used to emphasize risks and limitations of land use management.

2.2.8 GENERALIZED SOIL MAPS

When relevant features are abstracted to produce a smaller-scale map from a larger-scale map, that is generalization. For example, the soil survey of China done between 1979 and 1997 (C.F. Xi, personal communication) mapped urbanized areas at scales of 1:100 to 1:5000 and the adjacent lands at 1:10,000. This information was generalized and combined with landscapes mapped at 1:50,000 to produce a huge set of maps for the country. A correlation procedure and further generalization provided maps of 1:200,000. The objective of the decreed "Great Plan" was to produce a 1:1 M soil map of China and a set of derivative maps relating to potential and utilization of soil resources. Generalized maps at scales of 1:1, 1:2.5, and 1:4 M were planned for publication.

In the U.S., the agricultural areas have modern detailed maps published at scales from 1:12,000 to 1:24,000. This compilation of maps is called SSURGO (Soil Survey Geographic) database[9] and is being digitized and georeferenced at a scale of 1:24,000. Currently, SSURGO is transformed into an updated geodata format called Soil Data Mart. Each of the map units identified in SSURGO and Soil Data Mart can be composed of up to three components, some of which are unknown. Most modern published surveys include a generalized soil association map showing major physiographic areas at 1:150,000 to 1:300,000.

A nationally consistent set of generalized soil maps at a scale of 1:250,000 is called STATSGO (State Soil Geographic) database,[9] and the maps have been digitized. The composition of delineations was estimated from transects drawn across published SSURGO maps. A generalized version of STATSGO, known as NATSGO (National Soil Geographic) database is currently used as a base map for many national-level soil-based interpretive maps. Procedures for generalizing maps are probably less familiar than how to describe soils or do laboratory analyses for soil characterization.

2.2.9 SPECIAL SOIL RESOURCE INVENTORIES

Two rather unique surveys are the Canadian Biophysical Resource Inventory surveys and the Australian Land System surveys.

Jasper and Banff National Parks in Canada were mapped by a group of natural resource scientists including wildlife biologists, plant ecologists, soil scientists, geomorphologists, and parkland managers.[15] The initial landform delineations were modified based on suggestions of the specialists and the inventory mapped on air photos at a scale of 1:24,000. Biophysical surveys in Quebec and other provinces were at smaller map scales, but the integration of knowledge in the legend designs greatly enhanced the utility of these resource inventories for many users.

Innovative reconnaissance surveys were the land system schemes designed to evaluate sparsely populated terrain in Australia.[16] Landscape (terrain) provinces at 1:250,000 were successively divided into patterns, units, and components at larger scales. Air photos served as a basic tool for separating and delineating terrain taxa.

A modification of the land system approach was used in Victoria, Australia,[17] to evaluate land capability, mainly for agriculture. Classes of the highest category, land zones, are divided into land systems, land units, and land unit components. Components have a limited range of values appropriate for some features, but an indefinite range for other features not related to the sequence. It was stated that covariance of features in a sequence was more important than the range of values.

Special-purpose surveys of landscapes all use models of the landscape. Some are written down; others are carried in the minds of the surveyors. Recent attempts to reconstruct the mental models of surveys using statistics and geographic information systems (GIS) layers to provide environmental elements have been only partially successful, indicating the value of recording models for future users.[18,19] This textbook introduces numerous studies that successfully used digital spatial datasets and pedometric techniques to describe soil-landscapes.

A study of soil resource inventories[20] identified five kinds of information that would be necessary to predict soil performance in an on-site appraisal:

- The level of detail of information that would be required to evaluate soil resources for that objective
- The soil properties that would be critical for the projected land use
- The land use objective for which the soil resources are to be evaluated
- The degree of limitations that critical soil properties would impose on that use
- The effects of the geographic distribution of limiting soil conditions of the projected use

The information needed to predict soil performance from inventories is obviously the same as that required for on-site investigations. In addition, the study addressed three additional criteria: quality of the base map, including ground control; legibility of the map; and reliability of the recorded data of both the map and associated text. A sound basis for appraisal is provided in the report.[14]

2.3   SOIL CLASSIFICATION

At the First World Congress of Soil Science, W.W. Weir[21] made the point that because soil classification is wholly within the realm of thought and is governed by the laws of identity, it follows that we create the concepts indicated by the categories. He believed that we created soils as a universal concept, and that we did the same for soil series. He further commented that in the world of things, soils exist as real physical objects, but in classification we deal with them as thought entities called categories or classes. And to summarize this philosophy, Weir concluded that a category in soil classification, representing a general concept, may be defined, whereas an individual soil as a natural object may only be described.

Classification schemes readily handle mutually exclusive classes and systematically produce clean and tidy boxes of information, whereas nature has responded to a quite different set of guidelines. As difficult as defining a soil may seem, defining and modeling soil-landscapes also presents numerous challenges.

A classification is no better or no worse than the purpose for which it was designed. The adequacy of a system is judged by how well it satisfies its stated objectives or purpose.

A representative profile based on a set of field observations is the central concept for a soil. Insofar as each concept represents different genetic soils, they are the objects that are organized and arranged in a scheme of classification. The purpose is to show relationships among and between the many kinds of soils recognized by the designers of a system. Such a scheme reveals order in nature as perceived and described by the authors. There have been many schemes to classify soil profiles (pedons), and most are reasonably consistent with the intended purpose.

Because morphogenetic systems are derived from the basic premise of soil science about the soil-forming factors, it follows that the real objects are geographic bodies; however, no comprehensive system has yet been implemented to classify soil-landscape bodies over large areas.

2.3.1 HIERARCHICAL STRUCTURE

The sheer number of kinds of recognized soils is so large that a nested or hierarchical scheme is the most common way to accommodate them. Although concepts of individual soils at the field level can consistently be placed together into ever larger more inclusive groups, once a scheme is developed, it can only operate from the higher categories to the lower ones to separate the universe of soils into ever smaller, more detailed groups.[22]

All morphogenetic classifications have a genetic thread holding the categories together. The highest level is the most comprehensive, abstract, and is unstated; it is the universe of soils. The recorded highest category is an abstraction of the concepts of soils, usually different pathways of formation and evolution of soils. The soil features used to recognize each category are thought to be the result of processes, and they differ according to the concepts of how different kinds of soils develop. Classes are the subdivisions at each categorical level.

2.3.2 KEYS TO CLASSIFICATION SCHEMES

The human mind automatically classifies everything to simplify the complexity of the environment in which it lives, including ideas and thoughts. Keys are an expedient way to focus on the essential information needed to classify soils. A key is a tool of exclusion and is most efficient with yes–no decisions. Is this soil composed of dominantly mineral or organic materials? If mineral, exclude all further information about organic soils. Is this set of properties the result of soil-forming processes? On and on, one asks the questions looking for the yes–no answers and moving on through the keys searching for the first placement.

All soils are important in their geographic locations; however, in the structure of a key, soils are abstractions of mental models, and choices are made about importance. It is possible to note which constraints to soil development were thought to be more important to the designers of the system.

Keys may not tell us as much about soil-landscape models as we might hope for, but nevertheless, they are important because of the strong influence they have on soil surveyors using them to identify soils. They channel thinking and thought processes in rather rigid ways by focusing attention on selected soil properties.

2.3.3 IMPORTANCE OF SOIL SERIES

Soil series have been recognized for more than 100 years in the U.S.[23] The soil series were divided into soil types based on the overall texture of the profile generally associated with kinds of parent material or geologic deposit. Because soil series represented soil-landscapes, they had fairly wide ranges of properties, but very little information of actual ranges was available. Boundaries between soil series were not fixed, and as long as surveys were at widely scattered locations, there was little difficulty recognizing different series. When great soil groups were recognized, the soil series were placed in appropriate groups; however, the lack of precise definitions of boundaries meant that the placement of many series was uncertain.

The correlation of soil descriptions and data for correct placement was a major task of the soil survey. As new soils were detected, they were set up as provisional or tentative series, and as more information was obtained, they were finally recognized as official soil series. The survey program relied mainly on descriptions and properties of soil series to support the mapping and interpretation of soils throughout the country.

Are soil series soil-landscapes? Of course. Are soil series conceptual entities in a classification system? Of course. There has always been some ambiguity with the term *soil series* because it is used to name soil map units, is the name of an ideal mutually exclusive entity of classification, and is both singular and plural.

2.3.4 SOIL TAXONOMY

The adoption of *Soil Taxonomy*[9] was slow because lots of testing was required to make sure it satisfied its objectives. The soil series were the testing blocks; if classes at higher categorical levels split a series, either the criteria of a higher-level class was changed or the soil series was reexamined and modified accordingly.

Models of soils guided the development of *Soil Taxonomy*. The order level consists of soils whose properties are thought to result from major courses of development; the suborders are soils of the order classes whose additional properties are thought to be major controls of the current soil-forming processes; the great groups have additional properties that affect current processes; and so on through the definitions of the categories.[22] The subgroups reveal intergrades between other classes and also extragrades not related to other specific taxa; the family category provides information about the capacity for further change and serves as the bridge between the abstractions of higher-level genetic concepts and the series, the basic units of this system of soil classification.

There are assumptions that classes of the higher categories represent large land areas and that successively lower levels represent smaller land areas. Classes at each categorical level have central concepts of properties and assumed landscapes; consequently, the scheme reflects many qualitative aspects of soil-landscape models.

Because the classes are considered to be mutually exclusive with fixed boundaries, the massive job of adjusting the official descriptions of all soil series was undertaken. This step finally separated the soil series as landscape mapping units from the conceptual entities of classification used to name the map units. The thoughts of W.W. Weir became reality once again. At present, there is no taxonomy of soil-landscapes in widespread use.

2.3.5 DIAGNOSTIC HORIZONS AND FEATURES

Quantified diagnostics are essential to modern classification schemes. Emphasis is given to setting the boundary limits of classes in order to facilitate placement of descriptions and data into mutually exclusive classes. Often the central concept of a class is not clearly described and must be constructed from knowledge of the limits.

Crucial to applying class boundaries is the issue of boundary errors. Measurements are made on samples, whether depths and colors in the field, or lab measurements of physical, chemical, and biological properties. All measurements are approximations of true values, and so variability becomes important in classifying information. There are limits for depths, horizon thickness, and percent clay, sand, and coarse fragments. There are limits of colors, textures, cation exchange capacity, and soluble compounds. If every measurement is an approximation, how can we know with certainty the proper placement of soils whose properties are very near the arbitrary limits of classes?

Although studies demonstrate that a composite of four or more lateral samples in a horizon significantly reduces probable error of many properties,[24] seldom is this information reported for routine characterization data. With single samples the boundary errors of many properties are commonly ±2 to 6%, and even with composite sampling the errors are commonly ±2 to 3%. Guidelines generally have been qualitative, suggesting that good judgment should be used.

2.4 ACCOMPLISHMENTS

2.4.1 MAPS

Soil surveys in the U.S. dating from about 1980 are on digital base maps. The county-level maps or equivalent are the SSURGO maps, the 1:250,000 scale correlated set of maps are the STATSGO maps, and the national maps and its derivatives are the NATSGO maps. Because the maps are digitized, the publication scales or computer-generated ones may be at any scale appropriate and compatible with the databases.

The FAO-UNESCO (Food and Agricultural Organization of the United Nations) set of world maps[25] was published in the late 1970s at a scale of 1:5 M and were the only widely accepted maps covering the world. Russia and the U.S. also compiled world maps, but these were never as widely circulated or used as the FAO set of maps. As the availability of maps and databases increased, so did the production of maps for regions of the world. As map scales became smaller, soil patterns of larger areas reflected climatic zones and major physiographic regions, recalling again the zonal soil concepts of Dokuchaev and his colleagues.

2.4.2 DATABASES

The databases for early surveys consisted mainly of profile descriptions, a little characterization data, and general information about the use of soils in the survey area. The extent of the map units gave an idea of the dominance and distribution of kinds of soils in the survey.

In the 1980s, a new system in the U.S. to store and manage soil data was initiated. Patterned somewhat after the Canadian system, CANSIS, it was called NASIS (National Soil Information System)[9] and has evolved and grown into a complex, multifaceted system for maintaining map information, map unit information, soil description data, laboratory data, and numerous soil interpretation records for all map units. Standardization caused some state data and data from older formats to be excluded, as it was not possible to provide adequate correlations with data generated by different or updated technologies. The capability to extract or build soil-landscape models has not been emphasized and remains a challenge for the future.

In the late 1980s, an effort to develop an international standard database was started, and SOTER,[26] a soil terrain system, was tested in several parts of the world. Eventually, FAO provided support and ISRIC (International Soil Reference Information Center) in Wageningen became the caretaker. The European Soil Bureau adopted it for use within the European Union. SOTER databases relate more to classification taxa; however, some landscape parameters are included.

2.4.3 SOIL SURVEY STANDARDS

In a soil survey, standards are agreed-on items. The U.S. *Soil Survey Manual* of the NCSS is primarily a book of standards. Soil description terms are defined as are classes of many properties, such as particle size, texture, consistency, acidity, boundary thickness, and so forth.

Soil Taxonomy, another major book of standards, provides definitions of the various categories and taxa of the classification scheme used to identify and name soils and map units. A series of Keys to Soil Taxonomy as updates of the system were designed for field use; consequently, only the bare minimum of explanation was provided. A second edition of *Soil Taxonomy* was published in 2000 and provided updated definitions and concepts used in the soil survey.

A laboratory manual gives the details of characterization methods used in the U.S. by the NCSS, and another guide provides interpretations of lab data as applied in the soil survey. Most standards have been put on CDs and are also offered online from the Information Technology Center in Ft. Collins, CO.

The day-to-day operations of the National Soil Survey program are described and standards set forth in the *National Soil Survey Handbook*. It is a complex set of documents maintained by the soil survey staff of the Natural Resources Conservation Service, and it is also available online. It is unfortunate that a glossary of terms for all of the standards does not exist in one document; consequently, there may be ambiguity in the use and explanation of some terms and standards.

2.4.4 CORRELATION

The key to success of the U.S. soil survey program has been correlation. It is the process of applying standards to obtain and maintain consistency of concepts, of data collection and interpretation, of technologies used to support the program, and even in publishing the results.

Correlations are not foolproof; they are biased by participants as well as by the designers of standards. A lot of information about soil variability has been lost, misplaced, and perhaps misrepresented at times, although unintentionally. Soil-landscape modeling has not been subjected to standardization and correlation; thus, there are few records of coordinated or comprehensive pedogenic models in the U.S.

2.4.5 SOIL-LANDSCAPE MODELS

Pedologists think in terms of soil profiles, their horizonation, and the movement of water. Individual properties such as carbonates or redoximorphic mottles are usually visualized in the context of a profile. Pedologists have learned to respect and trust geomorphological features because the story of parent materials is there.

Geomorphology seems to respond as though the self-similarity of fractal theory determines the patterns at all scales of observation. Rivers do the same things over and over, alluvial fans build up in similar fashion everywhere, even in roadside ditches, and mass movement teases surficial soils down slope. Pedimentation is perhaps the dominant process that slowly molds the pedopshere into shapes and sizes that are recognized as soil-landscapes.[27] The stories of stepped land surfaces and landscape inversion are written in soils even though a page or two is missing here and there.[28]

Why do we know so little about the soil-landscape models of pedologists? Field soil surveyors develop working hypotheses from a collection of observations; they conceptualize relationships that exist in landscapes; they predict occurrences based on these models; and they test, evaluate, and refine the models sufficient to support reasonable maps of a landscape with delineations that can be related to conceptual central concepts of soils.[29–31] It is good science without the ability to control any of the variables of the experiments. Their task has been to recognize and delineate the patterns of soil formation expressed as soil-landscapes. Written records of their conceptual models were seldom required or shared.

How do you work in a multiproperty continuum whose properties are not necessarily coincident and find the stories that have been recorded? Daniels and Hammer[32] suggest beginning with stratigraphy, hydrology, and geomorphology to support what has been recorded in a soil profile.

2.4.6 MAP UNITS TEST MODELS

Field mapping is the constant testing of the correspondence of pedogenic models and landscape segments. It is a double relationship — one between a taxonomy and a landscape, and another between an identified landscape and a predicted response.

There are two uncertainties associated with this desire to predict the unknown. One is the correctness of prediction, and the other is the chance of being wrong in the prediction. There are trade-offs between the chances of being wrong (that chance associated with variability) and the degree of correctness (the variability of measurement). The experiments of soil surveyors have very little control; the variability in a landscape is fixed, so the task is to decipher it as well as possible given existing constraints. There is a risk for surveyors as well as users of their information.[33]

An evaluation of air photo interpretation (API) vs. field mapping[34] found that in a simple area, API correctly delineated soil series 66% vs. 84% for field mapping. Drainage class, parent material, and land use capability were also less for API. In a complex area, API was considerably less accurate for the above determinations. Studies of this kind confirmed that air photo interpretation is a wonderful tool to assist soil mapping, but it could not replace the fieldwork.

In a study to examine whether soil features could be combined in such a way that mappable soil bodies are the result,[35] point profile data for surface thickness, surface layer organic matter, sand in the subsoil, and clay in the subsoil in a 2-m grid system were used to determine pattern complexity. When the number of classes of each property was increased to increase class homogeneity, the pattern complexity also increased. Starting with categorical classes too narrowly defined resulted in a geographic pattern so complex that a simple pattern of slightly more heterogeneous units was missed entirely. It was possible to set statistical tolerance intervals on some properties and produce map units that were 75 or 90% pure with a 90% probability. The relevance of such statistical measures for decision makers is as yet unknown.

2.5 OPPORTUNITIES

Compared to the extensive coverage of qualitative soil-landscape mapping, there is little known about the mapping of quantitative soil-landscape models. The excitement in basic soil science is not simply in surveying the different kinds of variability in space and time, but also in determining the fundamental mechanisms for variability.[36] There are opportunities in soil survey and in soil classification.

2.5.1 SEARCH FOR UNIFYING CONCEPTS

There is a long-standing belief that the patterns we see at one scale are related in some manner to patterns at other scales. Pedologists have not solved the question of unifying laws with their numerous trials of making maps of the same area at different scales. Preliminary studies show that delineations of the same map unit in a survey have fractal dimensions that commonly increase as unit sizes get larger or boundaries become strongly convoluted.[37] If a distribution of fractal dimensions of a map unit were composed of several self-similarity relationships, it might suggest that processes over small areas are not the same as for large areas and might lead to new ways of understanding soil-landscapes.

As yet we do not know enough to test and evaluate this part of chaos theory to determine if it may be a unifying concept, or even if it is relevant to our understanding of soil-landscapes. Databases of digitized soil surveys contain data on the length of perimeters and the areas of map unit delineations. A fractal dimension is twice the value of the slope of a plot of log P (perimeter) vs. the log A (area) of delineations, that is, $P = A^{0.5D}$. It should be possible now to evaluate hundreds of soil-landscapes of interest, an exercise not previously possible.

2.5.2 STANDARDIZING LANDSCAPES

By most definitions, a landscape is what lies before us or in our line of sight. It seems worthwhile to develop a set of descriptions and eventual definitions for many soil-landscapes. A common scale in the U.S. could be 1:24,000, as one U.S. Geological Survey (USGS) topographic series exists at that scale and digitized soil surveys can be overlain at this scale.

2.5.3 GEOMETRY OF DELINEATIONS

We have many map products yet know very little about the cartographic features of the map units delineated on the maps. Geologists, cartographers, and even soil scientists have used perimeter–area ratios to evaluate roundness, dissection, or convolutions and to classify sizes and shapes of delineations.[5,38] It is thought that wherever a change in the factors, or degree of interaction, influenced the processes that resulted in this soil, there would be a constraint and likely a boundary separating differences among soils.

If the area of delineation is compared with the area of a circle having the same perimeter as the delineation, it can be interpreted as a measure or degree of constraint imposed in that landscape. This can be called a complexity index[38] and is equivalent to CI = $P^2/4$ ΠA, where P and A are the perimeter and area of a map unit delineation, respectively. The digitized soil survey data used to calculate fractal dimensions could also be used to calculate complexity indices.

2.5.4 LANDSCAPES OF SOIL SERIES

Local patterns of a soil series can be seen on maps of a county survey, but what is the pattern throughout the extent of a soil series? In most instances, we do not have good knowledge about the spatial extent of a soil series; however, with STATSGO the general limits could be estimated and soils surveys within these limits examined for patterns.

Plant ecologists and geographers have used a number of characteristics of spatial relationships, including the number of near neighbors and distances to and angles between these near neighbors.[39] Frequencies of these properties can be compared with theoretical random distributions.

Another technique is to produce a map of centroids and compare the distribution of empty and filled cells with those generated by various random generators. The question of interest is whether or not there is some scale at which the spatial distribution is similar to a random frequency. Other models consider landscape patterns as deterministic; however, the rates, locations, and times of occurrence are not predetermined and might produce features that are indistinguishable from random ones.

A preliminary test of five map units of glacial till- and outwash-derived soils evaluated three random frequencies for three cell sizes. Overall, the chi square test accepted 2 of 15 Poisson frequencies, 6 of 12 double Poisson frequencies, and 5 of 10 negative binomial frequencies. How do frequencies that can be generated randomly affect our pedogenic models of soil-landscapes? Do they suggest that at some scale processes affecting landscapes are indistinguishable from our assumptions of cause-and-effect relationships?

2.5.5 COMPOSITION OF DELINEATIONS

Most pedologists had been taught that although soils were parts of a continuum, they could not readily handle soil information that way. The concept of continua has always been a part of pedology, but not the techniques to apply to soil survey, soil classification, or making interpretations for use and management of soil resources. Statistical methods associated with agriculture dealt with discrete entities, and only the variability of measurements was treated as a limited continuum. The variability of soil properties, and even sets of properties, like profiles, was observed, but not described very well, and only occasionally was it quantified. This is rapidly changing.[40]

2.5.6 APPLICATION OF GEOSTATISTICS

The mathematics of geostatistics appear overwhelming to most pedologists whose careers have revolved around qualitative concepts and models. In glaciated areas it is difficult to grasp a 300-m range based on a kriging technique because other studies reveal spatial patterns within patterns within patterns, and a 300-m range seldom makes sense.

Rethinking a continuum is like seeing the visible portion of the electromagnetic spectrum with new eyes. The reds and yellows and blues actually grade into one another, and a wavelength can be used to identify points along that continuum. This book is about some aspects of what lies ahead.

To understand how soils will react to changes, it is essential that process models, whether simulating pedogenic evolution or other dynamic changes in biosphere components, collect the correct spatial and temporal distributions of soil attributes in soil-landscapes.[41] In detailed soil surveys there is little information about attribute distributions as continua; however, at smaller scales the information from detailed surveys has many of the features of spatially distributed attributes. The problem of data for modeling processes is therefore partly a matter of scale. Another concern has been the lack of techniques to obtain and present attribute distributions that are appropriate and adequate for modeling processes.

2.5.7 CRITERIA DEFINITIONS

Throughout *Soil Taxonomy* there are different definitions for properties of some differentiating criteria. This ambiguity for terms such as *humic*, *aquic*, and *content of organic carbon* is confusing for most users. The implication is that there are many more diagnostics in *Soil Taxonomy* than tacitly assumed, causing undue attention to definitions and specifying which criteria are actually being used. Cleaning up the definitions and reducing ambiguity should facilitate computer-assisted classification of soils.

2.5.8 Series as Taxa

When soil series were redefined to be in compliance with the class limits imposed by the hierarchy of *Soil Taxonomy* they no longer were landscape map units. They assumed the role of providing identity only to pedons. This facilitated comparisons among and between soil classification systems but was a step farther away from learning how to use geographic attributes in a comprehensive classi-fication of soil-landscapes.

The existence of repeating horizon variations over short distances prompted the creation of a variable-width pedon to include half of the distance of each variation with a maximum of 3.5 m. This concept is valid genetically; however, it has failed in application, so that the reasonable solution is to use a fixed-size pedon and define new kinds of complexes as standard map units in soil-landscapes.

2.5.9 Allowable Errors

There are many class limits in classification, but there are few rules or even guidelines to assist classifiers in being consistent in their decisions. It is often assumed that the mean values of properties are also the expected values and that the data can be placed in the correct classes. When this assumption is less certain, joint class memberships are possible, but guidelines for placement and for pre-sentation of such information have not been adequately implemented.

Allowable errors are not about mistakes; they are statistical expressions of measured variability. Acceptable and allowable are judgments, and they benefit science when they follow agreed-on standards. Field-determined textures have wider ranges of variability than laboratory determinations. Colors estimated with Munsell color charts have less accuracy than those measured with a spectropho-tometer. Soil-landscapes described as part of soil survey operations have lots of variability. Guidelines are needed for applying fuzzy set theory to situations where one or more attributes of a soil lie beyond class limits. Individual properties may be amenable to continuum representation, but what about the whole set that we call a soil pedon?

2.5.10 Soil-Landscape Anthology

Our literature is full of fragments of soil-landscape models. Hundreds of theses have been written about studies of soil-landscape segments, and many have never been published. We are awash in a sea of ignorance, flooded by unknown infor-mation. We are all guilty of this oversight, but it could be changed. There is no single source or repository of our knowledge of soil-landscapes in the U.S., either qualitative or quantitative. As technologies have changed, so have perceptions of soil genesis and the evolution of soils in their landscapes. Imagine a CD set containing the anthology of American soil-landscape models.

2.6 CONCLUSIONS

It is difficult to know what you do not know when information is very fragmented. I believe this to be the case for soil-landscape models; nevertheless, soil science has made a lot of progress during the past century, and here is what I conclude:

1. There is a solid background and legacy of soil-landscape models in the U.S. Many details are in unpublished research studies, and descriptive models are implied but seldom stated in the thousands of published soil surveys in the U.S. and elsewhere. Some reconstruction is possible.
2. People working in soil survey programs became so busy and pressured for products that knowledge of soils as landscapes became more and more dispersed. Real encouragement to document models and save the knowledge has never been implemented, but reconstruction of many models may be possible.
3. The U.S. is very fortunate to have had a continuous correlation effort in the National Cooperative Soil Survey. This meant defining standards for all aspects of soil survey and then following through with the application of the standards. Regardless of the glitches, errors, and any oversights, the results have provided remarkable consistency throughout the country. It is truly a major accomplishment.
4. The collective experience and wisdom of the NCSS concerning uncertainty in nature, in science, and in the human mind suggest that the truth of soil-landscapes cannot be discovered. However, there appear to be many potential pathways to understand relationships in the environments of which we are a part. How fortunate we are.

REFERENCES

1. FD Hole, JB Campbell. *Soil Landscape Analysis.* Rowman and Allanheld Publ., Totowa, NJ, 1985, pp. 42–82.
2. IA Krupenikov. *History of Soil Science.* Amerind Publ. Co., New Delhi, 1992, pp. 156–174 (translation of 1981 Russian edition).
3. CE Kellogg. *Development and Significance of the Great Soil Groups of the United States*, USDA Miscellaneous Publication 229. U.S. Government Printing Office, Washington, DC, 1936.
4. H Jenny. *Factors of Soil Formation: A System of Quantitative Pedology.* McGraw-Hill, New York, 1941, pp. 1–20.
5. VM Fridland. *Pattern of the Soil Cover.* Keter Publ. House, Jerusalem, 1976, pp. 15–41 (translated from 1972 Russian version).

6. AC Orvedal, MJ Edwards. General principles of technical grouping of soils. *Soil Sci Soc Am Proc* 6: 386–392, 1941.

7. TM Bushnell. To what extent should location, topography or physiography constitute a basis for differentiating soil into units or groups? *First Int Congr Soil Sci* (Washington, DC) IV: 158–163, 1927.

8. CF Marbut. The contribution of soil survey to soil science. *Soc Promot Agric Sci Proc* 41: 116–142, 1921.

9. Soil Survey Division, Natural Resources Conservation Service, U.S. Department of Agriculture. Technical references. Online access to *Soil Taxonomy*, Keys to Soil Taxonomy, Official Series descriptions, *Field Book for Descriptions, Soil Survey Manual, National Soil Survey Handbook*, lab methods, NASIS tools, national instructions, and technical notes. Available at http://soils.usda.gov/technical, accessed September 2, 2003.

10. AA Klingebiel, PH Montgomery. Land capability classification. In *USDA Handbook 210*, 1961 (reprinted 1973). Online access as part 622 of the *National Soil Survey Handbook*. Available at http://soils.usda.gov.

11. TM Bushnell. Some aspects of the soil catena concept. *Soil Sci Soc Am Proc* 7: 466–476, 1942.

12. SA Romero. A Study of Soil Variability at Mt. Pleasant. MS thesis, Cornell University, Ithaca, NY, 1944.

13. EA Norton. *Soil Conservation Survey Handbook*, USDA Miscellaneous Publication 352. USDA, Washington, DC, 1939.

14. Soil Resource Inventory Group. Soil resource inventories and development planning. In *SMSS Technical Monograph 1*. SCS, USDA, Washington, DC, 1981, pp. 271–364.

15. WD Holland, GM Coen, Eds. *Ecological (Biophysical) Land Classification of Banff and Jasper National Parks*, Vol. I, *Summary*, Alberta Institute of Pedology Publication M-83-2. University of Alberta, Edmonton, 1983.

16. GD Aitchison, K Grant. Terrain evaluation for engineering. In *Land Evaluation*, GA Stewart, Ed. Sydney: Macmillan of Australia, 1968, pp. 125–146.

17. FR Gibbons, JCFM Haans. Dutch and Victorian Approaches to Land Appraisal, Soil Survey Paper 11. Netherlands Soil Survey Institute, Wageningen, 1976, pp. 23–32.

18. EN Bui, A Loughhead, R Corner. Extracting soil-landscape rules from previous soil surveys. *Aust J Soil Res* 37: 495–508, 1999.

19. B Zhou, R Wang. Knowledge-based classification in automated mapping. *Pedosphere* 13: 209–218, 2003.

20. MG Cline. Objectives and rationale of the Cornell study of soil resource inventories. In *SMSS Technical Monograph 1*. SCS, USDA, Washington, DC, 1981, pp. 7–14.

21. WW Weir. What is the relative weight that should be given field and laboratory data in the definition of several categories in a comprehensive scheme of soil classification? *First Int Congr Soil Sci* (Washington, DC) IV: 113–121, 1927.

22. H Eswaran, T Rice, R Ahrens, BA Stewart, Eds. *Soil Classification: a Global Desk Reference*. CRC Press, Boca Raton, FL, 2003, pp. 27–42, 101–200.

23. RW Simonson. *Historical Aspects of Soil Survey and Soil Classification*. SSSA, Madison, WI, 1987.

24. LP Wilding, LR Drees. Spatial variability and pedology. In *Pedogenesis and Soil Taxonomy*. LP Wilding, GF Hall, NE Smeck, Eds. Elsevier Sci. Publ., Amsterdam, 1983, pp. 83–116.

25. Food and Agricultural Organization of the United Nations (FAO-UNESCO). *Soil Map of the World*, 1:5 M, Vol. 1, *Legend*. UNESCO, Paris, 1974.

26. FAO-ISRIC-ISSS. Global and national soils and terrain digital database (SOTER). In *Procedures* manual. ISRIC, Wageningen, 1995.

27. RV Ruhe. Geomorphic surfaces and the nature of soils. *Soil Sci* 82: 441–455, 1956.

28. VO Targulian. Soil as a biotic/abiotic natural system: a reactor, memory and regulator of biospheric interactions. *Eurasian Soil Sci* 29: 30–41, 1995.

29. RW Arnold. Multiple working hypothesis in soil genesis. *Soil Sci Soc Am Proc* 29: 717–724, 1965.

30. RW Scully, RW Arnold. Soil-geomorphic relationships in postglacial alluvium in New York. *Soil Sci Soc Am J* 43: 1014–1019, 1979.

31. WE Hanna, LA Daugherty, RW Arnold. Soil-geomorphic relationships in a first-order valley in central New York. *Soil Sci Soc Am J* 40: 89–92, 1975.

32. RB Daniels, RD Hammer. *Soil Geomorphology*. John Wiley & Sons, New York, 1992, pp. 1–23.

33. RW Arnold. Soil survey reliability: minimizing the consumer's risk. In *Data Reliability and Risk Assessment in Soil Interpretations*, SSSA Special Publication 47, WD Nettleton, A Hornsby, RB Brown, TL Coleman, Eds. SSSA, Madison, WI, 1996, pp. 13–20.

34. JA Pomerening, MG Cline. The accuracy of soil maps prepared by various methods that use aerial photograph interpretation. *Photogrammetric Eng* 19: 809–817, 1953.

35. IJ Jansen, RW Arnold. Defining ranges of soil characteristics. *Soil Sci Soc Am Proc* 40: 89–92, 1976.

36. G Sposito, RJ Reginato, Eds. *Opportunities in Basic Soil Science Research*. SSSA, Madison, WI, 1992, pp. 1–7.

37. RW Arnold. Fractal dimensions of some soil map units. *Trans 14th Int Congr Soil Sci* (Kyoto) V: 92–97, 1990.

38. PA Piech. Selected Concepts of Soil Map Delineations. MS thesis, Cornell University, Ithaca, NY, 1980.

39. RW Arnold. *Spatial Analysis of Soil Map Units*, Agronomy Mimeo 78-5. Cornell University, Ithaca, NY, 1978.

40. MJ Mausbach, LP Wilding, Eds. *Spatial Variabilities of Soils and Landforms*, SSSA Special Publication 28. SSSA, Madison, WI, 1991.

41. PA Burrough. The technologic paradox in soil survey: new methods and techniques of data capture and handling. In *Soil Survey: Perspectives and Strategies for the 21st Century*, ITC Publication 21, JA Zinck, Ed. ITC, Enschede, The Netherlands, 1993, pp. 15–22.

3 A Historical Perspective on Soil-Landscape Modeling

David J. Brown

CONTENTS

ABSTRACT

Over the past several decades, a number of new and powerful technologies have been applied to soil-landscape modeling: satellite remote sensing, geographic information systems (GISs), global positioning systems (GPSs), and digital elevation models (DEMs). However, the key soil-landscape theories underlying these applications were proposed and developed by scientists and surveyors from the early 19th to the mid-20th centuries. Soil-landscape modeling has its origins in 19th-century geological surveying before soil surveyors broke away to create their own discipline around the turn of the 20th century. The Russian geologist-geographer V.V. Dokuchaev introduced a climatic-geographical approach to soil mapping in the late 19th century, drawing on the ideas of the German scientist Alexander von Humboldt and merging geological and geographical ideas in the formulation of the five factors of soil formation known today. In the early 20th century, through the work of Dokuchaev disciple Glinka and the U.S. National Cooperative Soil Survey (NCSS) leader Curtis Fletcher Marbut, a predominantly climatic, zonal concept came to dominate soil-landscape modeling. Against this background, an unknown British scientist working in East Africa named Geoffrey Milne challenged the dominant zonal paradigm in pointing to dramatic changes in soil properties and formation from hilltop to valley bottom within a single climate zone. To address this regular and repeating soil-topography relationship, Milne coined the *catena* concept to describe a complex map unit with associated hillslope hydrologic processes of formation. The U.S. soil survey community distorted and confused Milne's catena, and only through the work of Robert Ruhe in the 1950s and 1960s was the concept saved from scientific obscurity. The dominant paradigm in the NCSS (U.S.) and much of the soil survey world for the past century has been Marbut's soil anatomy, based on a 19th-century biological metaphor with the primary objective of constructing a hierarchical, natural soil classification system. While geological surveying, Humboldt's plant geography, Dokuchaev's geology–geography synthesis, and the catena concept have all contributed to the development of modern, quantitative soil-landscape modeling, soil anatomy has hindered progress by diverting intellectual resources and prestige away from soil mapping.

3.1 INTRODUCTION

The scientific discipline of pedology grew out of soil surveys that were initiated in the late 19th and early 20th centuries to produce soil maps and related management interpretations. The fundamental problem in soil mapping has been the need to construct soil maps over large areas based upon a relatively limited number of soil observations. To solve this problem — to interpolate or extrapolate from a few soil observations to the Earth's surface — scientists and surveyors have constructed soil-landscape models: theoretical, empirical, graphical, verbal, qualitative, quantitative, explicit, and tacit. This chapter addresses the history of soil-landscape models, focusing on developments from the early 1800s to the 1950s.

Given the centrality of soil-landscape modeling for both soil surveys and pedology, it seems surprising that so little attention has been paid to the history of this science. Simonson[1–3] and Gardner[4] have written extensively on the history of soil survey activities in the U.S., but the bulk of their work addresses administrative issues, classification, and underlying soil profile concepts, with relatively little attention paid to how surveyors have actually drawn lines on a map and how soil-landscape theories have informed this work. Similarly, Taylor[5] discusses mapping practices only as part of a general history of Australian soil survey activities. Effland and Effland[6] chronicle the history of soil-geomorphology studies in the U.S., but stop short of demonstrating how this work has actually been applied to soil mapping. The explanation for the relative dearth of histories on soil mapping science is quite simple. Soil-landscape modeling has historically been left to the proverbial "field man," and soil surveyors' theories — until quite recently — have rarely been made explicit.[7,8] Accordingly, the *history* of the science of soil-landscape modeling has been as tacit as the science of soil-landscape modeling itself.

Tandarich[9,10] has traced the genealogy of soil investigations and made the case that pedology is fundamentally interdisciplinary with historical ties to geology, agricultural chemistry, biology, and geography. While undoubtedly true, this claim begs the question: How *specifically* have these disciplines contributed to the development of pedology and soil-landscape modeling? Given the internal diversity of disciplines like geology, geography, and biology, changes in those disciplines over time, and their relatively recent emergence from the primordial soup of natural history, it would be instructive to know what particular ideas, metaphors, methods, problems, questions, and answers have been inherited by the newer science of soil-landscape modeling from its more established relatives.

I make the case that four major scientific traditions, concepts, or metaphors have impacted the development of soil-landscape science as we know it today: (1) soil characterization in 19th-century geological surveys, (2) Alexander von Humboldt's early 19th-century plant geography, (3) Geoffrey Milne's soil catena concept, published in 1935, and (4) late-19th-century biological morphology.* The integration of Quaternary geology and geomorphology into soil-landscape modeling in the 1950s and 1960s,[6] while important, lies beyond the scope of this chapter. It would be a mistake to view these themes as completely distinct or hegemonic. Paradigm shifts in the history of science are like waves washing upon a beach, with a great deal of mixing from one wave to the next. Nonetheless, identifying and analyzing key ideas in the history of soil mapping can help distinguish their relative importance in contemporary soil-landscape modeling. In addition to sections on these theoretical concepts, I review the introduction of aerial photography in soil survey work, of great practical importance to developments in soil-landscape modeling. A brief discussion of contemporary relevance

---

* Morphology, in this instance, refers to the study of the form, structure, and development of biological organisms.

closes out the chapter. A timeline of important events, ideas, and publications with associated scientists is provided as a guide in Table 3.1.

3.2   GEOLOGISTS, GEOLOGICAL SURVEYS, AND SOIL MAPPING

The earliest soil surveyors, soil mappers, and what we now might call soil-landscape modelers were geologists. William Smith's 1815 geological map of England, Wales, and part of Scotland "that changed the world"[11,12] includes a claim in the legend to exhibit the "varieties of soil, according to the variations in the substrata" (Figure 3.1). Smith did not recognize soil profiles as they are known to soil surveyors today, nor did he devote appreciable attention to the study of unconsolidated surface materials. Rather, he constructed a seminal geologic map using fossils to identify strata — then assumed for his publication a relationship between soil variability and variability in the rocks below. The relevance of this map to the history of soil-landscape modeling is that (1) geologists, from the outset, produced maps; (2) many early geologists like William Smith were landscape scientists trained to read landscapes in the field and utilize surveying equipment[11,12]; and (3) most 19th-century geological surveys included soil investigations, albeit of variable quality.[4,14–16]

3.2.1   GEOLOGIC UNDERSTANDINGS OF SOIL–GEOLOGY RELATIONSHIPS

Simonson[17] and other U.S. soil survey workers/historians[4,18] have claimed that prior to the initiation of the NCSS (U.S.) program in 1899, geologists simply mapped surface geology and assumed a perfect correspondence between geology and soils. Early in the 19th century, this was for the most part true. Amos Eaton, who conducted some of the earliest studies of agricultural geology, wrote in 1818 that

> all soils, excepting what proceeds from decomposed animal and vegetable matter, are composed of the broken fragments of disintegrated rock. From this fact it is natural to infer that the soil of any district might be known by the rocks out of which it is formed. (Quoted in Gardner, 1998)

According to Gardner[4] (p. 8), this statement "was to characterize field studies in soils and geology for nearly a century." However, by the late 1800s most geologists had attained a relatively sophisticated understanding of the various relationships between rocks and soils. For example, the U.S. Geological Survey (USGS) scientist Israel Russell[19] studied precipitation and temperature influences on rock weathering and the production of secondary clays. Russell[19] (p. 27) observed that "the soil formed by the decay of a great variety of rocks is a red clay, which, in the more advanced stages of decomposition, is strikingly similar,

TABLE 3.1
Timeline Showing the Dates of Significant Publications, Ideas, or Applications with Associated Scientists

Decade Starting	Significant Publication, Idea, or Event with Associated Scientist
1730	*Systema Naturae*, 1st ed., 1735; **Carl von Linné or Linnaeus** (1707–1778), Swedish botanist, naturalist, and taxonomist
…	
1800	Scientific journey through the Americas, 1799–1804; **Alexander von Humboldt** (1769–1859), climatic theory of plant geography
1810	Paper on isothermal lines, 1817; **Alexander von Humboldt**
	Geologic map of England and Wales, 1815; **William Smith** (1769–1838)
1820	Scientific journey through Russia, 1829; **Alexander von Humboldt**
1830	State geological surveys initiated in U.S.
1840	
1850	*Origin of Species*, 1859; **Charles Robert Darwin** (1809–1882), proposed theory of biological evolution
1860	Birth of the U.S. Geological Survey (USGS), 1869
1870	
1880	Russian Chernozem, 1883; **Vasilii Vasilevich Dokuchaev** (1846–1903), Russian geologist-geographer, five factors of soil formation
1890	*Relations of Soil to Climate*, 1892; **Eugene W. Hilgard** (1833–1916), U.S. agricultural geologist/chemist
1900	Birth of the U.S. National Cooperative Soil Survey (NCSS), 1899
1910	*Treatise on Soil Science*, 1914; **Konstantine Dimitrievich Glinka** (1867–1927), Russian student of Dokuchaev
	Soil anatomy, ca. 1916–1920; **Curtis Fletcher Marbut** (1863–1935), NCSS (U.S.) leader, 1913–1933
1920	Experiments with aerial photography in 1923 and 1927
1930	1st county mapped with aerial photography in 1930; **Thomas M. Bushnell**, leader of Indiana survey, 1922–?; aerial photography adopted by NCSS in 1935
	Catena concept, 1935; **Geoffrey Milne** (1898–1942), British soil chemist, East Africa
1940	Debates over catena concept, 1942–1945; **Thomas M. Bushnell**
	Little soil survey activity due to WWII
1950	Soil Survey Division and Soil Conservation Survey (SCS) merged, reorganization
	Initiation of NCSS Soil-Geomorphology program in 1953; led by **Robert V. Ruhe**, U.S. Quaternary geologist and pedologist

Source: Compiled from a variety of references cited in this chapter.

both in appearance and in constituents, the world over." The warmer and more humid the climate, he argued, the greater the degree of weathering, with a consequent soil reddening due to the production of dehydrated ferric oxides. In a prominent 1904 treatise on *Rocks, Rock-Weathering and Soils*, the geologist George Perkins Merrill[20] (p. 360) stated:

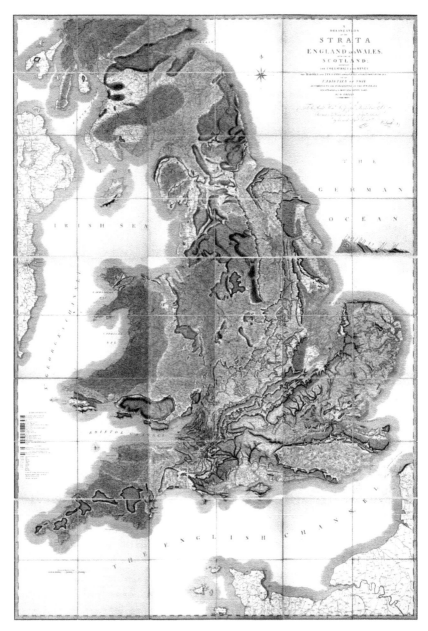

FIGURE 3.1 William Smith's 1815 geological "map that changed the world."[12] The legend indicates that the map expresses "varieties of soil according to the variations in the substrata," a common approach in the early 19th century. (Reprinted from Schneer, C.J., William "Strata" Smith on the Web, 2004, available at http://www.unh.edu/esci/wmsmith. html.[13]) (See color version on the accompanying CD.)

That, however, a rock contains all the desired materials, is no certain indication as to character of its decomposition product, since in this process of decomposition much desirable matter may have become lost. Nevertheless most soils retain what we may call inherited characteristics and a direct comparison whenever possible is by no means uninteresting.

Merrill discussed the fact that some soils formed on limestone were devoid of lime while, other soils formed on rocks other than limestone were found to have significant amounts of lime — a perplexing observation that was explained by reference to Hilgard's[21] theories on the influence of climate, precipitation, and profile leaching. The type and degree of soil inheritance from the rock below was acknowledged to be conditional on a number of factors.

For the most part, 19th-century geological surveyors published soil information in reports attached to geological maps.[4,22,23] Even where independent soil maps were published,[14] soil distributions were based heavily on geological maps.[15,16] This indicates that for the most part, geological surveyors used geologic relationships to *delineate* map units (just as surveyors today use topography and surface reflectance from aerial photographs). However, surface and subsoils within map units were described (particularly color and texture), and samples were frequently taken back to laboratories for mechanical and chemical analyses.[22–24] The development of a more sophisticated understanding of soil–geology relationships within geology followed decades of describing and analyzing soils on a wide variety of formations and deposits.

This discussion has focused for the most part on the use of geology to map soils, but it should also be noted that geologists use soils to map geology in areas where outcrops are lacking[25] and have done so for at least a century.[26,27] Just as soil surveyors came to use vegetation and topography for soil mapping because these landscape attributes were readily visible at the surface, so geologists have used soils to map geology because they can be readily sampled at the surface. "In regions which are not covered by glacial deposits ... or by thick sheets of transported materials ... the soils will usually indicate the nature of the underlying solid rocks"[27] (p. 281). At the dawn of the 20th century, geologists were aware that there was no exact correspondence between soils and geology, and on sloping lands the field geologist was instructed to be wary of colluvial soil transport that might obscure bedrock stratigraphy. But surveyors found that sandstones did tend to produce sandy soils, shales and marls produced clays, etc.[27] The study of soil-geology relationships, then, has long been of mutual benefit for both soil and geologic surveyors.

3.2.2 The Use of Geology in the U.S. National Cooperative Soil Survey

When the U.S. soil survey program was initiated in 1899 within a relatively new Department of Agriculture, the first nationally funded soil survey program in the world, the federal agency hired surveyors almost exclusively from agricul-

FIGURE 3.2 Curtis Fletcher Marbut (1863–1935), the geologist turned antigeologist who led the U.S. National Cooperative Soil Survey (NCSS) from 1913 to 1933 and promoted the soil anatomy concept. (From Krusekopf, H.H., Ed., *Life and Work of C.F. Marbut*, Soil Science Society of America, Columbia, MO, 1942.[29] With permission.)

tural colleges.[4,28] This unofficial policy held even when the "Ag" colleges were not held in high academic regard and there were few qualified candidates receiving degrees from such colleges[28] (p. 29). Many of the cooperating institutions at the state level were geological surveys,[14] but this seems to have been avoided when possible. For example, when the Bureau of Soils entered into a cooperative agreement for a reconnaissance soil survey with the Washington State Geological Survey rather than the Washington State Experiment Station in 1908 at the insistence of the state legislature, this caused some controversy within the NCSS[28] (p. 82). It was only later in the 20th century that geologically trained scientists like Mark Baldwin and James Thorp, who trained under Allen D. Hole at Earlham College, joined the soil survey.[10] Geologists and geologic surveying had a great deal to offer the new U.S. soil survey program initiated in 1899 — expertise in Quaternary geology, physiography, reading landscapes, and basic surveying skills in addition to decades of soil studies — but this expertise was largely ignored.

When Curtis Fletcher Marbut (Figure 3.2), leader of the NCSS (U.S.) from 1913 to 1933, joined the survey in 1910, it seemed to some soil surveyors that he "still regarded soil as an interesting geologic formation"[30] (p. 47). In his early years with the survey, Marbut was responsible for refining a systematic soil classification based on physiographic provinces and geologic parent material — a classification that became the soil survey standard for the second decade of the 20th century.[31] In a stunning intellectual reversal, the origins of which are not clear, sometime around 1916 Marbut switched from being the disliked geologist to an extreme antigeologist.[4] Like Saul struck down on the road to Damascus, the man who a few years earlier had designed a geological and physiographic classification scheme for soils in the U.S. now claimed that "these [soil] features have no harmonious relation whatever to the features of the parent geological material"[32] (p. 18). His rejection of geology was so complete and zealous that he was almost removed as leader of the soil survey.[33] It would take almost half a century — with the work of Robert Ruhe and the soil-geomorphology projects initiated in 1953[6] — for geology to once again make significant contributions to soil survey work in the U.S.

Having soil, geologic, and topographic surveying under one scientific and administrative roof in the U.S. — something that almost happened in the late 19th century — would likely have led to a very different history of soil-landscape modeling. In the late 1880s, the agricultural geologist/chemist E.W. Hilgard at the University of California and the USGS director John Wesley Powell made repeated attempts to get an "agricultural" or soil survey established within the USGS. Powell, who already incorporated topographic surveying into the USGS, was interested in expanding the size and relevance of his agency, and even considered moving the USGS into the new Department of Agriculture. Hilgard, in the end, turned down an opportunity to lead the new survey because he did not want to move to Washington, D.C., and the initiative subsequently failed.[34]

3.2.3 SOIL SURVEY OF SOUTHEASTERN ENGLAND

Hall and Russell,[35] two distinguished British agricultural chemists, conducted one of the most involved and scientifically rigorous studies of soil-geology relationships through the first half of the 20th century in their *Report on the Agriculture and Soils of Kent, Surrey, and Sussex*. More importantly, for this discussion, they also published an explicit account of how this map was constructed,[35] which provides a window into early-20th-century theories of soil-geology relationships.

The essential problem that Hall and Russell[35] faced was to find some means of correlating soils to existing or readily obtainable data.

> As it would be almost out of the question to construct a soil map on a basis of analysis only, examining for example field by field along the common boundary of two types of soil in order to draw that boundary, some guiding principle must be sought for, and this in the area in question has been provided by the Geological Survey. It was a matter of experience that within the district there was a general correlation between soils and geological outcrop, and at the outset a number of determinations were made to ascertain if the outcrop lines laid down on the geological map would also serve as boundary lines between two soil types. (pp. 185–186)

Detailed hard rock and drift (surface deposit) maps were obtained from the Geological Survey at a scale of 1 inch to 1 mile. Soils and subsoils at selected locations were analyzed for both particle size distribution and major element composition (e.g., silica, alumina, potash, etc.). Additionally, farmers were interviewed to ascertain the yields, land use history, and management problems associated with selected fields. For the time, the soil and geology data were exceptionally detailed and allowed for the rigorous comparison of the two.

Hall and Russell were well aware that climate, vegetation, and topography could also control soil variability[35] (pp. 182–186). Citing Tulaikoff,[36] they acknowledged that the Russian climatic approach might be important at the continental scale, but argued that it was not relevant for the study at hand given the relative uniformity of climate across southeastern England. Given the long cropping history in England, they believed that most native vegetation influences were not only difficult to discern, but the effects were likely to have been obliterated from the soil by human management. Hall and Russell were also clearly aware that geologic materials might be redistributed across the landscape and sampled to avoid such complications — explicitly avoiding steep slopes, hollows, and streambeds. The goal was to establish general soil relationships with geology that were intended to be interpreted "in the light of local conditions, such as climate, water-supply and drainage"[35] (p. 182).

By no means did Hall and Russell assume soil-geology relationships, but they examined and tested correlations in great detail. Some of the geological formations examined (relatively few) were found to have very tight and regular relationships with specific soil types. More commonly, the lithology of geological formations gradually changed according to the mode and pattern of sediment deposition preceding lithification. Nonetheless, Hall and Russell were still able to employ soil-geology relationships to characterize soil properties across these formations using predictable geographic patterns of lithological variation. For example, they found that the Hythe Beds were comprised of a clayey limestone to the east, then a calcareous sandstone further west, and an infertile (noncalcareous) sandstone at the western edge of the district. In the Lower Wealden formation with alternating clay and sand, they found the clays to be consistently low in alumina and potash and difficult to flocculate, translating into soils with a lack of structure irrespective of texture.

Hall and Russell demonstrated that geological survey information, used carefully, could assist in the construction of soil maps. Though they conducted one of the more explicit early studies of soil-geology relationships, these agricultural chemists were building on a century of work in geological surveys. For larger regions, as Hall and Russell acknowledged, climatic variables might play a more important role in controlling soil variability, a topic we turn to next.

3.3 VON HUMBOLDT, DOKUCHAEV, AND SOIL GEOGRAPHY

The richness of science no longer lies in the abundance of facts but in their linkage.

— **Attributed to Alexander von Humboldt (1769–1859)[37] (p. 151)**

FIGURE 3.3 Self-portrait of Alexander von Humboldt, pencil drawing of 1814. (Reprinted from Kellner, L., *Alexander von Humboldt*, Oxford University Press, London, 1963.)

Alexander von Humboldt (Figure 3.3) made major contributions to 19th-century natural history. Humboldt attended lectures given by the famous geologist A.G. Werner, but made his name largely on the basis of a 5-year scientific journey through South America and Mexico (1799–1804), which provided the observations for decades of scientific publishing. Major contributions from this work included, among many accomplishments, new insights into the Earth's magnetic field, volcanism, meteorology, astronomy, natural philosophy, and plant geography.[38,39]

Humboldt's development of plant geography bears the greatest relevance to the history of soil-landscape modeling. Whereas the 18th-century botanist Linnaeus focused on identifying, naming, and classifying individual plants, Humboldt was more interested in vegetation types such as a rain forest, grassland, or heath. With a strong experimentalist bent, he developed instruments to make meteorological measurements, *quantified* correlations between climate and vegetative forms, and used the understanding gained from these studies to delineate climate-vegetation zones.[39,40]

3.3.1 HUMBOLDT AND RUSSIAN SOIL GEOGRAPHY

At the height of Humboldt's international fame in 1829, he was invited as a guest of the Russian tsar to conduct a tour of mines in Siberia and the Urals, and used this trip as an opportunity to study the natural history of that continent. At the close of this trip, Humboldt convinced the Russian government to establish a network of meteorological stations.[38] Having given talks all through Russia and met most of the prominent Russian scientists of that period, Humboldt's ideas

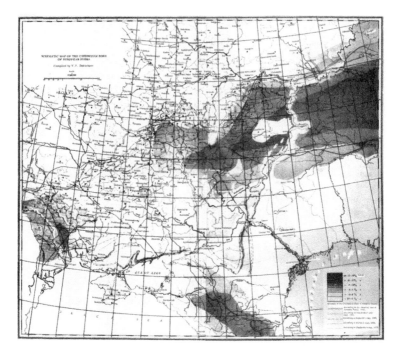

FIGURE 3.4 V.V. Dokuchaev's 1883 "Schematic Map of the Chernozem Zone of European Russia" with isohumus lines following and demonstrating a familiarity with Alexander von Humboldt's plant geography. (Reprinted from Dokuchaev, V.V., *Russian Chernozem*, Vol. 1, Kaner, N., Trans., Israel Program for Scientific Translations, Jerusalem, 1967.)

and methods would certainly have been included in the training of a young Russian scientist named Vasilii Vasilevich Dokuchaev (1846–1903), who is regarded as the father of modern pedology.[2] The clearest evidence of Humboldt's influence on Dokuchaev can be found in the design of the latter's first soil map. In 1817, Humboldt published a paper on isotherms, or geographic lines of constant temperature, and went on to promote all kinds of isolines for geographic representation and analysis.[41] Dokuchaev employed this same technique in his isohumus map (Figure 3.4) to illustrate the results of his seminal Russian Chernozem* study published in 1883.[42]

Dokuchaev[42] (p. 14) referred to his soil studies as "geologic–geographic investigations," and though he did not mention von Humboldt by name, "geography" in the late 19th century referred to Humboldtian science. Limited by funds and only able to make a few traverses through the large region in question, Dokuchaev was able to construct the first coherent map of Chernozem humus accumulation using Humboldt's geographic principle of *climatic correlation*,

---

* Chernozems are grassland soils with thick, dark organic surface horizons and typically calcareous subsoils.

predicting humus accumulation at unknown locations based upon the availability of climate data and soil-vegetation-climate relationships. The network of meteorological stations that Humboldt proposed in 1829 provided the data necessary for the construction of Dokuchaev's map of Russian Chernozems.

Dokuchaev did not believe that climate and vegetation were the only factors influencing soil formation. He clearly understood and wrote extensively about the importance of geology, geomorphology, topography, hydrology, and land surface age on soil formation and mapping. Dokuchaev[42] (p. 338) argued, for instance, that

> favorable rock composition is one of the most important prerequisites for the formation of Russian chernozem.... However, favorable parent rock is not the only prerequisite for chernozem formation, since chernozem results from the combined effects of climate, country age, vegetation, topography and parent rock.

Accordingly, though Dokuchaev mapped Chernozems in geographic belts, he also noted that the soil type was not really as continuous as represented, stating "the chernozem zone is interrupted by forest area, bogs, hilly areas, river valleys, sands and solonetses"[42] (p. 314). In sampling to construct the isohumus map, Dokuchaev carefully selected sites to avoid local variability — selecting soils on flat surfaces or in the middle of gentle slopes — but took care to note the limitations thus imposed on the final product.

The five factors of soil formation as proposed by Dokuchaev can be partitioned into (1) the geological and physiographic factors, which included parent rock, country age, and topography, and (2) geographic factors, which were comprised of Humboldt's climate and vegetation. The brilliance of Dokuchaev came first from hypothesizing a soil-climate relationship analogous to well-established vegetation-climate correlations. Dokuchaev then synthesized geologic and geographic perspectives into a single coherent theory describing the nature and distribution of soil humus accumulation, which was later expanded to encompass a wide variety of soils across Russia and the world.[15]

In contrast to the combined geological and geographical approach of Dokuchaev, his disciple Konstantine Dimitrievich Glinka (1867–1927) articulated a purely geographic or climatic theory of soil formation and distribution.[43] In the introduction to his 1914 treatise, Glinka wrote that topography, vegetation, and parent rock influenced soil variation by modifying soil microclimates (e.g., moisture regimes in lowlands, vegetation and shade, rock color and insolation, etc.), and could therefore be considered subsidiary climatic factors — unifying the fundamental control of climate on soil formation. He acknowledged only one parent material influence on soil formation, devoting just 5 pages in a 674-page book to a discussion of the association of rendzina soils with calcareous parent rocks[43] (p. 513–517). This prompted the Russian editor Polynov to insert a note in a later addition stating that "there is no doubt that quartzite, quartz sandstone, and quartz sand — as parent rocks — also exert a considerable influence on the corresponding soils"[43] (p. 517). Moreover, Glinka recast the importance of land surface age in biological terms, arguing that soils evolved over time with changing conditions, as did organisms. All together, this proved, according to Glinka, that soils were "geographically conditioned" like plants in Humboldt's geography[43] (pp. 4–15). "This provides sufficient justification for singling out the soils as a particular group of natural bodies, with which a special branch of science should be concerned"[43] (p. 5).

The U.S. soil survey leader Curtis Marbut brought Glinka's climatic soil geography ideas into the U.S. circa 1916–1920[4] and later translated Glinka's 1914 German text into English.[44] With a series of lectures in 1928, Marbut[45] (p. 19) argued:

> When we superpose over a soil map, maps of the various kinds of climatic forces, and the various kinds of natural vegetation, we find certain definite relationships. When, however, we superpose over a soil map of *mature* soils, a geological map, we find no relationship between the general broad, predominant characteristics of the soils and the characteristics of the geologic formations. In the same way when we superpose a topographic map over a map of mature soils we do not find a relationship.

Whereas Glinka justified a climate-dominated approach in terms of how other factors modified soil climate, Marbut argued that poorly drained and geologically influenced soils were simply immature expressions of mature climatic, zonal soil types. The Glinka–Marbut zonal soil concept came to dominate pedology in the early 20th century.

3.3.2 OTHER GEOGRAPHIC INFLUENCES IN PEDOLOGY

Dokuchaev was not the only late-19th-century scientist to recognize the importance of geography in soil formation and distribution. The agricultural geologist/chemist Eugene Hilgard[21] (p. 59) also independently came to recognize "that there must exist a more or less intimate relation between the soils of a region and the climatic conditions that prevail, or have prevailed therein"[21] (p. 9). Hilgard[46] also devoted considerable attention to soil-vegetation relationships, in particular examining — both in the field and in the laboratory — the relationships between tree species and the carbonate content of soils in Mississippi.

> It is needless to say that these presumptions were quickly submitted by me to the test of chemical analysis, which, while corroborating the general induction, yet soon showed the need of qualifications, corollaries, and conditions to be fulfilled, in order that the hypothesis might stand.[46] (p. 610)

Hilgard, like his late-19th-century Russian counterpart Dokuchaev, also identified and employed relationships between soils and geology or landforms where useful.[34,46] The fact that both Hilgard and Dokuchaev arrived at similar geologic-geographic theories of soil formation in the late 19th century suggests that they were influenced by the same developments in geological surveys, geography, and the 19th-century establishment of meteorological networks in both Russia and the U.S.

3.3.3 RUSSIAN SOIL MAPPING TECHNIQUES, CIRCA 1900

According to Dokuchaev's student Sibirtsev,[15] Russian soil surveyors at the turn of the 20th century constructed soil maps in two explicit stages: (1) data collection and (2) spatial interpolation. In addition to obtaining or constructing a topographic map with roads, hills, valleys, etc., soil surveyors were encouraged to gather as much information as possible about the soil-forming factors — particularly geology and vegetation. Using this information, traverse routes were selected to capture variability in soil-forming factors. In addition to describing and sampling natural soil exposures, pits and boreholes were excavated as topography or surface soil morphology changed.

> More detailed soil investigations necessitate a clear notion of the topography, and soils should be traced according to the forms and changes of the relief. In other words, the area under investigation should be subdivided into definite parts, for instance: flat hummock, plain, gentle slope, steep slope, low-lying land at the foot of the slope, closed depression, etc., and the soil should be sampled on every such part.[15] (p. 205)

Every profile described was located on the base map as accurately as possible. Combining field descriptions with laboratory characterization, all of the profiles examined were then classified in preparation for the construction of the final map.[15]

The final map was constructed by interpolating between known points on the map using what we would now call environmental correlation:

> At first glance, it may appear that this working stage involves many arbitrary choices, especially in drawing the boundaries of patches or strips. However, as already pointed out many times, every soil occupies a specific area and its occurrence is necessarily related to definite causes which must be determined by the investigator. Most often, the soil patches and strips correspond to the topographic features (patches on hummocks, hills, depressions or strips along slopes), which should be indicated on the cartographic bases, or else the investigator himself may mark them following a visual or instrumental survey.[15] (p. 328)

As a check, correlations between soil formers and soil types were studied carefully for regularity. In areas where regular relationships could not be worked out, soil surveyors were required to take a systematic approach, sampling on a regular grid if this could be afforded. At least ideally, Russian soil surveyors marked point profile locations (not polygons) on base maps, and employed and tested explicit soil factor correlations to draw map unit boundaries around and between those points. The topographic factors — both relief and landscape position — played a major role in local soil surveys.

3.3.4 Use of the Factorial Approach in the NCSS (U.S.)

Relative to their Russian counterparts, early-20th-century U.S. soil surveyors had little scientific guidance on how to assess the spatial variability of soils. In the 1904 *Instructions to Field Parties*[47] there were instructions on everything from alidade plane table techniques to laundry procedures — but absolutely no mention of how to delineate soils in the field. By 1906, a procedure for finding and mapping soil boundaries was established that required a large number of systematic borings and empirical interpolation between sampling locations.[48] In 1920, a publication titled *The Value of the Soil Survey* proclaimed that soil surveyors "carefully inspect every forty acres and show soil variations on the map as small as from five to ten acres."[49] And in his personal reminisces on soil survey work during the first four decades of the NCSS, Macy Lapham stated that soils were "examined systematically by means of frequent borings"[28] (p. 245). Individuals might have had personal theories that helped them identify changes in soil materials, but they were not part of any official procedure or science of soil surveying. "Every man will go about the work somewhat differently, and any discussion of the methods employed must necessarily be taken largely from a man's personal experience," a surveyor wrote in 1909[50] (p. 186).

It was not until 1914 that the *Instructions to Field Parties* included a small but explicit acknowledgment that soil surveyors might use landscape features to map soils:

> Often slight depressions or elevations, a change of the color of the surface material, or a change of the character of the surface or of the vegetation will indicate to the experienced soil man a change of soil conditions to be investigated or verified by an examination with his soil auger.[51] (pp. 69–70)

Physiography, depositional processes, and parent rock were employed in soil classification, but were not presented as a tool for map unit delineation. The NCSS did not publish another field manual until 1937, when the *Soil Survey Manual* included one paragraph in a 135-page document on topography and vegetation, "external features which assist in the sketching of boundaries after they have been located"[52] (p. 101). The British field manual[53] published in 1936 was only slightly more detailed in the discussion of field mapping, with the author listing changes in five circumstances that could indicate a change in soil properties: (1) lithology, (2) topography, (3) vegetation, (4) surface color, and (5) the sound and feel of the soil underfoot. How much of a change and of what type indicated a change in soil characteristics? "At this point the field man only may decide, and this is only possible after experience on the spot and cannot possibly be described here"[53] (p. 110).

Marbut's mimeographed translation of Glinka's treatise was published in 1927,[44] around the time that the factors of soil formation were first taught in U.S. soil science courses.[1] The first explicit discussion of the five factors of soil formation in a U.S. government publication can be found in the annual U.S. Department of Agriculture (USDA) Yearbook of Agriculture for 1938, titled *Soils and Men*,[54] and the first soil survey report to include an explicit discussion of all five factors of soil formation (climate, organisms, relief, parent material, and soil age) was issued in January 1940.[55] A year later, NCSS soil scientist James Thorp published a paper on the use of environmental factors in soil mapping,[56] and Hans Jenny proposed a pseudoexperimental approach to quantify Dokuchaev's factors of soil formation.[57] In the 1940s and 1950s, boilerplate discussions of the five soil-forming factors became a standard feature of U.S. county survey reports.

3.4 MILNE'S SOIL CATENA CONCEPT

Over large areas where local variation in topography were regularly repeated, a given colour on any map finally produced (on any but an impracticably large scale) would have to be interpreted as indicating the occurrence not of a single soil but of a sequence of soils occurring generally over the area, to be worked out on the actual ground in each instance according to topography and other local influences.

— **W.S. Martin, 1932, as reported by Geoffrey Milne[58] (p. 5)**

The inspiration for the soil catena concept came from W.S. Martin, a British soil chemist based in Uganda[58] (p. 5). However, it was the Tanganyika-based soil chemist Geoffrey Milne (1898–1942) (Figure 3.5) who formally defined, expanded, and promoted this soil-topography concept both within and beyond East Africa.[60–63] Despite Milne's relative isolation in Africa and the confusion that resulted when U.S. scientists reinterpreted the catena to fit existing soil-topography concepts,[54,64,65] the catena has become a central concept in soil-landscape modeling,[66,67] inspired a namesake journal *Catena*, and been the subject of numerous book chapters and reviews.[68–71]

3.4.1 THE ORIGIN AND DEFINITION OF THE CATENA CONCEPT

In 1928, Geoffrey Milne arrived at the East African Agricultural Research Station in Amani, Tanganyika, with training that included a B.Sc. and M.Sc. in chemistry and agricultural chemistry, respectively, and 8 years of lecturing experience in the latter subject.[59] The first and primary task assigned to Milne was to coordinate the construction of a soil map for the region.[72] Toward that end, in May 1932, a meeting of soil chemists from Uganda, Kenya, Tanganyika, and Zanzibar was convened at Amani to discuss, among other things, mapping the soils of East Africa.[58] A central problem for this project was the need to construct a map for two conflicting purposes: (1) a detailed (large-scale) map for agricultural advising and (2) a regional (small-scale) map for inclusion in both a world soil map and a British Association geographical project. Highlighting the scale problem, Martin brought a sequence of soil monoliths from Uganda to illustrate dramatic changes in profile morphology from hilltop to valley bottom and suggested that the group use complex map units to capture soil-hillslope variability. Mapping units based on soil-topography relationships arose from the very practical scale issues of the

FIGURE 3.5 Photograph of Geoffrey Milne who coined the soil catena concept to describe complex soil map units with regularly repeating soil-topography relationships. Milne also outlined the hydrologic processes responsible for the differentiation of soils on hillslopes, an early expression of what we now know as process geomorphology. (Reprinted from Milne, K., in *Geographers: Biobibliographical Studies*, Vol. 2, Freeman, T.W. and Pinchemel, P., Eds., Mansell, London, 1978, pp. 89–92.)

project at hand: "the soils of a large piece of country are to be mapped on a small piece of paper"[60] (p. 191).

Milne followed up on Martin's idea by coining the term *catena* (chain) to characterize a regularly repeating soil-topography sequence. This idea was first circulated in a memo to the other soil chemists in early 1933,[73] then published formally in a relatively obscure *Soil Research* paper.[60] Milne proposed two clearly distinct ideas: (1) the *fasc* (Latin for "bundle") as a taxonomic grouping intermediate between the series and great soil groups, and (2) the *catena* concept as a "composite unit of mapping"[60] (p. 193). The catena name itself was "intended to serve as a mnemonic, the succession of different soils corresponding to the links in a hanging chain" in a progression from one hilltop to the next[63] (p. 16).

The catena concept as first proposed in 1933 and published in *Soil Research* was a simple soil-topography map unit. However, in a paper presented at Oxford for the Third International Congress of Soil Science in conjunction with the presentation of the *Provisional Soil Map of East Africa* in 1935, Milne expanded his concept.[61] First, he suggested that there be two different types of catenas, those formed from uniform parent rock and those formed on more complex geology, where geology–topography relationships also contributed to soil–topography relationships. The original Buganda catena was included in the second category, with upland soils forming on ferricrete-capped hilltops and lowland soils forming on younger incised bedrock. Second, he provided a concise discussion of processes leading to catena formation:

> Soil differences are brought about by differences of drainage conditions, combined with some differential reassortment of eroded material and the accumulation at lower levels of soil constituents chemically leached from higher up the slope.[61] (p. 346)

With this single sentence, Milne proposed an entirely new dimension for soil-landscape modeling, what we now recognize as hillslope hydrology, process geomorphology, and landscape geochemistry. Milne followed with a letter to *Nature* discussing the influence of fluvial erosion and deposition on soil-landscape formation,[63] proposing that erosion be considered a soil-forming process (rather than a geologic process) if the removal and addition of materials was incremental and proceeded in parallel with other soil-forming processes. In so doing, Milne brought an experimental scientist's perspective to a natural historian's project, something Jenny[57] later proposed at the soil profile scale with his quantitative factorial approach.

While soil scientists in the 1930s and 1940s did not acknowledge the process dimension of the catena, Ruhe cited Milne's work extensively in his seminal "Elements of the Soil Landscape" paper and strongly objected to the "distortion" of the catena by the U.S. soil survey community[74] (p. 166). In a later review of a soil–geomorphology textbook based on the factorial approach, Ruhe[75] (p. 177) argued:

> [The catena concept] not only integrates the factors in explaining soil differences, but it also focuses on past history of the land surface, geohydrology, erosion, sediment transport, and pedogenic processes.

Milne's applications of the catena concept for soil studies in East Africa, published posthumously,[76] were very much in the soil-geomorphology spirit later articulated by Ruhe. The nine-unit soil-landscape model of Conacher and Dalrymple[77] was also based explicitly on the catena, providing a more detailed discussion of specific processes, discrete landscape units, and soil relations.

While the catena concept referred to an abstract soil–topography relationship with associated formation processes, Milne also proposed that a catena refer to a landscape soil field unit analogous to the vertical soil profile (Figure 3.6).

> To the geomorphologists, geologist or engineer, the profile of the ground would be the outline of my diagram. To the pedologist the profile is what he finds in depth at a selected point. The language of soil description lacks a suitable term having a cross-country dimension, and the want of it is felt as soon as soils are discussed in relation to the lie of the land.... To help in such discussions the word catena has been adopted ... to describe a topographic complex of soils such as is represented in my example.[62] (p. 549)

FIGURE 3.6 Geoffrey Milne's graphical two-dimensional representation of a catena, with the typical soils found at each landscape position identified by the numbers 1 to 7. This representation was used to highlight the role of erosion-deposition processes in soil-landscape formation and published in 1936. (Reprinted from Milne, G., *Nature*, 138, 548–549, 1936. With permission.)

The catena could either refer to a *specific* soil-landscape transect studied in the field, or if that transect were representative of a region, the catena would then acquire a symbolic status representing regional soil–topography relationships — the catena as map unit. In actual use, Milne and his assistant often employed the catena as a three-dimensional soil mantle over a small watershed,[76,78] an extension later articulated explicitly by Huggett.[79]

3.4.2 CHALLENGING THE ZONAL SOILS TRADITION

The idea that soil formation and spatial variability could be related to topography or relief was hardly a novel concept in the 1930s. Both Dokuchaev and his student Sibirtsev discussed soil-topography relationships in the late 19th century.[15,42] As discussed earlier, Russian soil surveyors were utililizing soil–topography relationships in detailed soil mapping by the end of the 19th century.[15] In the U.S., Indiana survey leader Thomas Bushnell pointed out the importance of topography for soil mapping in 1927.[80] The prominent British pedologist Gilbert W. Robinson wrote in his 1932 textbook[81] (p. 335):

> There are areas even in Britain, where relatively simple geology and topography under a uniform climate result in extensive tracts of soil which, if not actually constant in character, exhibit variation which can be easily related to topographical conditions.

That soil varied with topography was a widely accepted fact by the early 1930s, and the immediate acceptance of the catena concept was in large part due to the fact that Milne provided an interesting name for this commonly recognized phenomenon. But in several important ways, Milne's catena concept also challenged and expanded existing ideas regarding soil–landscape relationships.

Within the zonal framework, soil-topography relationships were acknowledged and described, but only as *subsidiary* to the primary climatic control. Both Dokuchaev[42] and Sibirtsev[15] discussed the topography-related effects of

CHART 1				THE NIKIFOROFF SYSTEM

KEY FOR FIELD CLASSIFICATION OF SOILS
IN A UNIT AREA

Soils in the zone of .... the determined by common morphological characters of the phytomorphic (well-drained) associates, or the typical normal soils of the zone.

A. Combination of soils in a given physiographic region	B. Associations differentiated on the basis of parent material (geological)	Associates		
		C_1. Oromorphic (eroded on locally arid)	C_2. Phytomorphic (well drained)	C_3. Hydromorphic (poorly drained or locally humid)
		D_1 D_2 etc. *Subdivided into phases as they occur	D_1 D_2 etc. *Subdivided into phases as they occur	D_1 D_2 etc. *Subdivided into phases as they occur
	B_1 B_2 B_3 B_4 B_5 etc.	* A phase is a variation or modification of an associate which differs in some characteristic but with other associate characteristics in common. Associates and phases, further differentiated according to textural classes.		

FIGURE 3.7 The hierarchical Nikiforoff field classification system with the oromorphic, phytomorphic, and hydromorphic associates for different landscape positions. Note that this association fits comfortably within the zonal soil concept. (From Ellis, J.H., *Sci. Agric.*, 12, 338–345, 1932. With permission.)

insolation, drainage, erosion-deposition processes, uneven surface age, and chemical leaching, but largely as side discussions to explain why "soils are not always normal in constitution"[42] (p. 343). For Marbut, soils on lower landscape positions were not even full-fledged soils, but immature siblings of upland soils within their assigned climate zone.[45] In 1915, the Russian soil geographer Neustruev proposed that soil-climatic zones consisted of zonal complexes.[82] This basic idea was developed further into a field classification system, brought into the U.S. by C.C. Nikiforoff in the early 1930s (Figure 3.7) and published by the Canadian soil scientist and surveyor J.H. Ellis.[83] Within this system, soils were *hierarchically* classified according to (1) climate zone, (2) physiography, (3) parent material, (4) relief/drainage, and (5) other factors, like local vegetation, wind, or stones. As can be seen in Figure 3.7, *soil associations* were defined as groups of soils with differing drainage on uniform parent material within the same climate and physiographic zones. Nikiforoff's soil association fit comfortably within the climatic zonal framework, as topography was considered the least important soil-forming factor,[83] a belief echoed by several U.S. soil survey workers.[54,80]

In challenging the hierarchical, climate-dominated approach, Milne pointed to soils "whose profiles and conditions of formation differ fundamentally" from hilltop to valley bottom, geographically associated soil sequences that cut across the great soil groups[61] (p. 345) — a radical proposition at the time.

> We are not entitled to classify a black clay as a youthful red loam, nor to represent it as red loam on the map when in its own very different character it occupies an important proportion of the land surface. The soils of the bottomlands constitute just as much of the truth about soil conditions in these parts of Uganda as do the red loams of the ridges or the murram soils met with in between.[60] (p. 193)

In both his catena papers and in previous work on the importance of geologic provenance,[58,84] Milne issued a direct challenge to the zonal soils concept, though that challenge was not fully appreciated at the time.

Milne also felt that zonal soil maps were insufficiently grounded in empirical data. Early in the *Soil Map of East Africa* project, he expressed dissatisfaction with Marbut's 1923 U.S. soil map, arguing that

> assumptions about soils from knowledge ... of climate, topography, and other external factors should henceforth be verified on the actual ground, at least for a few points in every area, before being put into the new map. This view would imply that the mapping of sample areas in some detail must come first, before the possibilities of generalizing by broad strokes could be estimated.[58] (p. 3)

With a large area to map and few resources, even before the first meeting of soil chemists Milne outlined a creative, systematic program for soil survey in East Africa based on (1) occasional soil traverses over large areas while on safari, (2) detailed studies in particular areas where there was a demand for soil analysis to solve an immediate management problem, (3) site descriptions, profile diagrams, and intact samples sent in by various colonial officers, and (4) systematic local surveys at representative locations. In systematic local surveys, Milne and his colleagues sampled profiles on a tight grid over approximately 100 acres (40.5 ha), to both characterize local variability and relate that variability to local conditions such as topography and drainage.[72]

For the final *Soil Map of East Africa* (Figure 3.8), Milne completely filled well-studied areas, used broken coloring for areas with incompletely verified soil information, and left unsampled areas blank, as he was unwilling to extrapolate soil-environment relationships into regions without ground truth.[63] Milne also devised an innovative "pajama striping" scheme for cartographically representing catenas, which was employed on the final map (Figure 3.9). This was the first regional-scale soil map with an explicit representation of uncertainty and within-map-unit variability, something that even today is rarely available in published soil survey maps. Underneath the abstract, theoretical catena concept, Milne pursued a rigorously empirical and explicit soil mapping project utilizing catenary principles.

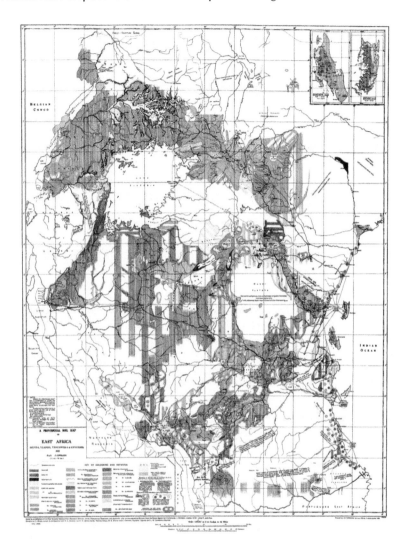

FIGURE 3.8 Geoffrey Milne's 1935 *Soil Map of East Africa*. Note the "pajama striping" for the catenas and large land areas left blank or partially blank where soil investigations were absent or incomplete. (Reprinted from Milne, G. et al., *A Provisional Soil Map of East Africa (Kenya, Uganda, Tanganyika and Zanzibar) with Explanatory Memoir*, Amani Memoir 31, East African Agricultural Research Station, Amani, Tangayika, 1936.) (See color version on the accompanying CD.)

3.4.3 CATENA CONCEPT VS. SOIL ASSOCIATION

When Milne first proposed the new catena term, W.S. Martin opposed the idea in part because he believed that the existing *suite* coined by G.W. Robinson would do as well.[73] Robinson's suite referred to a combined classification and mapping

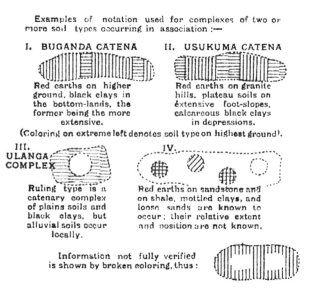

Examples of notation used for complexes of two or
more soil types occurring in association :—

I. BUGANDA CATENA   II. USUKUMA CATENA

Red earths on higher   Red earths on granite
ground, black clays in   hills, plateau soils on
the bottom-lands, the   extensive foot-slopes,
former being the more   calcareous black clays
extensive.   in depressions.

(Coloring on extreme left denotes soil type on highest ground).

III.
ULANGA
COMPLEX

IV.

Ruling type is a   Red earths on sandstone and
catenary complex   on shale, mottled clays, and
of plains soils and   loose sands are known to
black clays, but   occur; their relative extent
alluvial soils occur   and position are not known.
locally.

Information not fully verified
is shown by broken coloring, thus :

FIGURE 3.9 Map legend for the 1935 *Soil Map of East Africa* showing the graphical representation for compound map units. (Reprinted from Milne, G. et al., *A Provisional Soil Map of East Africa (Kenya, Uganda, Tanganyika and Zanzibar) with Explanatory Memoir*, Amani Memoir 31, East African Agricultural Research Station, Amani, Tangayika, 1936.)

term for a set of differing soils formed from a common parent material.[81] But Milne rejected the suite as inappropriate for use in the East Africa mapping project because (1) soils within a suite could vary for a number of reasons other than topography, (2) the uniform parent material requirement was not always met for soil-topography relationships in East Africa, and (3) the use of a term for both mapping and classification would result in confusion and ambiguity.[60] Unfortunately, the latter two problems were realized despite Milne's best intentions when U.S. survey workers reinterpreted the catena as equivalent to the existing soil association concept.

By 1938, the catena was redefined in the U.S. as "all the soils in a region developed from the same parent material but differing relief and in degree and character of profile development," what had been previously termed an *association* in the U.S. Unfortunately, this definition limited the utility of the catena as a mapping concept, as it depended "on the uniformity of the factors other than relief"[54] (p. 989). Milne's soil mapping and landscape formation model that cut across the zonal soil paradigm was reduced to a drainage-based, field classification unit similar to those already in use (e.g., Figure 3.7),[83,85] leading one prominent pedologist to suggest that Milne's concept be termed a *macrocatena*, with the U.S. version more aptly termed a *microcatena*[56] (p. 42).* In 1945, Bushnell[65] justified this new definition by publishing a handwritten note from Milne suggesting that it might be okay to limit the catena concept to the type I variety only (similar parent *rock*), though Milne was quite clear in the note that soil-topography sequences on different parent *materials* would still be considered a type I catena — a distinction lost on Bushnell. Since Milne passed away suddenly in 1942,[59] he was not in a position to contribute directly to the debate. Ironically, the U.S. redefinition meant that the original Ugandan soil-topography sequences no longer qualified as catenas.[86]

The catena came to be used in the U.S. primarily "as a means of facilitating the logical grouping of soil units and for remembering their characteristics and relationships"[54] (p. 989), or in other words, the catena was used as a field classification device. Following the publication of Jenny's *Factors of Soil Formation* in 1941,[57] Bushnell tried to reconcile the factorial approach with his own field keys and the catena concept by proposing a radically new soil classification based upon the five factors, with the catena as a *taxonomic* unit.[64,65] A debate ensued as to whether the catena should be considered a geographic association or taxonomic grouping of soils[54,64,65,87] — a confusing situation Milne expressly tried to avoid by coining the catena term instead of using the existing *suite*. As a result, by 1951 the catena was dropped from the official U.S. soil survey lexicon, and in yet another ironic twist, Milne's original catena — minus the explicit landscape process components — came to be known as an *association* comprised of "regularly geographically associated" soils[88] (pp. 302–306). Only in the tropics did soil surveyors continue to use the catena concept and term as originally defined.[68]

3.4.4 APPLICATIONS OF THE CATENA CONCEPT (OR ASSOCIATION)

The 1938 USDA yearbook, *Soils and Men*,[89] included a nominal soil association map in the sense of Milne's catena "with a characteristic pattern of distribution"[54] (p. 989). In map unit descriptions, however, soil-landscape relationships were incompletely defined and compiled from existing county soil survey reports that did not include soil associations.[89] The first county soil survey report to include the explicit use of soil associations (in the catena sense) was a Tennessee survey published in 1948, with the survey actually initiated in 1939.[90] A few county surveys initiated in 1938 and 1939 employed the association concept, and by 1952 this had become standard survey practice.[91] At the same time, U.S. soil surveyors began to publish their soil-landscape theories explicitly in the form of three-dimensional association diagrams. One of the earliest association diagrams resembling those of today (Figure 3.10) was published in a glossy Taylor County, IA, report released in 1954.[92] However, U.S. surveyors were still supposed to identify all soil boundaries in the field and sample every soil body. The association concept helped guide field sampling, but augmented rather than replaced traditional detailed survey protocols.[88,91]

---

* In contemporary terms, a toposequence might also be considered a microcatena, as all factors other than topography must be held constant, though in practice the toposequence term is often incorrectly employed as equivalent to the catena.

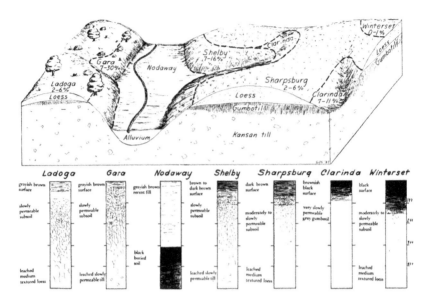

FIGURE 3.10 Three-dimensional soil association block diagram published in the 1954 survey of Taylor County, IA, report. This is one of the earliest three-dimensional association block diagrams, closely resembling diagrams found in contemporary U.S. soil survey reports. (Reprinted from Scholtes, W.H. et al., *Taylor County, Iowa, Soils*, Soil Survey Series 1947, No. 1, USDA-SCS, Washington, DC, 1954.)

In the 1940s, Australian soil surveyors used the catena concept to *replace* detailed soil mapping, allowing them to survey large land areas with limited resources:

> The procedure was to select small areas of 100–500 ac typifying representative landscape with characteristic topography, vegetation, microrelief, and surface drainage, and to carry out detailed surveys on them. From these was defined the array of soil types likely to be encountered in similar areas in the whole survey project and thus the common associations of soil types were formulated. These could be recognized in the landscape along the lines of traverse and the whole area could be mapped rapidly, in fact at about ten times the speed of the earlier detailed surveys.[5] (p. 16)

Australian soil surveyors used the term *association* rather than *catena*, as the later term had been redefined by the U.S. survey,[86] but the sense was very much the same. It was along these lines that in 1950 Kellogg proposed that tropical soils be mapped as associations.[93]

In 1946, approximately in parallel with the development of soil association mapping in the Australian soil survey, the first "land systems" survey was conducted by Christian and Stewart.[94] These surveys involved teams of scientists, usually including a pedologist, geologist, and botanist/ecologist, and set out to map the combined geological, soil, and vegetation landscape relationships (land systems) over large remote areas. The region surveyed was broken into smaller units with relatively consistent soil–geology–topography–vegetation relation-

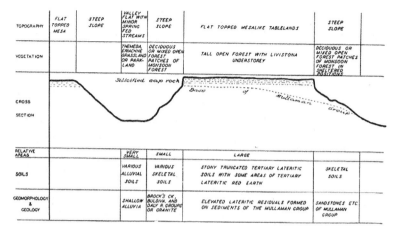

FIGURE 3.11 Mullaman land systems diagram from the survey of Katherin-Darwin in Australia. (Reprinted from Christian, C.S. and Stewart, G.A., *General Report on Survey of Katherin-Darwin Region, 1946*, Land Research Series, No. 1, CSIRO, Melbourne, 1953.)

ships, and a schematic diagram constructed to illustrate the relevant relationships for each of these land system units (Figure 3.11). These land systems diagrams were clearly analogous to Milne's catena diagram (Figure 3.6), though the objective was to map all landscape components (soils, vegetation, and geology), not just use vegetation, geology, and topography to map soils.

3.5 LINNAEUS, DARWIN, AND MARBUT'S SOIL ANATOMY

Biology was a new, ascendant, and rapidly growing field in the late 19th and early 20th centuries. Consequently, biological metaphors based on an evolutionary or life cycle concept of development were foundational in a number of 19th- and early-20th-century sciences, including Spencer's 19th-century sociology,[95,96] Clementian ecology,[97] and the geographic cycle of William Morris Davis.[98] As was discussed earlier, Marbut was inspired by Glinka's climatic, zonal soil geography. But Marbut also formulated a new concept in soil studies, *soil anatomy*,[18] built on an explicit biological metaphor.[99,100] In contrast to 19th-century geological surveying, Humboldtian science, or Milne's catena concept, this metaphor lacked an explicit geographic dimension.

3.5.1 THE ROOTS OF LATE-19TH-CENTURY MORPHOLOGY

As a boy growing up in rural Missouri, Curtis Marbut carried around *Gray's Manual*, a popular field guide for plant identification and classification.[101]* As

---

* Asa Gray was a prominent 19th-century Harvard botanist who played a central role in the development of plant taxonomy for the U.S.

with most natural historians of his era, Marbut was trained in a hierarchical taxonomy for flora and fauna, which descended from the natural system of 18th-century Swedish scientist Carl von Linné (1707–1778), or Linnaeus as he is known in English-speaking countries. Linnaeus published the first edition of his biological taxonomy *Systema Naturae* in 1735, in many ways defining the 18th-century project of collecting, naming, and classifying the natural world. Linnaeus sought not just to find a convenient and utilitarian classification system, but believed that a scientific classification system should represent a fundament order in nature. Taxonomy was, in his view, the highest calling of a scientist and revealed the very essence of nature.[102,103] Through the 18th and early 19th centuries, collectors were furiously finding, describing, naming, and ordering the natural world, but in the late 18th century interest turned to *comparative anatomy*, with the goal of establishing a more fundamental basis for a natural classification. In the 19th century attention turned from the examination of mature organisms to embryology and the study of life cycle development.[103,104]

In the late 19th century, what was then known as *morphology* (the study of comparative anatomy and embryology) shifted to encompass the evolution project proposed by Darwin's *Origin of Species*, first published in 1859. Morphologists had earlier pursued the idea that lower organisms in the order of life could be discerned in the early life stages of higher organisms. With a surge of interest in developing an evolution-based taxonomy, morphologists began comparative studies of embryonic development in the belief that the evolutionary history of organisms could be determined from their contemporary life cycle development — ontogeny recapitulates phylogeny (Figure 3.12).[103,104] Though the "ontogeny recapitulates phylogeny" theory has since been discredited, the relevant point for this discussion is that at the turn of the 20th century, biological taxonomy, anatomy, embryology, and evolution were inextricably linked in the field of morphology.

3.5.2 MARBUT'S SOIL ANATOMY

> Soil surveys have created a new branch of soil science — soil anatomy.
>
> — **Curtis Fletcher Marbut, 1921**[18] **(p. 141)**

NCSS staff scientist George Coffey, who studied and published on the genetic classification approach of Sibirtsev, was emphatic that the ideal classification should be based upon "differences in the soil itself" that *result* from the effects of climate and geology — not based directly on either climate (as he viewed the Russian system) or parent rock (the early U.S. system)[105] (p. 34). By 1916, Marbut[106] had adopted this ideal as his own, a mantra repeated through all of his later writings.[18,32,45,107] In practice, it has proved difficult to classify soils without reference to climate both in Marbut's time[4] and today,[108] but an involved discussion of soil classification lies beyond the scope of this chapter.

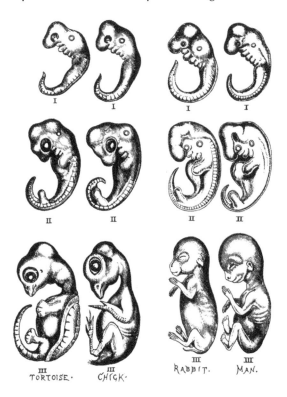

FIGURE 3.12 Biological morphology served as a metaphor for Curtis Marbut's 20th-century soil anatomy concept. In the late 19th century, Ernst Heinrich Haeckel argued that ontogeny, the development of the individual, recapitulated phylogeny, the evolutionary history of the species. Morphologists studied embryonic development as a tool for reconstructing evolutionary relationships. In the late 19th and early 20th centuries, biological metaphors were applied to many different disciplines, including sociology, geomorphology, ecology, and pedology. (George John Romanes, after Ernst Heinrich Haeckel, 1892, from Coleman, W., *Biology in the Nineteenth Century: Problems of Form, Function, and Transformation*, Cambridge University Press, Cambridge, U.K., 1977. With permission.)

To construct a scientific framework for the study and classification of the soil itself, Marbut turned to biology. Profile horizons were equivalent to animal limbs, and changes in the profile over time were another representation of the life cycle.[18,107] To understand Marbut's soil anatomy, we need only look at the diagram in Figure 3.12 and imagine different soil profiles in stages of progressive development replacing the tortoise, chick, rabbit, and man. As with biological morphology, Marbut's soil anatomy inextricably linked profile description, development studies, and classification. Following Linnaeus, Marbut insisted that the development of a hierarchical soil taxonomy was "fundamentally scientific in nature" and the highest calling for the survey[4,32] (p. 2). Consequently, from 1920 until his death in 1935, Marbut's single-minded objective was to develop a soil classification system comparable to biological taxonomy.[4]

Curtis Fletcher Marbut was a product of his time, a time in which biological metaphors were applied in a variety of fields. As a consequence of this choice of metaphors, however, Marbut came to view soil surveyors as the equivalent of 18th-century biological collectors sent out to the corners of the Earth to find, describe, and bring back samples to name and order in a natural taxonomic system. At the 1923 American Association of Soil Survey Workers meeting, the geographer P.S. Lovejoy discussed various utilitarian and theoretical justifications for soil survey work, to which Marbut responded:

> [Mr. Lovejoy] begins by stating that one point of view "regards soil surveys as primarily intended to do for soils what Gray, for instance, did for our native plants, etc." We who are doing the soil survey work wonder why he says that is one point of view, implying that there are others. We know that to be the point of view.[109] (p. 59)

The construction of maps and development of land use interpretations were to Marbut utilitarian and therefore secondary objectives for the survey.[4]

3.5.3 Marbut's Reign, 1913–1933

> There was a tendency to consider the field man as a mere mechanic who mapped the soil units as he encountered them, while the more interesting problems of soil development and of the classification and correlation of soils were left to the inspectors and the supervising and administrative officers in Washington.
>
> **— Macy Lapham, reflecting on soil survey work under Marbut, 1945[110] (p. 349)**

> Soil Survey — apparently intended to imply a 50-50 balance between the Soil and the Survey — It is about like Mutt's rabbit hash — 50-50 mixture of rabbit and horse — 1 horse and one rabbit. In our Association the soil study has grown to the size of a ton horse and the survey has shrunken to the size of a small cotton-tail.
>
> **— Thomas Bushnell, 1929[111] (p. 23)**

The quotations above from field surveyor Macy Lapham and the leader of the Indiana survey, Thomas Bushnell, clearly illustrate the decline in status of U.S. soil mapping in the 1920s and 1930s. Marbut, who led the U.S. soil survey program from 1913 to 1933, was cognizant of the principles of environmental correlation and even suggested that the development of an improved taxonomy system would facilitate the study of such correlations.[18] But he devoted little attention to this problem because mapping simply was not important within his soil anatomy paradigm.[4,33] In the early 1930s, survey staff pushing (unsuccessfully) for funds to begin using aerial photography complained that "the fundamental concern of the Soil Survey is the field study and classification of soils," despite the fact that the agency was obligated to produce soil maps. While most of the NCSS personnel (particularly at the state level) were engaged in mapping, the survey leadership assigned the highest priority and intellectual prestige to profile studies and classification. After Marbut died in 1935 and Charles Kellogg was put in charge of the soil survey,* mapping and interpretations began to receive more attention from Washington.[4] However, for the 20 years from 1915 to 1935 there were virtually no advances in soil mapping within the U.S. soil survey program.[4]

Marbut's impact on the intellectual development of soil studies in the U.S. lasted far beyond his reign as soil survey leader. In 1936, the Soil Science Society of America (SSSA) was created by merging the American Soil Survey Association into the Soils Section of the American Society of Agronomy. Section V of this new scientific organization was named Soil Genesis, Morphology, and Cartography, but the focus was described as follows:

> Studies in which the soil is considered primarily as an individual entity, to be dissected and classified on the basis of its inherent characteristics shall be presented before Section V.[112] (p. 506)

Mapping is not mentioned in this description, and the use of the word *dissected* in this formal subdisciplinary definition clearly reflects Marbut's biological metaphor. In 1951, Section V was renamed "Soil Genesis, Morphology, and Classification,"[113] which remained the section name until 1994, when Pedology was adopted.[114] Following Marbut, the *intellectual* focus of U.S. pedology for the better part of a century has been the dissection and classification of soil profiles — modeled on 19th-century biological morphology. The application of this metaphor has led to the development of a complex, hierarchical taxonomy analogous to biological taxonomy.[108] However, a negative consequence of this focus has been a relative lack of interest in the scientific advancement of soil-landscape modeling.

3.5.4 THREE-DIMENSIONAL SOIL BODIES

Beginning in the late 1920s and particularly in the 1930s to 1950s, a number of U.S. soil survey workers extended Marbut's soil anatomy metaphor to soil mapping with the theoretical construction of a three-dimensional soil body. The kernel of the three-dimensional soil body was first proposed during Marbut's tenure when Bushnell[80] (p. 158) argued:

> An area of any soil type is a body of three dimensions and as such can not be completely defined on the basis of a two dimensional vertical profile. The vertical profile does reveal most soil characters because, by definition of a type, the layers are uniform in horizontal directions, but it fails to show the surface form, or topography which may well be regarded as a true soil character, in even the most scientific classifications.

---

* Kellogg was acting chief from 1933 to 1935, but it is unclear to what extent he was free to implement change with Marbut looking over his shoulder.

By Marbut's definition, soils had to be classified according to their characteristics — not external factors such as climate, geology, or topography. With an ingenious twist of logic, Bushnell proposed that topography be considered a *soil* characteristic so that classification might better address local variability. Soil surveyors, Bushnell argued, mapped variations in topography and landscape position as much as changes in profile horizons, due to the practical limitations on the ability to sample profiles. In considering topography a soil characteristic, Bushnell reconciled Marbut's soil classification theory with the actual practices of soil mapping. While Bushnell employed this idea for his Indiana soil keys,[85] it failed to gain traction in the survey as a whole as long as Marbut remained in charge.

Kellogg, in his 1937 *Soil Survey Manual*, embraced the idea that soils were three-dimensional bodies with both "internal features," like horizons, and "external features," such as topography and vegetation[52] (p. 101). For taxonomy purposes, these bodies were eventually termed pedons and polypedons.[115] Francis Hole, the influential pedologist and geographer at the University of Wisconsin, extended the three-dimensional soil body concept still further with his approach to soil-landscape analysis.[116] In 1953, Hole[117] first proposed that external soil features like slope, landscape position, drainage, etc., be described in a manner analogous to the description of profile features (Figure 3.13). Later, he coined the term *soilscape* to describe three-dimensional soil-landscape bodies that occupied specific habitats, much like biological organisms.[116]

While the three-dimensional soil bodies of Bushnell, Kellogg, and Hole might look superficially like Milne's catena concept[60] or Dokuchaev's soil geography,[42] there are two fundamental differences: (1) pedons, polypedons, and soilscapes are fundamentally discrete, whereas Dokuchaev and Milne saw soil-landscapes as spatially continuous; and (2) Milne and Dokuchaev viewed environmental factors as *separate from* but *correlated with* soil properties, whereas NCSS workers saw slope, vegetation, etc., as *part of* soil units to be described. The differences between these scientific traditions are subtle but important.

3.6 THE INTRODUCTION OF AERIAL PHOTOGRAPHY, 1927–1952

Aerial reconnaissance in World War I (1914–1919) led to the development of aerial photography, and soon after the USGS began using and refining this new tool for topographic survey work.[118] The first account of the use of air photos and aerial observation for soil survey came in the early 1920s,[119] but it was not until the late 1920s that Bushnell and NCSS scientist Mark Baldwin began systematic experiments on the use of aerial photography for soil surveys. After working with only a few scattered air photos obtained at minimal cost from other agencies, Bushnell[111] (p. 28) proclaimed: "I am unwilling to map another square mile without the aid of aerial photos." He reported running a small experiment, with two different experienced surveyors covering the same area with alidade plane

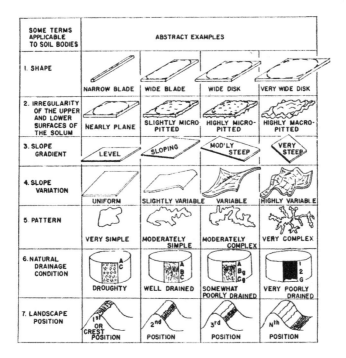

FIGURE 3.13 Francis Hole's scheme for describing soil bodies as three-dimensional landscape entities. Note the similarity of this ordinal approach and contemporary schemes for describing soil profile features like structure and acidity. (From Hole, F.D., *Soil Sci. Soc. Am. Proc.*, 18, 131–135, 1953. With permission.)

table traverses and producing two radically different maps. Soil maps produced with air photos, Bushnell argued, offered both significantly better spatial precision and time savings, as there was no need to construct a base map. In 1930, Jennings County was completely mapped using commercially obtained air photos, with four other counties completed by 1931.[120] The state of Michigan had also begun to use aerial photography, and reported the first use of a mirror stereoscope for terrain mapping.[121] Despite the pleadings of Bushnell and many others,[122] despite a detailed cost analysis in Indiana demonstrating that reduced field costs more than balanced photo purchase expenses,[123] and despite obvious benefits in spatial accuracy, the NCSS as a whole did not adopt this technique until around 1935, when inexpensive photographs became available from the new Soil Conservation Service, which used aerial photography to map soil erosion and land use.[4,110,123–126] Even as late as 1937, the use of aerial photography was considered a secondary technique with plane tabling still the basic mapping approach.[52]

Initially, aerial photography was used simply as a direct substitute for the plane table survey — a base map and field locator.[111,120] As early as 1932, however, soil surveyors in Michigan began using air photos to identify soil external features like vegetation, surface reflectance (for bare soils in agricultural areas), and, using a mirror stereoscope, topography. Soil boundaries were thus identified and slope phases determined over particularly difficult, heavily vegetated terrain where ground transect surveys would have been arduous.[121] The practice of using air photo *interpretation* for rugged areas continued into the early 1950s.

> In rough or heavily wooded country, the field worker has great difficulty in observing soil boundaries throughout their course. In such areas, the soils are identified, and their boundaries that coincide with land-form boundaries, drainage lines, or vegetation pattern are projected from the line of traverse as far as they can be seen and beyond that left hanging. Then in the office the stereoscope is used to connect the boundaries not connected in the field. Thus the scientist is able to "observe these soil boundaries throughout their course" without having actually seen them on the ground.[127] (p. 741)

According to official NCSS policy into the early 1950s, in situations where changes in vegetation, surface soil, and topography *could* be observed on the ground, they *were* to be observed on the ground. Only in rough or densely vegetated terrain or in reconnaissance surveys was air photo interpretation to be relied upon.[88,128,129] Every single map unit identified was to be sampled and characterized in the field using an auger or soil pit. Despite the introduction of aerial photography in 1935 for base map compilation, the person-days required to map a square mile did not decline, though maps became considerably more detailed.[91] In contrast, the Australian Land Systems Surveys of the 1940s employed air photo interpretation of topography and vegetation to systematically and extensively identify and map soil–vegetation–geology associations. Land systems surveyors were the first to map extensively by interpreting changed in topography and vegetation using stereo aerial photography, with field transects employed only to ground truth photos.[94] Mapping by means of extensive stereo air photo interpretation is now standard practice in the U.S. soil survey program.[130]

3.7 SUMMARY AND CONTEMPORARY RELEVANCE

3.7.1 HISTORICAL SUMMARY

Nineteenth-century geologists and geological surveyors were the first soil-landscape modelers. Within this tradition, maps were delineated according to geologic formation, lithology, and type of surficial deposit. Soil information was usually provided in an accompanying map report, though sometimes independent soil maps were constructed based upon the geologic map. At the start of the 19th century, most geologists believed that soil properties at the Earth's surface directly correlated with the rock or deposits from which they formed. Geologists did, however, examine soils in both the field and lab in the course of their survey work, leading to a more sophisticated understanding of contingent soil–geology relationships by the end of the century.

To the geological tradition of soil-landscape modeling, the late-19th-century naturalist V.V. Dokuchaev added the climatic plant geography of Alexander von Humboldt, creating a new field of soil geologic-geographic investigations that we now know as the five factors of soil formation: the geologic/physiographic factors of parent rock, topography, and land surface age together with the geographic factors of climate and vegetation. In the early 20th century, Dokuchaev's disciple Glinka and the U.S. soil survey leader Marbut emphasized the geographic-climatic dimension to the neglect of geology and physiography, and soil-landscape modeling became dominated by what was known as the Russian zonal concept.

Though topography had been used by surveyors from the late 1800s forward, an unknown soil chemist working in East Africa, Geoffrey Milne, asserted a more prominent theoretical role for this factor of soil formation with the publication in 1935 of his catena concept. The catena concept posited that topography-associated changes in soil properties and formation cut across zonal soil groups. The catena was therefore proposed as a compound map unit that represented repeating soil-topography relationships for an area rather than a dominant soil type. Milne also discussed the role of geology, hillslope hydrology, and mass transport in shaping soil-topography patterns. In the U.S., however, soil surveyors used the catena term to label an existing soil *association* concept — a field classification of soils in one climate zone on uniform parent material with differing drainage, what one soil scientist termed a *microcatena*[56] (p. 42). The association term was conversely employed in the sense of Milne's catena — though without the explicit formation processes. Despite this confusion in the soil survey community, the original catena concept was resurrected by Robert Ruhe in 1961,[74] and in the late 20th century has been cited as the fundamental theory for both quantitative soil-landscape modeling[67] and soil geomorphology.[66]

In 1920, Curtis Marbut introduced a new concept for soil surveys, what he called soil anatomy[18] (p. 141). Marbut's soil anatomy was explicitly modeled on biology, in particular the 19th-century biological subdiscipline of morphology, which encompassed comparative anatomy, embryology, evolution, and taxonomy. According to Marbut, the primary objective for the U.S. soil survey program was *not* to produce soil maps, but rather to construct a hierarchical, natural (or genetic) classification comparable to biological taxonomy. Unlike the scientific paradigms and traditions discussed previously, soil anatomy as originally formulated lacked a strong spatial or geographic dimension. After Marbut, U.S. soil scientists extended soil anatomy with the concept of soils as discrete, three-dimensional bodies with internal (horizons) and external (slope and vegetation) features.

Pedologists like to state that they study "soils as natural bodies"[131] (p. 3), with the belief that this concept embraces Dokuchaev, Glinka, Marbut, and Jenny.[9] As the historical analysis presented in this chapter demonstrates, there are fundamental differences between Milne's, Marbut's, Glinka's, and Dokuchaev's natural body concepts, differences that do not indicate the progressive development of a central idea. These different soil concepts, though not always made explicit, continue to shape questions, methods, and priorities in contemporary soil studies and soil-landscape modeling.

3.7.2 Contemporary Relevance

> The great new thing in professional science in the first half of the 19th century was Humboldtian science, the accurate, measured study of widespread but inter-connected real phenomena in order to find a definite law and a dynamical cause.[40] (p. 105)

The core concept in contemporary soil-landscape modeling — what we now call environmental correlation[132] — can be traced back to the early-19th-century quantitative plant geography and climatology of Alexander von Humboldt. Dokuchaev applied this concept to soils and extended this correlative approach to include not just climate and vegetation, but also geologic and physiographic factors — the combined environmental factors that are most commonly employed today. In many ways, Geoffrey Milne reclaimed the geology-geography environmental correlation of Dokuchaev that had been lost with the climatic focus of the early 20th century, but he also shifted the theoretical focus of soil-landscape modeling from continental-scale problems to hillslope-scale variability — the theoretical basis for contemporary digital soil-terrain modeling.[67] The hydrologic process dimension of Milne's catena concept also provides a link to contemporary process geomorphology in the Gilbert tradition.[99] Contemporary soil-landscape modeling, which grew out of a rich mapping tradition in geologic surveying, could benefit from a closer relationship to academic geologists and 21st-century geological surveys. There are many ideas, paradigms, and traditions from the past that still have great relevance for cutting-edge soil-landscape modeling today.

The development of GIS tools has undoubtedly played a major role in the revival of soil-landscape modeling, but it would be wrong to conclude that the relative lack of interest in prior years was due solely to the lack of spatial databases and modern computing power. Humboldt was quantifying climate-vegetation relationships in the early 19th century. Dokuchaev produced a regional, quantitative map of soil humus content in 1883.[42] Hilgard[21,46] studied soil-vegetation and soil-climate correlations in the late 19th century. Milne examined soil-topography relationships in pre-WWII East Africa lacking even the aid of reasonable topographic maps. With the ready availability of stereo aerial photography from the mid-1930s forward and relatively generous support for soil surveys from the 1930s to the 1970s, there has been ample opportunity for soil scientists to expand and refine soil-landscape modeling theories and techniques. However, academics and survey leaders have often been content to let field surveyors work out soil-landscape models on their own.[8] This is the root cause behind the general lack of explicit, published, and tested soil-landscape models. Prior to the introduction of GIS, soil-landscape modeling was not a high academic priority.

While 19th-century geological surveys, Humboldt's plant geography, Dokuchaev's factorial model, and Milne's catena concept have all contributed to the development of modern soil-landscape modeling, Marbut's soil anatomy has undoubtedly hindered the development of this science. Under Marbut's leadership, priority and prestige were given to the development of soil classification and related soil profile genesis studies. There was a short reversal of this imbalance in the 1930s with Marbut's passing, competition from a rival Soil Conservation Service mapping project, and the demands of New Deal social welfare programs.[4] However, from the 1950s forward, the preoccupation of the U.S. soil survey program and U.S. pedology community generally has been the construction and ongoing revision of what we now know as *Soil Taxonomy*.[108] Not surprisingly, U.S. soil surveying did not change appreciably from the 1950s to the end of the 20th century. Under the leadership of Robert Ruhe, a series of soil-geomorphology investigations were initiated in 1953, but the results of these studies have never been fully integrated into soil mapping.[6] Developments in quantitative soil-landscape modeling have been led overwhelmingly by European and Australian scientists.[133–135]

Perhaps the most important lesson to take from the history of soil-landscape modeling is that the most important ideas have come from outside the scientific mainstream. Both V.V. Dokuchaev in Russia and E.W. Hilgard in California were far removed from the centers of late-19th-century geologic science, and Geoffrey Milne was even more isolated from soil survey developments in pre-WW II Tanganyika. Environmental correlation was an idea borrowed from plant geography, not soil investigations. And Milne's hillslope processes were more commonly discussed in the context of early-20th-century land management than in the soil formation literature. The U.S. soil survey began using aerial photography almost 15 years after the USGS, 5 years after successful experimentation in Indiana, and only then due to the availability of inexpensive photos from the rival Soil Conservation Service. More recently, U.S. geologist Robert Ruhe and Australian scientists like Bruce Butler were largely responsible for drawing attention to the importance of Quaternary geology and geomorphology in soil-landscape modeling.[6,136] To the extent that history can be used as a guide for the future, 21st-century developments in soil-landscape modeling might well spring from the far corners of the globe and scientific fields other than soil science and pedology.

REFERENCES

1. RW Simonson. Early teaching in USA of Dokuchaiev factors of soil formation. *Soil Sci Soc Am J* 61: 11–16, 1997.
2. RW Simonson. *Historical Aspects of Soil Survey and Soil Classification*. Soil Science Society of America, Madison, WI, 1987.
3. RW Simonson. The U.S. soil survey: contributions to soil science and its application. *Geoderma* 48: 1–16, 1991.
4. DR Gardner. *The National Cooperative Soil Survey of the United States*. USDA, NRCS, Res. Econ. Soc. Sci. Div. Soil Surv. Div., Washington, DC, 1998.
5. JK Taylor. *The Development of Soil Survey and Field Pedology in Australia*, 1927–1967. CSIRO, Melbourne, Australia, 1970.
6. ABW Effland, WR Effland. Soil geomorphology studies in the United States soil survey program. *Agric Hist* 66: 189–212, 1992.

7. BD Hudson. The soil survey as paradigm-based science. *Soil Sci Soc Am J* 56: 836–841, 1992.
8. RW Simonson. Historical aspects of soil survey and soil classification. Part VII. 1961–1970. *Soil Surv Horiz* 28: 77–84, 1986.
9. JP Tandarich. The intellectual background for the factors of soil formation. In *Factors of Soil Formation: A Fiftieth Anniversary Retrospective*, SSSA Special Publication 33, R Amundson, J Harden, M Singer, Eds. SSSA, Madison, WI, 1994, pp. 1–13.
10. JP Tandarich. The development of pedologic thought: some people involved. *Phys Geogr* 9: 162–174, 1988.
11. JM Eyles. William Smith: some aspects of his life and work. In *Toward a History of Geology*, CJ Schneer, Ed. MIT Press, Cambridge, MA, 1969, pp. 142–158.
12. S Winchester. *The Map That Changed the World: William Smith and the Birth of Modern Geology*. Harper Collins, New York, 2001.
13. CJ Schneer. William "Strata" Smith on the Web, 2004. Available at http://www.unh.edu/esci/wmsmith.html.
14. JP Tandarich. Wisconsin agricultural geologists: ahead of their time. *Geosci Wis* 18: 21–26, 2001.
15. NM Sibirtsev. *Selected Works*, Vol. 1, N Kaner, Trans. Israel Program for Scientific Translations, Jerusalem, 1966.
16. GN Coffey. The development of soil survey work in the United States with a brief reference to foreign countries. *Proc Am Soc Agron* 3: 115–129, 1911.
17. RW Simonson. Historical aspects of soil survey and soil classification. Part I. 1899–1910. *Soil Surv Horiz* 27: 3–11, 1986.
18. CF Marbut. The contribution of soil survey to soil science. *Soc Promot Agric Sci Proc* 41: 116–142, 1921.
19. IC Russell. *Subaerial Decay of Rocks*, Bulletin 52. USGS, Washington, DC, 1889.
20. GP Merrill. *A Treatise on Rocks, Rock-Weathering and Soils*. The Macmillan Co., New York, 1904.
21. EW Hilgard. *A Report on the Relations of Soil to Climate*, Bulletin 3. USDA Weather Bureau, Washington, DC, 1892.
22. RT Hill. *Present Knowledge of the Geology of Texas*, Bulletin 45. USGS, Washinton, DC, 1887.
23. JS Flett. *The First Hundred Years of the Geological Survey of Great Britain*. HMSO, London, 1937.
24. DD Owen. *Report of a Geological Survey of Wisconsin, Iowa and Minnesota; and Incidentally of a Portion of Nebraska Territory*. Lippincott, Grambo, Philadelphia, 1852.
25. JW Barnes. *Basic Geological Mapping*. The Open University Press, Milton Keyes, U.K., 1981.
26. FH Lahee. *Field Geology*. Hill Publishing Co., London, 1916.
27. J Geikie. *Structural and Field Geology*, 4th ed. D. Van Nostrand, New York, 1920.
28. MH Lapham. *Crisscross Trails*. Willis E. Berg, Berkeley, CA, 1949.
29. HH Krusekopf, Ed. *Life and Work of C.F. Marbut*. Soil Science Society of America, Columbia, MO, 1942.
30. TD Rice. C.F. Marbut. In *Life and Work of C.F. Marbut*, HH Krusekopf, Ed. Soil Science Society of America, Columbia, MO, 1942, pp. 46–48.
31. CF Marbut, HH Bennett, JE Lapham, MH Lapham. *Soils of the United States*, Bulletin 96. USDA Bureau of Soils, Washington, DC, 1913.

32. CF Marbut. A scheme for soil classification. In *Proceedings of the 1st International Congress of Soil Science, June 13–22, 1927*, Vol. 4. American Organizing Committee, Washington, DC, 1928, pp. 1–31.

33. RS Smith. Dr. C.F. Marbut's contribution to soil survey. In *Life and Work of C.F. Marbut*, HH Krusekopf, Ed. Soil Science Society of America, Columbia, MO, 1942, pp. 50–53.

34. R Amundson, DH Yaalon. E.W. Hilgard and John Wesley Powell: efforts for a joint agricultural and geological survey. *Soil Sci Soc Am J* 59: 4–13, 1995.

35. AD Hall, EJ Russell. Soil surveys and soil analysis. *J Agric Sci* 4: 182–223, 1911.

36. NM Tulaikoff. The genetic classification of soils. *J Agric Sci* 3: 80–85, 1909.

37. LK Nyhart. Teaching community via biology in late-nineteenth-century Germany. *Osiris* 17: 141–170, 2002.

38. L Kellner. *Alexander von Humboldt*. Oxford University Press, London, 1963.

39. M Nicolson. Alexander von Humboldt, Humboldtian science and the origins of the study of vegetation. *Hist Sci* 25: 167–194, 1987.

40. SF Cannon. *Science in Culture: The Early Victorian Period*. Science History Publications, New York, 1978.

41. TL Hankins. Blood, dirt and nomograms. *Isis* 90: 50–80, 1999.

42. VV Dokuchaev. *Russian Chernozem*, Vol. 1, N Kaner, Trans. Israel Program for Scientific Translations, Jerusalem, 1967.

43. KD Glinka. *Treatise on Soil Science* (Pochvovedenie, original title), 4th ed., A Gourevitch, Trans. Israel Program for Scientific Translations, Jerusalem, 1963.

44. KD Glinka. *The Great Soil Groups of the World and Their Development*, CF Marbut, Trans. Edwards Brothers, Ann Arbor, MI, 1927.

45. CF Marbut. *Soils: Their Genesis and Classification*, 2nd ed. Soil Science Society of America, Madison, WI, 1951.

46. EW Hilgard. Soil studies and soil maps. *Overland Mon* 18: 607–616, 1891.

47. FE Bonsteel, WT Carter, OL Ayers. *Soil Survey of the Syracuse Area, New York*, Series 1903. USDA Bureau of Soils, Washington, DC, 1904.

48. USDA Bureau of Soils. *Soil Survey Field Book*. USDA, Washington, DC, 1906.

49. WJ Gieb. *The Value of the Soil Survey*. American Association of Soil Survey Workers, Madison, WI, 1922.

50. EL Worthen. Methods of soil surveying. *Proc Am Soc Agron* 1: 185–191, 1909.

51. USDA Bureau of Soils. *Instructions to Field Parties*. USDA, Washington, DC, 1914.

52. CE Kellogg. *Soil Survey Manual*, Miscellaneous Publication 274. USDA, Washington, DC, 1937.

53. GR Clarke. *The Study of Soil in the Field*. Clarendon Press, Oxford, 1936.

54. M Baldwin, CE Kellogg, J Thorp. Soil classification. In *Soils and Men*, USDA Committee on Soils, Ed. U.S. Government Printing Office, Washington, DC, 1938, pp. 979–1001.

55. JA Ewell, HL Dean, F Rudolph, EW Tigges. *Soil Survey of Cerro Gordo County, Iowa*, Series 1935, No. 13. USDA Bureau of Plant Industry, Washington, DC, 1940.

56. J Thorp. The influence of environment on soil formation. *Soil Sci Soc Am Proc* 6: 39–46, 1941.

57. H Jenny. *Factors of Soil Formation*. McGraw-Hill Book Co., New York, 1941.

58. G Milne, Ed. *Proceedings of a Conference of East African Soil Chemists, Amani, Tanganyika Territory, May 21–26, 1932*. Technical Conferences of the East African Dependencies, Government Printer, Nairobi, Kenya, 1932.

59. K Milne, G. Milne. In *Geographers: Biobibliographical Studies*, Vol. 2, TW Freeman, P Pinchemel, Eds. Mansell, London, 1978, pp. 89–92.

60. G Milne. Some suggested units of classification and mapping, particularly for East African soils. *Soil Res* 4: 183–198, 1935.

61. G Milne. Composite units for the mapping of complex soil associations. In *Transactions of the Third International Congress of Soil Science, Oxford, England*, Vol. 1. T. Murby & Co., London, 1935, pp. 345–347.

62. G Milne. Normal erosion as a factor in soil profile development. *Nature* 138: 548–549, 1936.

63. G Milne, VA Beckley, GH Gethin Jones, WS Martin, G Griffith. *A Provisional Soil Map of East Africa (Kenya, Uganda, Tanganyika and Zanzibar) with Explanatory Memoir*, Amani Memoir 31. East African Agricultural Research Station, Amani, Tangayika, 1936.

64. TM Bushnell. Some aspects of the soil catena concept. *Soil Sci Soc Am Proc* 7: 466–476, 1942.

65. TM Bushnell. The catena caldron. *Soil Sci Soc Am Proc* 10: 335–340, 1945.

66. GF Hall, CG Olson. Predicting variability of soils from landscape models. In *Spatial Variabilities of Soils and Landforms*, Vol. 28, MJ Mausbach, LP Wilding, Eds. Soil Science Society of America, Madison, WI, 1991, pp. 9–23.

67. ID Moore, PE Gessler, GA Nielsen, GA Peterson. Soil attribute prediction using terrain analysis. *Soil Sci Soc Am J* 57: 443–452, 1993.

68. A Young. *Tropical Soils and Soil Survey*. Cambridge Press, Cambridge, U.K., 1976.

69. PW Birkeland. *Soils and Geomorphology*, 3rd ed. Oxford University Press, New York, 1999.

70. M Sommer, E Schlichting. Archetypes of catenas in respect to matter: a concept for structuring and grouping catenas. *Geoderma* 76: 1–33, 1997.

71. AE Scheidegger. The catena principle in geomorphology. *Z Geomorph* 30: 257–273, 1986.

72. Geoffrey Milne to the East African Directors of Agriculture, Imperial Bureau of Soils, April 24, 1930. MSS.Brit.Emp.s.457, File 14, Items 11–14. Bodleian Library of Commonwealth and African Studies at Rhodes House, Oxford.

73. W.S. Martin to Geoffrey Milne, April 1933. MSS.Brit.Emp.s.457, File 14, Items 32–33. Bodleian Library of Commonwealth and African Studies at Rhodes House, Oxford.

74. RV Ruhe. Elements of the soil landscape. *Trans 7th Int Congr Soil Sci* 4: 165–169, 1961.

75. RV Ruhe. Review of "Pedology, Weathering and Geomorphological Research," by P.W. Birkeland. *Geoderma* 14: 176–177, 1975.

76. G Milne. A soil reconnaissance journey through parts of Tanganyika Territory. *J Ecol* 35: 192–265, 1947.

77. AJ Conacher, JB Dalrymple. The nine-unit landsurface model: an approach to pedogeomorphic research. *Geoderma* 18: 1–154, 1977.

78. WE Calton. The use of the catena in Tanganyika Territory. *Soils Fert* 15: 369–371, 1952.

79. RJ Huggett. Soil landscape systems: model of soil genesis. *Geoderma* 13: 1–22, 1975.

80. TM Bushnell. To what extent should location, topography or physiography constitute a basis for differentiating soil into units or groups. In *Proceedings of the 1st International Congress of Soil Science, June 13–22, 1927*, Vol. 4. American Organizing Committee, Washington, DC, 1928, pp. 158–163.

81. GW Robinson. *Soils, Their Origin, Constitution, and Classification*, 1st ed. Thomas Murby & Co., London, 1932.

82. AN Gennadiyev, KR Olson. Pedological cooperation between Russia and the USA, past to present. *Soil Sci Soc Am J* 62: 1153–1161, 1998.

83. JH Ellis. A field classification of soils for use in the soil survey. *Sci Agric* 12: 338–345, 1932.

84. G Milne. The Extent to Which Volcanic Rocks Give Rise to Distinctive Soils, Examined from East African Occurances, 1934. Typescript, MSS.Brit.Emp.s.457, File 13, Items 13–51. Bodleian Library of Commonwealth and African Studies at Rhodes House, Oxford.

85. TM Bushnell. Taxonomic considerations in soil correlation. *Bull Am Soil Survey Assoc* XV: 110–114, 1934.

86. G ap Griffith. Catena. *Soils Fert* 15: 169–170, 1952.

87. E Winters. Interpretative soil classifications: genetic groupings. *Soil Sci* 67: 131–139, 1949.

88. Soil Survey Staff. *Soil Survey Manual*. USDA, Washington, DC, 1951.

89. Soil Survey Division. Soils of the United States. In *Soils and Men*, USDA Committee on Soils, Ed. U.S. Government Printing Office, Washington, DC, 1938, pp. 1019–1161.

90. ME Austin, MH Galltin, JN Odom, F Rudolph, ME Swann, SR Bacon, MG Cline, AJ Vessel. *Soil Survey of Claiborne County, Tennessee*, Series 1939, No. 5. USDA-BPISAE Division of Soil Survey, Washington, DC, 1948.

91. RW Simonson. Lessons from the first half century of soil survey. II. Mapping of soils. *Soil Sci* 74: 323–330, 1952.

92. WH Scholtes, GD Smith, FF Riecken. *Taylor County, Iowa, Soils*, Soil Survey Series 1947, No. 1. USDA-SCS, Washington, DC, 1954.

93. CE Kellogg. World food production: the role of the photogrammetrist. *Photogramm Eng* 16: 94–100, 1950.

94. CS Christian, GA Stewart. *General Report on Survey of Katherin-Darwin Region, 1946*, Land Research Series, No. 1. CSIRO, Melbourne, 1953.

95. H Spencer. The social organism. *Westminster Rev* 73: 51–68, 1860.

96. VL Hilts. Towards the social organism: Herbert Spencer and William B. Carpenter on the analogical method. In *The Natural Sciences and the Social Sciences*, IB Cohen, Ed. Kluwer Academic Publishers, Dordrecht, The Netherlands, 1994, pp. 275–303.

97. FE Clements. *Plant Succession: An Analysis of the Development of Vegetation*. Carnegie Institution of Washington, Washington, DC, 1916.

98. WM Davis. *Geographical Essays*. Ginn & Co., Boston, 1909.

99. CE Thorn. *An Introduction to Theoretical Geomorphology*. Unwin Hyman, Boston, 1988.

100. D Worster. *Nature's Economy: A History of Ecological Ideas*, Studies in Environment and History, AW Crosby, Ed. Cambridge University Press, Cambridge, U.K., 1994.

101. LM Moomaw. Curtis Fletcher Marbut. In *Life and Work of C.F. Marbut*, HH Krusekopf, Ed. Soil Science Society of America, Columbia, MO, 1942, pp. 11–27.

102. L Korerner. *Linnaeus: Nature and Nation*. Harvard University Press, Cambridge, MA, 1999.

103. LF Farber. *Finding Order in Nature: The Naturalist Tradition from Linnaeus to E.O. Wilson*, Johns Hopkins Introductory Studies in the History of Science, MT Green, S Kingsland, Eds. Johns Hopkins University Press, Baltimore, MD, 2000.

104. W Coleman. *Biology in the Nineteenth Century: Problems of Form, Function, and Transformation*. Cambridge University Press, Cambridge, U.K., 1977.

105. GN Coffey. *A Study of the Soils of the United States*, Bulletin 85. USDA Bureau of Soils, Washington, DC, 1912.

106. CF Marbut. Report of the committee on soil classification. *J Am Soc Agron* 8: 387–388, 1916.

107. CF Marbut. Soil classification. *Bull Am Assoc Soil Survey Workers* 3: 24–33, 1922.

108. Soil Survey Staff. *Soil Taxonomy: A Basic System of Soil Classification for Making and Interpreting Soil Surveys*, 2nd ed. USDA-NRCS, Washington, DC, 1999.

109. PS Lovejoy. The soil survey: present and future. *Bull Am Assoc Soil Survey Workers* 4: 53–59, 1923.

110. MH Lapham. The soil survey from the horse-and-buggy day to the modern age of the flying machine. *Soil Sci Soc Am Proc* 10: 344–350, 1945.

111. TM Bushnell. Aerial photography and soil survey. *Bull Am Soil Survey Assoc* 10: 23–28, 1929.

112. Minutes of joint meeting. *Soil Sci Soc Am Proc* 1: 505–526, 1937.

113. Report of the fifteenth annual business meeting of the Soil Science Society of America. *Soil Sci Soc Am Proc* 16: 91–99, 1952.

114. Reports of SSSA divisions and committees, 1994. *Soil Sci Soc Am J* 59: 588–592, 1995.

115. RW Simonson, DR Gardner. Concept and function of the pedon. *Trans 8th Int Congr Soil Sci* 4: 127–131, 1960.

116. FD Hole. Approach to landscape analysis with emphasis on soils. *Geoderma* 21: 1–23, 1978.

117. FD Hole. Suggested terminology for describing soils as three-dimensional bodies. *Soil Sci Soc Am Proc* 18: 131–135, 1953.

118. HC Davey. Brief outline of aerial photographic work in the U.S. Geological Survey. *Photogramm Eng* 1: 3, 1935.

119. WB Cobb. Possibilities of the airplane in soil survey work. *Bull Am Soil Survey Assoc* 4: 77–80, 1923.

120. TM Bushnell. A new technique in soil mapping. *Bull Am Soil Survey Assoc* 13: 74–81, 1932.

121. CE Millar. The use of aerial photographs in the Michigan land economic survey. *Bull Am Soil Survey Assoc* 13: 82–85, 1932.

122. M Baldwin, WJ Geib, LH Smith, WE Hearn, AT Sweet. Committee on survey reports, maps and technique of mapping. *Bull Am Soil Survey Assoc* 13: 94–95, 1932.

123. TM Bushnell. Use of aerial photography for Indiana land use studies. *Photogramm Eng* 17: 725–738, 1951.

124. CW Collier. Use of aerial maps in soil conservation studies. *Photogramm Eng* 2: 21–26, 1936.

125. MF Miller. Progress of the soil survey of the United States since 1899. *Soil Sci Soc Am Proc* 14: 1–13, 1949.

126. MS Wright. The application of aerial photography to land use problems. *Soil Sci Soc Am Proc* 1: 357–360, 1936.

127. JD Rourke, ME Austin. The use of air-photos for soil classification and mapping in the field. *Photogramm Eng* 17: 738–747, 1951.

128. RW Simonson. Use of aerial photographs in soil surveys. *Photogramm Eng* 16: 308–311, 1950.

129. M Baldwin, HM Smith, HW Whitlock. The use of aerial photographs in soil mapping. *Photogramm Eng* 13: 532–536, 1947.
130. Soil Division Survey Staff. *Soil Survey Manual.* USDA, Washington, DC, 1993.
131. SW Buol, FD Hole, RJ McCracken, RJ Southard. *Soil Genesis and Classification*, 4th ed. Iowa State University Press, Ames, 1997.
132. NJ McKenzie, PJ Ryan. Spatial prediction of soil properties using environmental correlation. *Geoderma* 89: 67–94, 1999.
133. PHT Beckett, R Webster. Soil variability: a review. *Soils Fert* 34: 1–15, 1971.
134. PA Burrough. Soil variability: a late 20th century view. *Soils Fert* 56: 529–562, 1993.
135. AB McBratney, MLM Santos, B Minasny. On digital soil mapping. *Geoderma* 117: 3–52, 2003.
136. BE Butler. Periodic Phenomena in Landscapes as a Basis for Soil Studies, Soil Publication 14. CSIRO, Melbourne, 1959.

4 Geomorphological Soil-Landscape Models

Carolyn G. Olson

CONTENTS

ABSTRACT

This chapter presents soil formation concepts and resulting soil-landscape models with an emphasis on geomorphology entailing: (i) soil factorial models including Jenny's soil-forming factors and Runge's energy model for soil development; (ii) system dynamics and process models including Simonson's model for horizon differentiation based on additions, removals, transfers, and transformations; mass-balance modeling for gains and losses of substances in soil; process–response models considering both intrinsic and extrinsic thresholds and complex geomorphic response; and (iii) geomorphically based landscape models. Milne introduced the concept of the catena, and numerous other hillslope models have been developed highlighting the relationships among topography, landforms, hydrology, and soil formation considering the three-dimensional nature of soil-landscapes. In little over a century we have seen a significant shift from the dominance of conceptual models in soils and geomorphology to spatially explicit quantifiable soil-landscape models that use regression analysis and multivariate methods including spatial point systems, networks, continuous distributions, partitioning, trend analysis, and simulation modeling.

4.1 INTRODUCTION

Observation and conceptualization have been the mainstay of earth science disciplines for centuries. The earth sciences have a history rich in conceptual models based on these empirical observations. There are two codependent disciplines important to our core understanding of earth sciences: soil science and geomorphology. In order to assess the state of our current progress in environmental soil-landscape modeling, it is important to have an appreciation of the historical basis for our present position. In turn, this assessment paves the way for predicting potentials, thus gaining the benefit of analyzing future trends. The objective of this chapter is to provide a context for progress by examining the contributions of a few scientists and their philosophy of soils, geomorphology, and soil-landscape modeling.

Soil science and geomorphology, the parental disciplines of soil-geomorphology and soil-landscape modeling, use the scientific method as the traditional empirical approach to a study. Dijkerman[1] provides a detailed summary of the scientific approach with relevant soil science examples. He states that there are seven stages in the scientific method: (1) selection of the system to study, (2) measurement of properties, (3) ordering and condensing of data, (4) development of hypotheses, (5) testing hypotheses with data gathered, (6) structuring confirmed hypotheses into scientific laws, and (7) using scientific laws to predict unknown phenomena. It is important to emphasize that this scientific method has been the framework by which we approach our science and will continue to be for some time.

Although the scientific method has been the foundation for scientific investigations, it does not explain how scientific advances come about. Many scientists have presented their views on how science progresses. A few ideas relevant to the development of soil-landscape modeling are presented.

Osterkamp and Hupp[2] maintain that geomorphology, soil science, and related disciplines are complex composites of the basic sciences of physics, chemistry, and biology. They suggest that the basic sciences advance through paradigm definitions and replacements (see Kuhn[3,4]), but integrative disciplines such as geomorphology, soil science, or soil geomorphology cannot because of their complexity. Instead, these disciplines have developed by principles derived from the basic sciences. As we shall see, two common principles that appear consistently and have influenced soil scientists and geomorphologists through time are evolution, drawn largely from the biologic sciences, and equilibrium, from chemistry. Various permutations of these principles that have enjoyed popularity include process-response modeling and thresholds, dynamic equilibrium, and punctuated equilibrium.

Why model? Models enable us to study complex systems that otherwise might be considered intractable. Often they are an opportunity to forecast or predict interrelationships among environmental parameters. What makes a good model? Kirkby[5] lists several characteristics of a good model. Among these are an explicit physical basis, simplicity, generality, richness, and potential for scaling up or down. Most of these criteria are straightforward. Soil-landscape models are based in the physical world. Simplicity refers to the central concept of the model, not necessarily to the complexity of rigorous mathematical computations or relationships among parameters within the model. While most earth processes are complex, sometimes a simpler representative model is best. Simple functional systems can be calibrated and tested while complex or sophisticated models are difficult to validate. Oreskes[6] cautions that the more complex a model, the more difficult it is to prove invalid. Kirkby's scaling criterion provides a means of manipulating the model at various levels, from the microscopic to global scales. Generality refers to the ability of a model to be applicable when translated to a new environment, and richness refers to the information gained from applying the model.[5]

The advent of fast, inexpensive computing has allowed us to study and model even more complex issues that ordinarily cannot be studied by traditional scientific methods (e.g., global climate response to CO_2 or soil ecosystems). Recently, quantitative techniques for predicting the spatial and temporal distributions of properties to account for conceptual pedologic models are making use of variability theory. The introduction and increased use of geostatistics and modern statistical techniques in geology helped advance this particular approach in soil science, which has been referred to as pedometrics.[7] The most recent advances in pedometrics include the introduction of nonlinear geostatistical methods.[8] Some researchers combine these newer methods with multivariate methods to produce hybrid models; see, for example, Hoosebeck.[9] McBratney et al.[8] indicate that hybrid methods will be particularly powerful at catchment and regional scales. Several chapters in this book explore these newest approaches and examine current pedometric tools and methods.

4.2 CONCEPTUAL MODELS

4.2.1 SOIL FORMATION CONCEPTS AND EVOLUTION

The first widely recognized attempts to describe soil processes were based largely on climate and vegetation by V.V. Dokuchaev and the Russian school of soil scientists. Strzemski's translation[10] suggests that in the 1860s and 1870s, Dokuchaev described soil formation resulting from the combined activity of climate, organisms, relief, and parent rock. Climate and vegetation were considered of prime importance and formed the basis for Dokuchaev's "zonal" classification. Refinements were added by other well-known scientists, including Silbertsev (azonal and intrazonal soil concepts) and Glinka. Language barriers hindered wide or rapid circulation of their ideas until the turn of the century. These basic ideas were carried forward from the Russian school by Glinka, who translated them into German, and by Marbut, whose work is described below, to the English-speaking world.[11,12] Hilgard,[13] a U.S. geologist, first published material in Mississippi describing parent material, topography, and time as the most important elements of soil formation. This work emphasized the significance that regional climate variation and parent material have on vegetation and soil formation. Strzemski,[10] Arnold,[14] Smith,[15] and Tandarich and Sprecher[16] provide detailed summaries of the concepts developed by these and other early scientists.

Classification and mapping of soils, although not specifically central to this chapter, are intricately woven with the history of pedologic concepts and deserve brief mention here. C.F. Marbut,[12] strongly influenced by William Morris Davis, his former professor, adapted a form of Davis's erosion cycle to soil science. One of the major elements of Marbut's soil classification system for the U.S. Department of Agriculture was the concept of "mature soil" and the cyclic nature of soils, developing from youth to senility.[12,17] Evolution of the soil was a central concept. As in the Davisian scheme, Marbut's ideas did not consider process. In fact, Marbut seriously opposed the Russian genetics school, insisting that soils be classified by characteristics, not by reference to their genesis. Reference to this statement appears in the introduction to Marbut's 1927 speech to the First International Congress of Soil Science,[12] in which he outlines his scheme for classifying soils. Marbut was not entirely successful in selling his approach during his tenure as chief of the Soils Service. C.E. Kellogg, Marbut's successor, enlarged the nomenclature and saw to it that the details of Marbut's system were incorporated into *Soils and Men*, the USDA yearbook.[18] However, genesis was retained in that 1938 classification. Kellogg also renamed Marbut's mature soil, calling it normal soil. Within the context of soil survey, Byers et al.[19] discussed five principal factors of soil formation: parent material, climate, biological activity, relief, and time.

4.2.2 Factor Models

Hans Jenny was among the first to apply a mathematical approach to the conceptual model of soil development formulated by the Russian school. Jenny[20] developed his "fundamental equation of soil-forming factors" in such a manner that it has become a well-known conceptual or empirical deterministic model. Similar to the Russian school, he proposed that soils and soil properties were a function of five factors, climate (cl), organisms (o), topography or relief (r), parent material (p), and time (t), adding unspecified parameters, indicated by a series of dots, to allow for locally or regionally significant factors such as atmospheric dust additions.

$$S = f(cl, o, r, p, t...) \tag{4.1}$$

Jenny overcame the issue of solving this equation by solving for one factor at a time. Only one factor was allowed to vary while holding the others constant. For example, Jenny[20] suggested that if climate, organisms, topography, and parent material remained constant, a soil profile was a product of time alone.

$$S = f(time)_{cl,o,r,p} \tag{4.2}$$

This equation implies that soils change through time when related to the other variables and was termed a chronofunction.[20, 21] Similarly, climo-, bio-, topo-, and lithofunctions were developed. The system is somewhat analogous to Davis's erosion cycle and Darwinian evolution.

The importance of the time dimension in the study of soils may be considered a bias[2] or an appreciation of an essential component.[22] More likely, of all the factors, time may be considered the most independent of the variables in Jenny's equation. The emphasis on time led to the concept of chronosequences for soils. Birkeland[23] provides an excellent discussion of the applications that geologists have made using the chronosequence and chronofunction concept. Ecologists use chronosequences extensively because they appeal to the ecologists' views on succession,[24–26] an obvious analogy to Darwinian evolution. Purists would argue that by definition, there are very few proven chronosequences in nature (R.V. Ruhe, personal communication, late 1970s) because more than one soil factor usually varies with time. The five factors are not actually independent variables. However, if one understands the constraints, the basic construct of the five factors of soil formation remains an extremely valuable conceptual model for soil formation used by soil scientists, geologists, ecologists, and many others.

Runge[27] suggested that soils are too complex to be described suitably by Jenny's soil-forming factors. While he appeared to agree with Simonson's[28] emphasis on process, he stated that specific processes controlling soil development must be identified independently rather than as a combination of processes in balance. Consequently, Runge[27] discusses "energy models" for soil development in terms of potential energy vectors. Following refinements from his earlier studies with colleagues,[29–31] he examined soils in terms of columns through which water, the primary energy source for increasing or decreasing order, flowed. In the resulting energy model, Runge[27] stated that soil development (S) is a function of organic matter production (o), the amount of water available for leaching (w), and time (t):

$$S = f(o, w, t) \tag{4.3}$$

Organic matter was a renewing vector, and the amount of water available for leaching was the developing vector. Amounts of water are determined by rainfall intensity and duration and produce infiltration, runoff, and runon. Runge used this relationship to suggest that climate and relief were embedded in Equation 4.3, expressed by w. This is basically Jenny's 1941 equation with factors ranked or stratified differently. It is similar to the broader-based three-state-factor model[32] for an open system at the ecosystem level:

$$l, s, v, a = f(L_o, P_x, t) \tag{4.4}$$

where l is ecosystem properties, s is soil properties, v is vegetation properties, a is animal properties, L_o is the initial state of the system, P_x is the external flux potentials, and t is the age of the system. Clearly, Runge's energy models borrowed heavily from both Jenny and Simonson. Because of this, the general criticism applied to Jenny's five-factor model is also applicable to the Runge model; both are just as difficult to quantify mathematically.

Smeck et al.[33] discuss Runge's energy model concept in the context of the dynamic soil system and expand on the concepts from thermodynamics: equilibrium and steady state. In their version, potential vectors are referred to as energy fluxes. They suggested that horizon differentiation is driven by energy and material fluxes. Smeck et al.[33] identify fluxes occurring in several surface and near-surface vector directions, but seem to underemphasize lateral or horizontal contributions within the subsoil.

4.3 SYSTEM DYNAMICS AND PROCESS MODELS

Attempts to explain the physical, chemical, and biological reactions that transform rock and other parent materials into soil and soil horizons resulted in the development of models to describe processes. Net changes were related in terms of soil genesis, suggesting that the basic pedologic model was a genetic model. Huggett[34] considers these types of models very different from the factor approach ascribed to by Jenny. Huggett describes these models as a basis for the systems approach to pedogenic models, rather than the factorial approach. Systems approaches are generally process oriented and accommodate driving forces or fluxes of pedogenic processes to soil system dynamics.

Huggett[34] suggests that the systems approach is primarily internal to the soil, as it describes processes occurring within the soil. The factorial approach is described as external because factors can be studied individually as functions or taken together as multiple functions. Huggett[34] suggested that verbal models were exemplified by factorial and functional approaches, and mathematical models by systems approaches. Even as a group, these functions do not adequately illustrate the dynamic processes occurring in the soil.[34] In addition, he suggests that the systems approach is more versatile because it relates driving forces such as solar radiation and precipitation to soil system dynamics, whereas the factorial approach only correlates soil-forming factors to soil morphologic properties.

4.3.1 SOIL DEVELOPMENT MODELS

Simonson[28] developed a model for horizon differentiation based on additions, removals, transfers, and transformations. In this model, combinations of processes act in concert and the balance among them is the key to ultimate characteristics and properties of the soil that we characterize. At the time, this concept differed from the traditional thought that different soils were produced from different genetic processes and pathways. Simonson stated that the changes occurring during the production of soil horizonation were dependent on processes such as hydration, oxidation, solution, leaching, precipitation, and mixing. These simpler or more basic reactions common to all soils[28] were, in turn, controlled by Jenny's

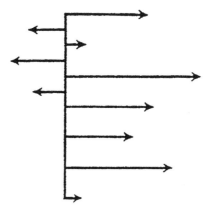

FIGURE 4.1 Schematic of processes, each represented by a vector of differing degrees of importance. (Modified from Simonson, R.W., *Soil Sci. Soc. Am. Proc.*, 23, 152–156, 1959. With permission.)

well-known factors of soil formation: climate, organisms, parent material, and relief. In Simonson's model, the emphasis was placed on the operation of processes in combination, some having a positive and some a negative influence on horizon differentiation. A diagram with arrows of different lengths cleverly and simply illustrated this combination of processes (Figure 4.1). Each length represented the relative importance of a single process to the entire system. Hall[35] suggests that Simonson's model,[28] although a form of a three-dimensional concept, was a limited model too simplistic and compartmentalized to succeed. Hall may have been correct, as Simonson's model has not received the attention that Jenny's state-factor approach has, particularly outside the soil science community.

T.C. Chamberlain,[36] credited with the development of multiple working hypotheses, completely changed the way geologists apply the scientific method when examining a phenomenon. Multiple working hypotheses permit the simultaneous examination of several explanations for an occurrence. The final conclusions are often a synthesis of the strongest concepts from several of the most likely hypotheses. Arnold[37] borrowed from Chamberlain to illustrate sequential models for soil development by applying multiple working hypotheses to explain observed relationships. Using four possible scenarios for the development of silt coats on ped faces of present-day Mollisols, Arnold integrated geomorphic processes with soil genetic processes and rates to produce evolutionary sequences of soil profile development. He points out that the advantage of using multiple working hypotheses is that they allow the investigator to collect scattered data and integrate them with genetic concepts into unifying ideas.

4.3.2 MASS BALANCE MODELING

Nikiforoff and Drosdoff[38] present a quantitative model for gains and losses of substances in soil. Their model seems to be among the first to attempt a mass balance approach in soils. They clearly state that calculating a balance between total loss and total gain can only be accomplished on a volume basis. A silt loam soil from Oregon was used to illustrate the model. Others, particularly those with chemistry backgrounds, have contributed similar mass balance and equilibrium approaches. Barshad[39] presents an excellent, detailed discussion of methods for calculating soil development based on a similar approach to net change utilizing gains and losses of primary and secondary constitutents in soil. Kirkby[40] appears to be one of the first to attempt a slope-compatible soil mass-balance model. Kirkby recognized the need to integrate both soil and slope development and, more importantly, to place them on similar timescales to more accurately examine their interaction.

Brimhall and Dietrich,[41] Chadwick et al.,[42] and Brimhall et al.[43] address mass balance in soils from an engineering geology view and add stress and strain, parameters for deformation, to the basic concepts of a mass balance model developed by Barshad.[39] Similar to that of Nikiforoff and Drosdoff,[38] this approach also presents a template for performing mass balance calculations on a volumetric basis. This model and versions of it have been applied to many recent studies (see, e.g., references 44 and 45) to assess mineral weathering and soil property changes on a landscape scale.

4.3.3 PROCESS-RESPONSE MODELS

In parallel with models developed in geomorphology (see, e.g., references 46 through 48), the concept of thresholds has been applied to soil process-response models (see, e.g., references 49 and 50). Hack[51,52] used dynamic equilibrium to explain erosional landscapes. Schumm[53,54] defines both intrinsic and extrinsic thresholds and complex geomorphic response with examples. In soil science, intrinsic thresholds are commonly described but not necessarily acknowledged as models (e.g., clay illuviation in the presence of salts).

Building on Simonson's 1959 model, Yaalon[22] described several categorical processes for soils that have been modeled: (1) feedback mechanisms (see, e.g., references 55 and 56) where systems are dampened or amplified by negative or positive changes, (2) dynamic equilibrium or steady state, and (3) self-terminating processes where gains exceed losses or losses exceed gains. Chadwick and Chorover[57] point out that feedback mechanisms control how and when thresholds occur. For example, when reactants are depleted, feedback stops and processes are terminated. In this case, feedback mechanisms are not separate from the self-terminating processes described by Yaalon.[22] Chadwick and Chorover[57] thoroughly discuss the concept of pedogenic thresholds and provide an extensive list of references.

4.3.4 SOIL-LANDSCAPE MODELING

A number of early scientists made significant contributions to conceptual models for geomorphic research in the latter half of the 19th century. Among these were

FIGURE 4.2 Hillslope profile evolution. On the left is the Penckian process of slope denudation, and on the right is the Davisian scheme. Shaded areas indicate the endpoint of denudation. (From Carson, M.A. and Kirkby, M.J., *Hillslope Form and Process*, Cambridge University Press, New York, 1972; and Davis, W.M., *Geol. Soc. Am. Bull.*, 43, 399–440, 1932. With permission.)

John Wesley Powell, who developed the concept of erosion of landscapes to base level and the polycyclical nature of landscape erosion and deposition[58,59]; G.K. Gilbert, one of the first to relate process to the origin of the landscape[60]; and W.M. Davis. Of these, William Morris Davis exerted the most substantial and enduring influence on the geologic community at the turn of the 19th century,[61] in part because of his prominence as a Harvard professor. Davis was also a strong Darwinian evolutionist in his approach to geology. Davis extracted ideas from his contemporaries, Powell and G.K. Gilbert as well as Darwin, and integrated them into his now-classic theory, the "cycle of erosion."[61] Landscapes progressed through a series of stages from youth to maturity to old age in a time-dependent model. Landscapes eroded by downwearing. Although this evolutionary model long influenced geomorphology and is still quoted today, it is critically lacking in several arenas. Its greatest shortcomings lie first with the inability to explain process and slope geometry, and secondarily with its two major assumptions: (1) uplift and denudation are mutually exclusive, and (2) streams have two phases of activity, rapid incision and then dormancy.[61]

Walther Penck, a German, is best known for his work on parallel retreat of slopes, or backwearing.[62] His model suggested that rapid incision produced steeper slopes than slower downcutting. Slope denudation produces slope retreat without changing the slope angle. Penck's model is in contrast to Davis's model and was severely criticized by Davis's supporters for many years. Figure 4.2 illustrates the relative differences between the slope evolution models of Penck and Davis.

Perhaps one of the underlying reasons for the integration of soil science and geomorphology in the U.S. is the early influence of William Morris Davis, who taught C.F. Marbut. As discussed earlier, Marbut proposed a soil classification with elements that included the concept of soil maturity and the development of soils from youth to senility.[12,17] These elements create a link to Davis's cycle of erosion that is too strong to dismiss as coincidence.

Geomorphically based landscape models such as those of Davis and Penck are often used to separate the Earth's surface into discrete elements or components. Those models alone are not sufficient to describe processes that occur in soils because processes are not independent. Processes occurring on one portion of the landscape influence soil processes on other parts of the landscape, a concept well understood by Milne.

Milne introduced the concept of the catena, a descriptive model in which soil characteristics vary with distance and topography and repeat relative to each other under the same conditions.[63–65] The catena concept was a fundamental advance in soil modeling, as it explained the pattern of soils on the landscape. This hillslope model recognized that processes of erosion and the particular position on the hillslope directly affect soil properties. Two types of catenas were described. The first was a catena in which soils on a hillslope developed on the same parent material. Changes in soil properties along the slope curvature were attributed to subsurface drainage and erosion and redeposition. The second type was a catena in which the hillslope had more than one parent material. Patterns were again predictable, but increased in complexity due to stratigraphic differences. Milne's model was adopted by soil scientists in the U.S. and appeared in government documents as early as 1938.[66] Today the term *catena* has a more limited definition among soil scientists and is loosely considered a hydrologic toposequence of related soils.

4.3.4.1 Hillslope Models

Hillslopes, one of the most ubiquitous landforms, have garnered much attention since Milne first introduced the catena concept and, as such, deserve a separate section for discussion. A broad understanding of hillslopes and slope-forms began to evolve from the early landscape models of Davis and Penck. Both continued to write about slope development after formulating their landscape evolution models. In the mid-20th century, publication of early two-dimensional hillslope conceptual models became prevalent. Kirk Bryan, a prominent physiographer, reexamined the work of Penck with his work on slope retreat in the western U.S.[67,68] Wood,[69] also working in arid environments, provided additional elements with the slope cycle: the waxing slope, the free face, the debris slope, and the pediment or waning slope. King[70] accepted Wood's slope elements and called them the "uniformitarian nature of hillslopes," a nod to Hutton, or "the fully developed hillslope," as redefined by Ruhe.[71,72] King compared slope elements and the effects of water movement and mass movement as part of his fully developed hillslope. Ruhe[71,72] presented a two-dimensional conceptual "fully developed hillslope" model of five units that compared well with Wood's and King's (Figure 4.3).

Conacher and Dalrymple[73] described a nine-unit hillslope model. This latter model, though based on a largely temperate climate regime, attempted to integrate slope profile components with material and water movement.

One of the first U.S. geomorphologists to work at the regional level of the soil-landscape was Robert Ruhe. From the 1950s to the 1980s, Ruhe and his students began producing conceptual models of the soil-landscape that emphasized process and quantification at scales varying from the hillslope to the drainage basin size (see, e.g., references 71, 72, and 74 through 81). His work and that of his student's on open and closed hillslope systems reflected Milne's catena concept that processes on the landscape influence soil processes on other portions

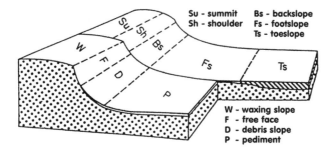

Su - summit   Bs - backslope
Sh - shoulder Fs - footslope
              Ts - toeslope

W - waxing slope
F - free face
D - debris slope
P - pediment

FIGURE 4.3 The elements of a hillslope. (Modified by Ruhe, R.V., *Trans. 7th Int. Congr. Soil Sci.* (Madison, WI), 23, 165–170, 1960. Foreground from Wood, A, *Proc. Geol. Assoc.*, 53, 128–138, 1942; and King, L.C., *Trans. Edinburgh Geol. Soc.*, 17, 81–102, 1957. Modified from Hall, G.F. and Olson, C.G., in *Spatial Variabilities of Soils and Landforms*, SSSA Special Publication 28, SSSA, Madison, WI, 1991, pp. 9–24. With permission.)

of the landscape[75,76] (see figures in Ruhe,[72] pp. 111, 114, 115, 117, 118). Ruhe's work emphasized the surficial processes related to landscape evolution and understanding the subsurface. This followed on the heels of a break from the traditionally descriptive sciences to the quantification and process approach in drainage basin analysis during the late 1940s (see, e.g., references 82 through 86). Linear statistical methods such as regression and analysis of variance were introduced in the early 1950s (see, e.g., references 84 and 85) to analyze erosional landforms. Within the decade, multivariate methods became available in geology, largely through the efforts of W.C. Krumbein (see, e.g., references 87 and 88), who had begun applying quantitative techniques to geology in the 1930s. By the 1960s, geomorphology began to be heavily influenced by the changes taking place in related disciplines such as geography, where spatial analysis was rapidly developing. This led to studies in quantitative geomorphology, including spatial point systems, networks, continuous distributions, partitioning, trend analysis, and simulation (see, e.g., references 40 and 89 through 91). Oddly, aside from drainage basin analysis and some morphometric analysis, statistical methods and spatial modeling of geomorphic studies did not develop as rapidly within the geomorphic and soil science communities as in the geography community. Conceptual models still played a significant role. Chorley[90] suggested that the slow acceptance of these methods in geomorphology may have been related to geomorphologists' preoccupation with time, a dimension not well served by those statistical methods. Among other reasons, he suggested that insufficient mathematical skills and poor communication among geographers and geomorphologists may have played a role. The most likely cause seems to be the lack of mathematical training among geomorphologists of that era, a condition that still may persist today. One of the few examples that used analysis of variance in soil geomorphology was a study by Daniels et al.[92] This study attempted a quantitative technique to describe erosional and depositional surfaces in nearly level areas. As earlier pointed out by Chorley,[90] there seems to be little follow-up even today.

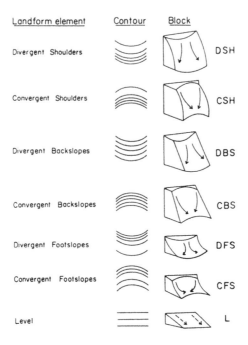

Landform element	Contour	Block	
Divergent Shoulders			DSH
Convergent Shoulders			CSH
Divergent Backslopes			DBS
Convergent Backslopes			CBS
Divergent Footslopes			DFS
Convergent Footslopes			CFS
Level			L

FIGURE 4.4 Hillslope elements, curvature, and flow lines on seven landscape positions. (From Pennock, D.J. et al., *Geoderma*, 40, 297–315, 1987. With permission.)

A natural progression in slope development studies was the realization of the key linkage of water movement to geomorphic and pedologic processes. The planar curvature of slopes became an important factor integrated into conceptual hillslope models. Hall and Olson[93] illustrate a succession of conceptual models that integrate hydrologic characteristics, particularly lateral flow, with landscape parameters. A few are summarized here. Aandahl[94] related slope curvature to differences in fertility status and soil morphological properties. In plot studies, Troeh[95] used quantitative descriptions of three-dimensional landform parameters to illustrate four basic convex–concave combinations. Ruhe and Walker[75] and Walker and Ruhe[76] were among the first to develop three-dimensional conceptual models for open and closed systems on hillslopes. Ruhe and Walker[75] describe a nine-unit geometry and Ruhe[72] (see Figure 6.1, p. 100) illustrated these changes in slope curvature with a matrix of nine basic forms, varying three components: (1) slope gradient, (2) slope length, and (3) slope width. Pennock et al.[96] added surface flow to this basic matrix of slope curvature (Figure 4.4).

Building on Troeh's work, Huggett[97] added surface flow to three-dimensional slope shapes (Figure 4.5). In a significant step forward, he fits this model to a larger idealized area, defined as a "soil-landscape system or valley basin." The surface boundaries of this system are similar to the drainage divide of a watershed, but he adds another dimension, the weathering depth, in order to include subsur-

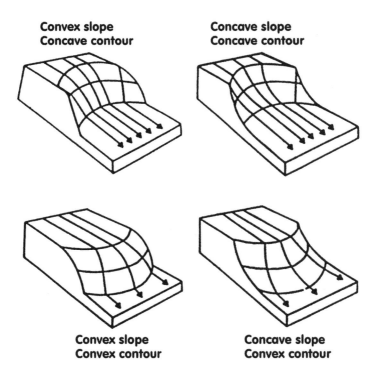

FIGURE 4.5 Four basic slope shapes with surface flow lines. (From Huggett, R.J., *Geoderma*, 13, 1–22, 1975. With permission.)

face flow. This neatly ties slope curvature models to hydrologic, geomorphic, and pedologic models. Material moves on and through soils, to different soils at different hillslope positions, as well as throughout the entire watershed. In a carefully documented watershed, Lanyon and Hall[98,99] developed predictive maps of landscape instability based on a number of soil and climate properties, including slope curvature and soil moisture. With these and many similar conceptual models, we begin to see first the shift toward a three-dimensional approach to a single hillslope, and then a progression to larger, regional models that incorporate soils and soil processes.

In the 1970s and 1980s, while still largely conceptual, increasing quantification of portions of these models became the norm as computational methods improved. Dijkerman[1] observed that most of our working models had been verbal in nature until the availability of computers became commonplace. Emphasis shifted toward mathematical models in soils in the early 1940s. Lateral movement of moisture in soil had been demonstrated early on using mathematical models[100] and was observed and measured by many (see, e.g., references 101 through 104). One of the first examples of merging geomorphic conceptual models and quantitative statistical programming was published by Pennock et al.[96] They developed a quantitative landform classification based on landform elements and slope morphology. Their method also provided a three-dimensional aspect to that of earlier work (see, e.g., references 71 and 105). Surface morphology, i.e., plan curvature, profile curvature, and gradient were the primary landform elements. Much like other researchers (see, e.g., references 87, 95, 97, and 99), Pennock et al.[96] expressed the idea that slope plan and profile curvature could be related to soil moisture content. Their major contribution[96] was the quantification of these relationships. Numerous researchers have expanded on these relationships, which have become especially useful as relational datasets to those working in a geographic information systems (GIS) framework.

4.3.4.2 Quantification and Explicit Soil-Landscape Models: Spatial Analysis

The development of techniques for quantitative spatial prediction began early, with the work of Speight[106] and Beven and Kirkby[107] paving the way for quantitative landform and watershed studies. Speight illustrated a numerical approach to land system classifications to reduce subjectivity to a minimum. Quantitative comparison of landscapes in different areas was shown to be possible. Beven and Kirkby[107] combined channel network topology and associated contributing areas with parametric basin models to predict hydrologic response in basins that were not gauged. With the advent of GIS-based digital elevation modeling (DEM) and modern computing facilities capable of large-scale calculations, the early work of these researchers has been greatly facilitated. Many researchers have expanded on their ideas (see, e.g., references 108 through 113). These researchers have taken advantage of the nonrandom variability of terrain attributes and carried out randomization studies to produce reliable relationships among other environmental parameters. For example, hillslope summits are always summits, but sampling of these can be done randomly and quantified.

4.4 CONCLUSIONS

Rapid and inexpensive computing has greatly improved our ability to study and model complex systems. In little over a century we have seen a significant shift from the dominance of conceptual models in soils and geomorphology to explicit quantifiable models. We have learned that soils are an integral part of the landscape, that soils, dependent on their hillslope positions, are predictable. We have defined landscapes in three dimensions and begun to include the hydrology. Our limitations in modeling soils and landscapes still lie with our understanding of the processes in their development. With these limits firmly in mind and the increased computational speed and complexity of hardware systems available today, we can focus on producing more functionally representative models of the soil-landscape system.

ACKNOWLEDGMENTS

The author thanks several individuals for their critical review of this document, including Robert Ahrens, Thomas Fenton, Lyle Steffen, and an anonymous reviewer.

REFERENCES

1. JC Dijkerman. Pedology as a science: the role of data, models, and theories in the study of natural soil systems. *Geoderma* 11: 73–93, 1974.
2. WR Osterkamp, CR Hupp. The evolution of geomorphology, ecology, and other composite sciences. In *The Scientific Nature of Geomorphology, Proceedings of the 27th Binghamton Symposium in Geomorphology*, BL Rhoads, CE Thorn, Eds. John Wiley & Sons, New York, 1996, pp. 415–442.
3. TS Kuhn. *The Structure of Scientific Revolutions*, 2nd ed. The University of Chicago Press, Chicago, 1970, 210 pp.
4. TS Kuhn. *The Essential Tension*. The University of Chicago Press, Chicago, 1977, 366 pp.
5. MJ Kirkby. A role for theoretical models in geomorphology. In *The Scientific Nature of Geomorphology, Proceedings of the 27th Binghamton Symposium in Geomorphology*, BL Rhoads, CE Thorn, Eds. John Wiley & Sons, New York, 1996, pp. 257–272.
6. N Oreskes. Why believe a computer? Models, measures, and meaning in the natural world. In *The Earth around Us, Maintaining a Livable Planet*, JS Schneiderman, Ed. W.H. Freeman & Co., New York, 2000, pp. 70–82.
7. R Webster. The development of pedometrics. *Geoderma* 6: 1–15, 1994.
8. AB McBratney, IOA Odeh, TFA Bishop, MS Dunbar, TM Shatar. An overview of pedometric techniques for use in soil survey. *Geoderma* 97: 293–327, 2000.
9. MR Hoosebeck. Incorporating scale into spatio-temporal variability: applications to soil quality and yield data. *Geoderma* 85: 113–131, 1998.
10. M Strzemski. *Ideas underlying Soil Systematics*, 1971 trans., Polish ed. TT73-54013. Foreign Scientific Publications Department of the National Center for Scientific, Technical and Economic Information, Warsaw, 1975, 541 pp.
11. CF Marbut. Soil classification. *Am Soil Surv Assoc Bull* III: 24–32, 1922.
12. CF Marbut. A scheme for soil classification. In *Proceedings and Papers of the First International Congress of Soil Science*, Commission 5, Vol. I, 1928, pp. 1–31.
13. EW Hilgard. *The Relations of Soil to Climate*, Bulletin 3. USDA Weather Bureau, Washington, DC, 1982, pp. 1–59.
14. RW Arnold. Concepts of soils and pedology. In *Pedogenesis and Soil Taxonomy 1: Concepts and Interactions*, LP Wilding, NE Smeck, GF Hall, Eds. Elsevier Science Publishers, New York, 1983, pp. 1–21.
15. GD Smith. Historical development of soil taxonomy: background. In *Pedogenesis and Soil Taxonomy 1: Concepts and Interactions*, LP Wilding, NE Smeck, GF Hall, Eds. Elsevier Science Publishers, Amsterdam, The Netherlands, 1983, pp. 23–49.

16. JP Tandarich, SW Sprecher. The intellectual background for factors of soil formation. In *Factors of Soil Formation: A 50th Anniversary Retrospective*, Soil Science Society of America Special Publication 33, R Amundsen, J Harden, MJ Singer, Eds. SSSA, Madison, WI, 1994, pp. 1–13.

17. CF Marbut. Soil of the U.S. In *Atlas of American Agriculture*, Part III, OE Baker, Ed. USDA, U.S. Government Printing Office, Washington, DC, 1935, pp. 12–15.

18. M Baldwin, CE Kellogg, J Thorp. Soil classification. In *Soils and Men*, U.S. Department of Agriculture Yearbook. USDA, Washington, DC, 1938, pp. 979–1001.

19. HG Byers, CE Kellogg, MS Anderson, J Thorp. Formation of soil. In *Soils and Men*, U.S. Department of Agriculture Yearbook. USDA, Washington, DC, 1938, pp. 948–978.

20. H Jenny. *Factors of Soil Formation*. McGraw-Hill, New York, 1941, 281 pp.

21. H Jenny. *The Soil Resource: Origin and Behavior*. Springer-Verlag, New York, 1980, 377 pp.

22. D Yaalon. Climate, time and soil development. In *Pedogenesis and Soil Taxonomy 1: Concepts and Interactions*, LP Wilding, NE Smeck, GF Hall, Eds. Elsevier Science Publishers, Amsterdam, The Netherlands, 1983, pp. 233–251.

23. PW Birkeland. *Soils and Geomorphology*. Oxford University Press, New York, 1999, 430 pp.

24. PM Vitousek. Factors controlling ecosystem structure and function, In *Factors of Soil Formation: A Fiftieth Anniversary Retrospective*, Soil Science Society of America Special Publication 33, R Amundson, J Harden, M Singer, Eds. SSSA, Madison, WI, 1994, pp. 87–97.

25. K Van Cleve, FS Chapin III, PW Flanagan, CT Dyrness, LA Viereck. Element cycling in taiga forests: state-factor control. *Bioscience* 41: 78–88, 1991.

26. PM Vitousek, LR Walker. Colonization, succession, and resource availability: ecosystem-level interactions. In *Colonization, Succession, and Stability*, A Gray, M Crawley, PJ Edwards, Eds. Blackwell Science, Oxford, 1987, pp. 207–223.

27. ECA Runge. Soil development sequences and energy models. *Soil Sci* 115: 183–193, 1973.

28. RW Simonson. Outline of a generalized theory of soil genesis. *Soil Sci Soc Am Proc* 23: 152–156, 1959.

29. KC Hinkley, ECA Runge, EJ Pedersen. Effect of soil on vegetation in NE Illinois. *Am Soc Agron* 137, 1970.

30. NE Smeck, ECA Runge. Factors influencing profile development exhibited by some hydromorphic soils in Illinois. In *Pseudogley and Gleys*, Trans. Comm. V and VI Int'l. Soc. Soil Sci., E Schlichting, U Schwertmann, Eds. Chemie Verlag, 1971, pp. 169–179.

31. TM Ballagh, ECA Runge. Clay-rich horizons over limestone: illuvial or residual? *Soil Sci Soc Am Proc* 34: 534–536, 1970.

32. H Jenny. Derivation of state factor equations of soils and ecosystems. *Soil Sci Soc Am Proc* 25: 385–388, 1961.

33. NE Smeck, ECA Runge, EE Mackintosh. Dynamics and genetic modeling of soil systems. In *Pedogenesis and Soil Taxonomy 1: Concepts and Interactions*, LP Wilding, NE Smeck, GF Hall, Eds. Elsevier Science Publishers, Amsterdam, The Netherlands, 1983, pp. 51–81.

34. RJ Huggett. Conceptual models in pedogenesis: a discussion. *Geoderma* 16: 261–262, 1976.

35. GF Hall. Pedology and geomorphology. In *Pedogenesis and Soil Taxonomy 1: Concepts and Interactions*, LP Wilding, NE Smeck, GF Hall, Eds. Elsevier Science Publishers, Amsterdam, The Netherlands, 1983, pp. 117–140.

36. TC Chamberlain. The method of multiple working hypotheses. *J Geol* 5: 837–848, 1897.

37. RW Arnold. Multiple working hypothesis in soil genesis. *Soil Sci Soc Am Proc* 29: 717–724, 1965.

38. CC Nikiforoff, M Drosdoff. Genesis of a claypan soil. *Soil Sci* 55: 459–483, 1943.

39. I Barshad. Chemistry of soil development. In *Chemistry of the Soil*, 2nd ed., American Chemical Society Monograph 160, FE Bear, Ed. Reinhold Pub. Corp., New York, 1964, pp. 1–70.

40. MJ Kirkby. Soil development models as a component of slope models. *Earth Surf Processes* 2: 203–230, 1977.

41. GH Brimhall, WE Dietrich. Constitutive mass balance relations between chemical composition, volume, density, porosity, and strain in metasomatic hydrochemical systems: results on weathering and pedogenesis. *Geochim Cosmochim Acta* 51: 567–587, 1987.

42. OA Chadwick, GH Brimhall, and DM Hendricks. From a black box to a grey box: a mass balance interpretation of pedogenesis. *Geomorphology* 3: 369–390, 1990.

43. GH Brimhall, OA Chadwick, CJ Lewis, W Compston, IS Williams, KH Danti, WE Dietrich, ME Power, DM Hendricks, J Bratt. Deformational mass transport and invasive processes in soil evolution. *Science* 255: 695–702, 1992.

44. OA Chadwick, CG Olson, DM Hendricks, EF Kelly, RT Gavenda. Quantifying climatic effects on mineral weathering and neoformation in Hawaii. *Proc 15th Int Soil Sci Congr* 8a: 94–105, 1994.

45. OA Chadwick, RT Gavenda, EF Kelly, K Ziegler, CG Olson, WC Elliot, DM Hendricks. The impact of climate on the biogeochemical functioning of volcanic soils. *Chem Geol* 202: 195–223, 2003.

46. SA Schumm. Geomorphic thresholds: the concept and its applications. *Trans Inst Br Geogr* 4: 485–515, 1979.

47. WB Bull. Geomorphic thresholds as defined by ratios. In *Thresholds in Geomorphology*, D Coates, J Vitek, Eds. Allen and Unwin Ltd., London, 1980, pp. 259–263.

48. WB Bull, *Geomorphic Responses to Climatic Change*. Oxford University Press, New York, 1991, 326 pp.

49. DR Muhs. Intrinsic thresholds in soil systems. *Phys Geogr* 5: 99–110, 1984.

50. EV McDonald, FB Pierson, GN Flerchinger, LD McFadden. Application of a soil-water balance model to evaluate the influence of Holocene climate change on calcic soils, Mojave Desert, California. *Geoderma* 74: 167–192, 1996.

51. JT Hack. Interpretation of erosional topography in humid temperate regions. *Am J Sci* 258A: 80–97, 1960.

52. JT Hack. Dynamic equilibrium and landscape evolution. In *Theories of Landform Development*, Proceedings of the 6th Annual Binghamton Symposium, SUNY, Binghamton, NY, WN Melhorn, RC Flemal, Eds., 1975, pp. 87–102.

53. SA Schumm. Geomorphic thresholds and complex response of drainage systems. In *Fluvial Geomorphology*, Proceedings of the 4th Annual Geomorphology Symposium Series, State University, Binghamton, New York, M Morisawa, Ed., 1973, pp. 299–310.

54. SA Schumm. *The Fluvial System*. John Wiley & Sons, New York, 1977, 338 pp.

55. J Torrent, WD Nettleton. Feedback processes in soil genesis. *Geoderma* 20: 281–287, 1978.
56. DA Reid, RC Graham, RJ Southard, C Amrhein. Slickspot soil genesis in the Carrizo Plain, California. *Soil Sci Soc Am J* 57: 162–168, 1993.
57. OA Chadwick, J Chorover. The chemistry of pedogenic thresholds. *Geoderma* 100: 321–353, 2001.
58. JW Powell. *Report on the Geology of the Eastern Portion of the Uinta Mountains.* U.S. Government Printing Office, Washington, DC, 1876.
59. JW Powell. *Exploration of the Colorado River of the West (1869–1872).* U.S. Government Printing Office, Washington, DC, 1875.
60. GK Gilbert. *Report on the Geology of the Henry Mountains*, U.S. Geographical and Geological Survey of the Rocky Mountain Region. U.S. Government Printing Office, Washington, DC, 1880.
61. Davis, W.M. The geographical cycle. *Geogr J* 14A: 481–503, 1899.
62. W Penck. *Morphological Analysis of Landforms*, KC Boswell, H Czech, Trans. Macmillan Co., London, 1924.
63. G Milne. Some suggested units of classification and mapping particularly for East African Soils. *Soil Res* 4: 1935.
64. G Milne. Composite units for the mapping of complex soil associations. *Trans 3rd Int Congr Soil Sci* 1: 345–347, 1935.
65. G Milne. Normal erosion as a factor in soil profile development. *Nature* 138: 548–549, 1936.
66. USDA, *Soils and Men*, U.S. Department of Agriculture Yearbook. U.S. Government Printing Office, Washington, DC, 1938.
67. K Bryan. Gully gravure, a method of slope retreat. *J Geomorphol* 3: 89–107, 1940.
68. K Bryan. The retreat of slopes. *Ann Assoc Am Geogr* 30: 254, 1940.
69. A Wood. The development of hillside slopes. *Proc Geol Assoc* 53: 128–138, 1942.
70. LC King. The uniformitarian nature of hillslopes. *Trans Edinburgh Geol Soc* 17: 81–102, 1957.
71. RV Ruhe. Elements of the soil landscape. *Trans 7th Int Congr Soil Sci* (Madison, WI) 23: 165–170, 1960.
72. RV Ruhe. *Geomorphology, Geomorphic Processes and Surficial Geology.* Houghton Mifflin Co., Boston, 1975, 246 pp.
73. AJ Conacher, JB Dalrymple. The nine unit landsurface model: an approach to pedogeomorphic research. *Geoderma* 18: 154, 1977.
74. RV Ruhe, RB Daniels, JG Caddy. *Landscape Evolution and Soil Formation in Southwestern Iowa*, USDA Tech Bull 1149. USDA, Washington, DC, 1967, 242 pp.
75. RV Ruhe, PH Walker. Hillslope models and soil formation. I. Open systems. *Trans 9th Int Congr Soil Sci* (Adelaide, Australia) IV: 551–560, 1968.
76. PH Walker, RV Ruhe. Hillslope models and soil formation. II. Closed systems. *Trans 9th Int Congr Soil Sci* (Adelaide, Australia) IV: 561–567, 1968.
77. RV Ruhe. *Quaternary Landscapes in Iowa.* Iowa State University Press, Ames, 1969.
78. WJ Vreeken. Stratigraphy, sedimentology, and moisture contents in a small loess watershed in Tama County, Iowa. *Iowa Acad Sci Proc* 75: 225–233, 1968.
79. HJ Kleiss. Loess distribution along the Illinois soil-development sequence. *Soil Sci* 115: 194–198, 1970.

80. HJ Kleiss. Hillslope sedimentation and soil formation in northeastern Iowa. *Soil Sci Soc Am Proc* 34: 281–290, 1973.

81. RV Ruhe, CG Olson. Soil welding. *Soil Sci* 130: 132–139, 1980.

82. RE Horton. Drainage basin characteristics. *Trans Am Geophys Union* 13: 350–361, 1932.

83. RE Horton. Erosional development of streams and their drainage basins: hydrophysical approach to quantitative morphology. *Geol Soc Am Bull* 56: 275–370, 1945.

84. AN Strahler. Equilibrium theory of erosional slopes approached by frequency distribution analysis. *Am J Sci* 248: 673–696, 1950.

85. AN Strahler. Equilibrium theory of erosional slopes approached by frequency distribution analysis. Part II. *Am J Sci* 248: 800–814, 1950.

86. AN Strahler. Hypsometric (area-altitude) analysis of erosional topography. *Geol Soc Am Bull* 63: 1117–1142, 1952.

87. WC Krumbein. The "sorting out" of geological variables illustrated by regression analysis of factors controlling beach firmness. *J Sediment Petrol* 29: 575–587, 1959.

88. WC Krumbein, FA Graybill. *An Introduction to Statistical Models in Geology.* McGraw-Hill, New York, 1965, 475 pp.

89. AE Scheidegger. Mathematical models of slope development. *Geol Soc Am Bull* 72: 37–59, 1961.

90. RJ Chorley, Ed. *Spatial Analysis in Geomorphology.* Harper & Row, New York, 1972, 393 pp.

91. MA Carson, MJ Kirkby. *Hillslope Form and Process.* Cambridge University Press, New York, 1972, 475 pp.

92. RB Daniels, LA Nelson, EE Gamble. A method of characterizing nearly level surfaces. *Z Geomorph* 14: 175–185, 1970.

93. GF Hall, CG Olson. Predicting variability of soils from landscape models. In *Spatial Variabilities of Soils and Landforms*, Soil Science Society of America Special Publication 28. SSSA, Madison, WI, 1991, pp. 9–24.

94. AR Aandahl. The characterization of slope positions and their influence on the total nitrogen content of a few virgin soils in western Iowa. *Soil Sci Soc Am Proc* 13: 449–454, 1948.

95. FR Troeh. Landform parameters correlated to soil drainage. *Soil Sci Soc Am Proc* 28: 808–812, 1964.

96. DJ Pennock, BJ Zebarth, E deJong. Landform classification and soil distribution in hummocky terrain, Saskatchewan, Canada. *Geoderma* 40: 297–315, 1987.

97. RJ Huggett. Soil landscape systems: a model of soil genesis. *Geoderma* 13: 1–22, 1975.

98. LE Lanyon, GF Hall. Land-surface morphology. 1. Evaluation of a small drainage basin in eastern Ohio. *Soil Sci* 136: 291–299, 1983.

99. LE Lanyon, GF Hall. Land-surface morphology. 2. Predicting potential landscape instability in eastern Ohio. *Soil Sci* 136: 382–386, 1983.

100. D Zaslavsky, AS Rogowski. Hydrologic and morphologic implications of anisotropy and infiltration in soil profile development. *Soil Sci Soc Am Proc* 33: 594–599, 1969.

101. RZ Whipkey. Subsurface storm flow from forested slopes. *Bull Int Assoc Sci Hydrol* 20: 276–288, 1965.

102. RZ Whipkey. Storm runoff from forested catchments by subsurface routes. In *Floods and Their Computation*, Proc. Symp. Int. Assoc. Sci. Hydrol., Leningrad, USSR, August 1967, Vol. 2. UNESCO/IASH, Belgium, 1969, pp. 773–779.

103. DR Weyman. Measurements of the downslope flow of water in a soil. *J Hydrol* 20: 276–288, 1973.

104. RZ Whipkey, MJ Kirkby. Flow within the soil. In *Hillslope Hydrology*, MJ Kirkby, Ed. John Wiley & Sons, New York, 1978, pp. 121–144.

105. JB Dalrymple, RJ Blong, AJ Conacher. A hypothetical nine unit landsurface model. *Z Geomorph* 12: 60–76, 1968.

106. JG Speight. Parametric description of land form. In *Land Evaluation*, GA Steward, Ed. Macmillan & Co., Melbourne, Australia, 1968, pp. 239–250.

107. KJ Beven, MJ Kirkby. A physically based, variable, contributing area model of basin hydrology. *Hydrol Sci Bull* 24: 43–69, 1979.

108. ID Moore, EM O'Loughlin, GJ Burch. A contour-based topographic model for hydrological and ecological applications. *Earth Surf Processes Landforms* 13: 305–320, 1988.

109. ID Moore, AR Ladson, R Grayson. Digital terrain modelling: a review of hydrological, geomorphological, and biological applications. *Hydrol Processes* 5: 3–30, 1991.

110. ID Moore, PE Gessler, GA Nielsen, GA Peterson. Soil attribute prediction using terrain analysis. *Soil Sci Soc Am J* 57: 443–452, 1993.

111. K McSweeney, PE Gessler, BK Slater, RD Hammer, JC Bell, GW Petersen. Towards a new framework for modeling the soil-landscape continuum. In *Factors of Soil Formation: A Fiftieth Anniversary Retrospective*, Soil Science Society of America Special Publication 33, R Amundson, J Harden, M Singer, Eds. SSSA, Madison, WI, 1994, pp. 127–145.

112. PE Gessler, ID Moore, NJ McKenzie, PJ Ryan. Soil-landscape modeling and spatial prediction of soil attributes. *Int J Geogr Inf Syst* 9: 421–432, 1995.

113. PE Gessler, OA Chadwick, F Chamran, L Althouse, K Holmes. Modeling soil-landscape and ecosystem properties using terrain analysis. *Soil Sci Soc Am J* 64: 2046–2056, 2000.

114. WM Davis. Piedmont benchland and Primärumpfe. *Geol Soc Am Bull* 43: 399–440, 1932.

Section II

Collection of Soil-Landscape Datasets

5 The Impact of Emerging Geographic Information Technology on Soil-Landscape Modeling

Sabine Grunwald and Sanjay Lamsal

CONTENTS

ABSTRACT

We provide an overview of historic and emerging soil mapping technologies, soil mapping paradigms, and data management. Geographic information technology (GIT), including the emergence of global positioning systems, geophysical mapping techniques such as electromagnetic induction and ground-penetrating radar, soil sensors such as profile cone penetrometers, soil spectroscopy, and remote sensors, are introduced and critically discussed. Special attention is given to the limitations of each emerging technology and the impact on environmental soil-landscape modeling. Geographic and soil information systems have served as integrators to manage and analyze soil and other environmental datasets. The Web-based distribution of soil-landscape data through the Internet has enabled global data sharing.

5.1 WHAT IS EMERGING?

5.1.1 SOIL MAPPING TECHNOLOGIES

Since the beginning of the systematic study of soils, various concepts have been adopted to map soils and their properties. Emerging soil mapping techniques have had a major impact on how we map and describe soils, resulting in a paradigm shift in pedology and soil science. Early soil surveyors used soil augers, spades, pencils, and notebooks. These were later complemented by laboratory measurements. The tacit knowledge of the soil surveyor to comprehend a soil-landscape was the basis for characterizing the spatial distribution of different soils. Soil surveying based on soil taxonomies has been focused on the mapping of morphological, physical, chemical, and biological properties, with less emphasis on pedogenesis. The use of aerial photographs for soil mapping, which began during the late 1920s and early 1930s, greatly increased the precision of plotting soil boundaries. Constraints on these hard-copy soil maps were imposed by the experience of the soil surveyor and available budgets. For example, in the U.S., for a standard soil survey at a map scale of 1:24,000, an average of one auger boring per 14 ha (40 acres) has been used.

The heterogeneous nature of soils across a landscape has long been recognized, and different concepts and sampling designs have been employed to model the variation. At large (field) scale, high-intensity soil surveys can be conducted. For example, Lark[1] used intensive grid sampling (20-m intervals) and mapped soil texture and soil depth to identify seven map units across a 6-ha field. At small (coarse) scale, such high-density sampling is not feasible due to limited budgets and time constraints. Alternative methods are in need to improve existing soil surveys to provide high-quality and high-resolution soil maps. With the advent of global positioning systems (GPS), geophysical soil mapping techniques, soil sensors, and soil spectroscopy, it has become possible to map larger areas rapidly with higher sampling densities, resembling more exhaustive datasets for fine-scale mapping. Recent technological advances offer new opportunities for spatially explicit mapping of heterogeneous soil patterns. However, limitations still persist in terms of accurate and precise mapping of soil properties. In this chapter we will discuss a variety of emerging geographic information technologies (GIT) and their potentials and limitations in the characterization of soil-landscapes.

5.1.2 SOIL MAPPING PARADIGMS

The perceptions of soils and soil mapping paradigms have changed with time. Early soil mapping was influenced by the use of soils for farming, ranching, and forestry. In recent years the effort has shifted to acquire more quantitative and accurate site-specific soil data. Bouma et al.[2] pointed out that soil data are important for precision agriculture (PA), but current soil survey data do not satisfy precision agriculture requirements, including an appropriate level of detail and soil property information. The PA industry explored various high-density soil mapping and geostatistical techniques to produce fine-scale soil property maps.[3] Likewise, environmental assessment studies require detailed information about the spatial and temporal distributions of soil properties. The U.S. Department of Agriculture (USDA) Natural Resources Conservation Service (NRCS) and its predecessors, with local and state agencies and land grant universities, have been generating soil information in the U.S. for over 100 years. Although originally focused on the agricultural use of soil data, the mission of NRCS is now much broader, i.e., "to help people conserve, improve and sustain our natural resources and the environment."[4] Though the need for high-quality and high-resolution soil property maps has been recognized, we are still struggling to meet the demands of farmers, environmental scientists, the forest industry, and others.

In order to reduce soil information to a manageable form, soils are classified at various levels of hierarchy (taxonomic units). The crisp geographic data model is a well-adopted model for soil mapping at the landscape scale.[5] This *entity-based approach*[6] or *object view*[7,8] defines the real world as an arrangement of discrete, well-defined objects that are characterized by their geometrical and topological properties and by their nonspatial attributes (Figure 5.1). It is seldom, however, that a soilscape can be adequately described and defined from one pedon.[9] Therefore, the relationships of soils with observable landscape features (topography, vegetation, parent material, etc.) are used to infer the variability of

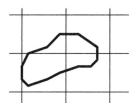

FIGURE 5.1 Representations of soils: (1) polygon, entity or object view, and (2) raster, field view.

soil properties, thereby interpolating the individual pedon properties/observations over the study area.[10] Based on such observed properties, similar soils are grouped together and translated into soil classes.[9] Soil properties are considered homogeneous within the map unit and sharp breaks in soil properties are assumed, forming the polygon boundaries. This soil mapping concept has been questioned extensively.[11–13] Crisp soil mapping is practical; however, it ignores the continuous spatial variation in both soil-forming processes and in the resulting soils. Burrough et al.[14] consider it a double-crisp model because the identified soil groups are supposed to be crisply delineated in both taxonomic space (the space defined by the soil properties) and geographic space (defined by the map unit boundary). Bie and Beckett[15] demonstrated that even with aerial photo identification (API), different surveyors choose radically different boundary spacing, and not every important change in soil properties occurred at physiographically distinct locations.

The raster-based or *field view* geographic data model[6,7] evolved as an effort to address the digital representation of continuously varying environmental properties such as soils (Figure 5.1). Commonly, geostatistical techniques[16–18] are employed to create continuous raster-based maps that model the spatial autocorrelation of soil properties measured at georeferenced locations. Typically, deterministic or stochastic methods are employed to characterize soil variation along with the prediction uncertainty.[19,20] The spatial discretization units in continuous maps are pixels or voxels whose size depends on the underlying spatial variability, sampling density, and survey design. McBratney et al.[21] suggested three levels of resolutions that are of interest: >2 km, 2 km to 20 m, and <20 m, which correspond to global/national, catchment/landscape, and local extents, respectively.

An alternative approach to map soils, considering its continuous nature, is provided by fuzzy sets, first introduced by Zadeh.[22] Fuzziness, a concomitant of complexity, is a type of imprecision characterizing classes that for various reasons cannot or do not have sharply defined boundaries or units. Numerous authors[23–28] describe the fundamental principles, operations, and applications of fuzzy sets to soils. Fuzzy methods allow the matching of individuals to be determined on a continuous scale instead of a Boolean binary or an integer scale. Contrary to crisp sets that assign membership values of 0 and 1, a fuzzy set is a class that admits the possibility of partial membership. Hence, fuzzy sets are generalizations of crisp sets to situations where the class boundaries are not or cannot be sharply defined. Though numerous applications of fuzzy set theory to soil-landscape modeling exist,[25,27–29] it has not been widely adopted. Reasons are seen in the fact that users prefer to have one crisp output map (e.g., bulk density map), rather than complex fuzzy output maps, which are more challenging to interpret. Defuzzification has been suggested to backtransform the fuzzy output into crisp soil data.[30] However, such a procedure opposes the paradigm of fuzzy set theory, i.e., the fact that our world has many shades of gray rather than being black and white. There has been extensive criticism about the development of membership functions for fuzzy application using soft input or subjective expert knowledge.

Past studies have clearly demonstrated that there is no single, overall soil classification and mapping paradigm that can be used uncritically at all locations and at all levels of resolution.[14] Regardless of the approach employed, soil mapping must take into account prevailing processes of soil formation, differences in lithology, landform, drainage, and others for delineating soil properties or classes. Continuous maps focus on specific soil properties, contrary to crisp maps, where soil properties are lumped into soil classes. In this regard, continuous soil maps are better suited to soil investigations focusing on specific purposes (so-called technical soil surveys), while crisp maps provide an overall picture of the soil variation across the landscape. Few studies[31,32] have shown that crisp soil mapping can provide predictions that are equally as good as continuous soil maps obtained by geostatistical techniques. Rogowski[33] compared taxonomic soil survey data to continuous soil data developed using geostatistical methods at farm and watershed scales. Results showed that the variability of bulk density and hydraulic conductivity can be estimated by comparing regularized variograms of measured and published values. These findings provide avenues of incorporating crisp soil data with raster-based soil datasets, reconciling the two contrasting geographic data models.

5.1.3 DATA MANAGEMENT

Historically, hard-copy soil maps were produced that show crisp soil map units as polygons and associated legends. With the advent of geographic information systems (GIS) storage, management, analysis, and display of soil data have changed tremendously. The emergence of GIS has had a pivotal impact on soil surveying. In the early 1980s, vector-based GIS emerged, which were used to represent crisp soil map units, while raster-based GIS were developed mainly for remote sensing applications. Vector and raster-based GIS eventually merged to become hybrid GIS, providing both geographic data models within one software package.

Spatial data are stored in the form of database management systems (DBMS).[6] Modern DBMS use many methods for efficiently storing and retrieving data, but all are based on three fundamental means of organizing information, which also reflect the logical models used for real-world structures known as the hierarchical, network, and relational schemata. Hierarchical systems of data organization are adopted in soil taxonomies (e.g., *U.S. Soil Taxonomy*).[106] Keys enable ease of access and retrieval of data. A disadvantage of hierarchical databases is large index files, which have to be maintained, and certain attribute values may have to be repeated many times, leading to data redundancy, which increases storage and access costs. The network database structure, though similar to the hierarchical database structure, avoids redundancy and linkage problems. Most commonly used in GIS are relational database (RDB) structures. An example of the RDB used in the Soil Survey Geographic (SSURGO) database is given in Figure 5.2. The data are stored in simple records, known as tuples, which are sets of

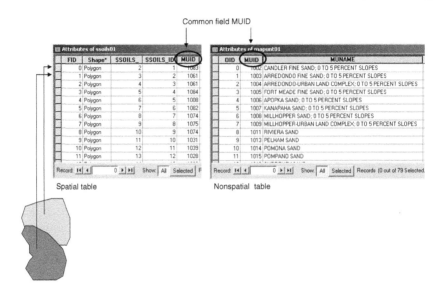

FIGURE 5.2 Spatial and nonspatial tables from SSURGO. The tables are stored in RDB format. Attribute fields from different tables can be associated using common fields (e.g., map unit identification (MUID) attribute field).

fields each containing an attribute. Tuples are grouped together in tables known as relations. Each table or relation is usually a separate file. Identification codes are used as unique keys to identify the records in each file. Data are extracted from a relational database defining the relation that is appropriate for the query. Advantages of RDBs are their flexible structure, ease to update, and nonredundancy. Disadvantages occur where the relationships between tables are complex and a number of joins are needed to produce a soil map. Such operations are error prone. Recently, a fourth structure has been introduced to GIS, which is called object orientation structure.[6] Object database management systems (ODBMS) integrate database capabilities with object programming language capabilities. An ODBMS makes database objects appear as programming language objects in one or more existing programming languages (e.g., C++, Java). Object database management systems extend the object programming language with transparently persistent data, concurrency control, data recovery, associative queries, and other database capabilities. Object-oriented databases use abstract data types, which add flexibility to a DBMS; however, there are two constraints that limit their application. First, the market adoption of object-oriented databases has been limited, despite the availability of such products for several years. Second, the Structured Query Language (SQL) is tightly coupled with the relational database model. SQL is a declarative language where users only specify the desired result rather than the means of production.[34]

Burrough[35] argues that GIS facilitate flexible weighting of properties in accordance with a specific purpose. Hence, the boundaries in technical groupings often cut across taxonomic subdivisions. Dudal[36] points out that there is no need anymore for data aggregation into taxonomic units to efficiently manage spatial datasets. Databases that contain geo-referenced soil property data enable users to select and group clusters of relevant soil property data by function of demand.

Emerging geographic information technology enables us to integrate geospatial datasets of soils and other resources. It facilitates universal data sharing across the Internet as outlined in the National Spatial Data Infrastructure (NSDI), which is a concept defined as the technologies, policies, and people necessary to promote sharing of geospatial data throughout all levels of government, the private and nonprofit sectors, and the academic community. The Federal Geographic Data Committee (FGDC) sponsors a decentralized system of servers called the Geospatial Data Clearinghouse, which is a collection of over 250 data servers that have geographic data primarily for use in GIS, image processing systems, and other modeling software. Such a framework enables users around the globe to access and share digital soil datasets.

5.2 EMERGING SOIL-LANDSCAPE MAPPING TECHNOLOGIES

5.2.1 GLOBAL POSITIONING SYSTEMS

Early soil surveyors recognized the importance of mapping soils in the context of geographic space; i.e., it mattered where (at which location) a soil observation was made and how it related to adjacent soils, topography, land use, and other properties. Reference systems such as the U.S. Public Land Survey System featuring relative georeferencing based on nested grids such as townships, ranges, and sections improved the capabilities to reference a soil sample to a specified geographic location. However, the uncertainty of georeferencing was relatively high. The introduction of GPS to soil mapping was pivotal, providing accurate and precise geographic coordinates of soil observations. To characterize soil-landscapes, each soil sampling location is georeferenced using a GPS based on x (easting) and y (northing) coordinates. Global positioning systems combined with GIS provide a powerful toolset to build soil information systems (SIS).

Global positioning systems consist of a satellite segment, control segment, and user segment.[37] The GPS receiver in the user segment records the radio signals broadcasted by each satellite, which are processed to obtain its distance (range) from the satellite.[38] The distance of the receiver from a minimum of three (more is even better) satellites defines a unique position in two- and three-dimensional space, respectively. An alternative to such autonomous GPS is a differential GPS (DGPS), which entails establishing a base station with a true coordinate location determined using high-accuracy survey methods. The GPS-measured position for the base station is compared to its predetermined location to define an error vector, which is used to correct the position measured by GPS units elsewhere, either in real time (real-time DGPS) or after the data are collected (post-processing DGPS).

Each dilution of precision — positional, vertical, horizontal, timing, and geometric — and environmental conditions contribute to the overall final positional accuracy of the GPS measurement. The current Standard Performance Service (SPS) published by the U.S. government is ±37 m horizontal and 77 m vertical 95% of the time.[37] With selective availability (SA) turned off, civilian use of GPS is often much better and regularly achieves a horizontal accuracy of about 20 m, while having a vertical accuracy of about 50 m. Even handheld GPS units facilitate relatively accurate mapping of geographic coordinates, while DGPS are capable of mapping geographic locations with submeter accuracy.

5.2.2 GEOPHYSICAL TECHNIQUES

For conservation, political, ethical, and cultural reasons, destructive auger-based soil sampling is not feasible at all sites[39] and prohibitive for spatially dense sampling.[40] The lack of sensitive tools to detect subtle shifts among soil properties limits the spatial delineation of soil variability.[41] As an alternative, geophysical methods that are noninvasive and reliable can complement the soil mapping effort.[42]

5.2.2.1 Electromagnetic Induction

Electromagnetic induction (EM) facilitates mapping the electrical conductivity (EC) of soils. Electrical conductivity correlates to soil properties affecting crop productivity, including soil texture, cation exchange capacity (CEC), drainage conditions, organic matter, salinity, and other subsoil characteristics. Other applications combine EC data with sparser georeferenced point soil observations to produce thematic soil maps. Electrical conductivity surveys are also used to prescreen a soil-landscape. Maps are then used to identify heterogeneous areas that are targeted for more detailed auger-based soil mapping.

Traditionally, the soil paste method[43] has been used to assess soil EC, but more recently commercial devices have become available to measure and assess bulk soil EC rapidly and economically across sites. Georeferenced *in situ* estimates of EC are now being made at the field scale using both direct contact sensors to measure resistance and noncontact sensors based upon EM technology.[44] In EM surveys, a sensor in the device measures the electromagnetic field induced by current inserted into the soil. The strength of this secondary electromagnetic field is directly proportional to the apparent electrical conductivity (ECa) of the soil. Hartsock et al.[45] summarized the variation in soil electrical conductivity as being attributed to multiple factors: amount and connectivity of soil water, bulk density, soil structure, water potential, timing of measurement, soil aggregation (e.g., cementing agents such as clay and organic matter, soil structure), electrolytes in soil water (e.g., exchangeable ions), soil temperature, the conductivity of the mineral phase (e.g., types and quantities of minerals, degree of isomorphic substitution, exchangeable ions), and more. Despite the multiple causes of EC variability, bulk soil ECa measurements have been related to individual factors that limit soil use and productivity, such as salinity,[46] moisture, clay content, calcium and magnesium content, depth to bedrock and fragipan,[45] and soil horizons.[47] Yoder et al.[48] predicted the movement of agrochemicals, and Sudduth et al.[49] predicted depth to topsoil using EM. We like to point out that EC provides surrogate information, and often there are multiple causes generating variable fields of EC measurements. However, EM surveys facilitate rapid mapping of soil-landscapes, generating exhaustive datasets. Hence, EC is a powerful tool to map soil variation across fields. Generally, the use of EM has been most successful in areas having reasonably homogeneous subsurface properties with a minimal sequence of dissimilar subsurface layer.[50] EM mapping coupled with traditional soil sampling can provide a compromise between soft and exhaustive, and hard (quantitative) and sparse soil data, respectively. Two instruments widely used for measuring bulk soil EC *in situ* are the EM38 meter (Geonics Limited, Mississauga, Ontario) and the Veris System (Veris Technologies, Salina, KS). The EC meters respond to the average bulk soil EC between the surface and a maximum depth of about 1 m (Veris) or 1.5 m (EM38).

5.2.2.2 Ground-Penetrating Radar

Ground-penetrating radar (GPR) is one of the geophysical exploration and subsurface delineation techniques that has proven to aid greatly in site characterization and mapping. The GPR method has found widespread acceptance for shallow subsurface mapping because it can detect shallow underground discontinuity and heterogeneity,[51] covering a wide area in a short period with high spatial resolution.[52]

Ground-penetrating radar information is acquired by reflecting radar waves off subsurface features. The ground-penetrating radar antenna is pulled along the ground by hand or behind a vehicle. The radio waves are propagated in distinct pulses from a surface antenna, reflected off subsurface features, and detected back at the source by a receiving antenna. The travel time of the energy pulses and the velocity change of radar waves can accurately trace the distance or depth to the subsurface feature. The electrical and magnetic properties of rocks, soils, and fluids (natural materials) control the speed of propagation of radar waves and their amplitudes. At radar frequencies, electrical properties are dominantly controlled by rock or soil density, chemistry, state (liquid/gas/solid), distribution (pore space connectivity), and content of water. Ground-penetrating radar measures differences in the dielectric constant of subsurface features. Dielectric constants range from 1 (air), 4 (dry sand), 5.5 (dry limestone), 6 (wet sandstone), 11 (till), 23.5 (wet sandy soil), 27 (wet clay), 64 (organic soil), to 81 (water). Ground-penetrating radar waves can reach depths up to 30 m in low-conductivity materials such as dry sand or granite. Clays, shale, and other high-conductivity materials may attenuate or absorb GPR signals, greatly decreasing the depth of penetration to 1 m or less. The resolution is controlled by the wavelength and polarization of the electromagnetic energy, the contrast in electromagnetic properties, and the size, shape, and orientation geometry of the target. The resolution increases with increasing frequency (decreasing wavelength), but at the expense of depth of investigation.

Although earlier GPR systems recorded the raw subsurface reflections on paper printouts, currently used GPR systems are equipped to record the reflections in digital format. Conyers and Goodman[39] discussed some of the field and post-acquisition techniques to acquire, process, and interpret GPR data. Typically, radar antennas are moved along the ground and two-dimensional profiles of a large number of periodic reflections are created, producing profiles of subsurface features and stratigraphy. In case data are acquired in a series of transects, and the reflections are correlated and processed, it is possible to create three-dimensional models of subsurface features. Ground-penetrating radar provides higher resolution of subsurface features than EM, but it is more depth restricted.[50]

Gish et al.[53] used georeferenced GPR data in concert with EM data to identify subsurface restricting layers. These data were coupled with hydrological models in a GIS to determine potential flow pathways from topographic maps of a subsurface restricting layer. Doolittle and Collins[50] compared EM and GPR techniques in areas of karst and found GPR effective in determining the thickness of surface layers and locations of buried solution features, while the presence of multiple, contrasting soil horizons and layers weakened relationships and created nonunique interpretation for EM. In Florida, GPR is used extensively to update soil surveys.[42] Ground-penetrating radar is most successful in sand-rich soils (e.g., Florida soils), but it fails in silt-rich material due to confounding impact of soil moisture and other factors. Successful applications of GPR include identification of subsurface features and cavities,[50] identification of soil layers and subsurface flow channels,[53] archaeological investigations,[39,54] fractures and faults in geological formations,[51] and more. The integrated use of GPR and EM, supported by ground truth verification, increases the confidence of subsurface feature interpretations.[50] However, many GPR applications are qualitative in nature, highlighting subsurface variability. The true nature of the variability is often confounded by many interrelated soil properties. Therefore, it is challenging to derive quantitative relationships between soil properties and GPR measurements.

5.2.3 SENSORS

5.2.3.1 Soil Sensors

Soils impose resistance to penetration by virtue of texture, structure, porosity, water content, cementing agents, and compaction.[55] A profile cone penetrometer (PCP) is an instrument in the form of a cylindrical rod with a cone-shaped tip designed for penetrating soils and for measuring penetration resistance expressed as cone index (CI). Two standards for PCP applications exist: the American Society for Testing and Materials[56] and the American Society of Agricultural Engineers[57] standards, which differ mainly in the cone apex angle of 60° and 30°, respectively. Truck or all-terrain vehicle (ATV)-mounted constant-rate PCPs provide more reliable CI data than handheld push penetrometers (Figure 5.3). Clustered cone index profiles collected on a grid with 10-m spacing on a 2.73-ha site in southern

FIGURE 5.3 Truck-mounted PCP system. (Courtesy of Soil Science Department, University of Wisconsin, Madison.)

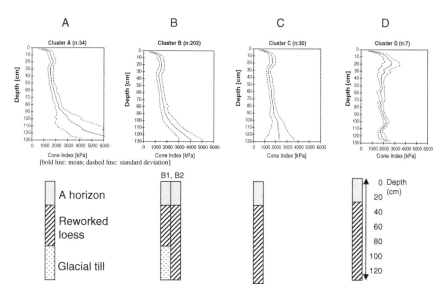

FIGURE 5.4 Cone index curves ($n = 273$) were clustered into four groups and related to soil horizons/materials. Data were collected at a 2.73-ha site in southern Wisconsin on a grid with 10-m spacing.

Wisconsin are shown in Figure 5.4. The identified four clusters were related to soil materials. A detailed description of the study can be found in Grunwald et al.[55]

Among the soil characteristics that influence penetration resistance are soil texture, porosity, structure, water content, cementing agents, organic matter, and compaction. Numerous authors related measured CI to soil properties. Rooney and Lowery[58] mapped soil horizons. Correlations between particle size and penetration resistance were presented by Kasim et al.,[59] Kurup et al.,[60] and Puppala et al.,[61] where coarse-textured soils showed greater penetration resistance than fine-textured soils. Water content and organic matter content are inversely related to penetration resistance,[62,63] while cementing agents like carbonate, silica, hydrous silicate, and hydrous iron oxide[61] increase penetration resistance. Lowery and Schuler[64] showed that penetration resistance and bulk density increased with increasing level of compaction. Grunwald et al.[65] derived pedotransfer functions relating cone index to soil texture, bulk density, and soil moisture content. Grunwald et al.[55] used a profile cone penetrometer to distinguish between glacial till and reworked loess soil materials in a glaciated landscape in southern Wisconsin. The CI data combined with soil property and topographic data were used to reconstruct the soil-landscape using clustering and multidimensional kriging.

The advantage of rapid nonintrusive mapping using PCP is limited by the fact that penetration resistance is the response signal to soil-factor combinations. Hence, CI represents a surrogate of soil properties such as soil moisture, soil texture, and bulk density in varying quantities. Profile cone penetrometer data can be used in combination with other soil sensors or traditional soil mapping techniques. A combination of tip and sleeve measurements and a soil moisture probe with sparse soil observations were explored by Soil and Topography Information, LLC (Madison, WI) to make inferences on soil horizons and soil texture. A color video camera was integrated in a PCP to record soil profiles while measuring CI[66] (Figure 5.5). Image processing techniques can be used to derive numerous soil properties (e.g., soil color, soil texture, soil structure). For example, Schulze et al.[67] found that soil color is closely related to soil organic matter. The integration of multiple soil sensors and traditional soil mapping techniques will provide new and exciting avenues to generate high-resolution and quality choropleth soil maps in the near future.

5.2.3.2 Soil Spectroscopy

Conventional soil extraction methods of phosphorus, nitrate, carbonates, metals, and others are labor intensive, expensive, and time consuming.[68] In addition, dense sampling is required to adequately characterize spatial variability of a soil-landscape, making broad-scale quantitative evaluation difficult.[69] As reflectance and emittance behavior of soil is highly dependent on its biochemical and physical fabric,[70] the analysis of such characteristics can provide information on soil properties. Soil spectroscopy refers to the use of sensing instruments[71] to measure the absorption, emission, or scattering of electromagnetic radiation from soil to qualitatively or quantitatively study soil properties. Diffuse reflectance spectroscopy is increasingly used for the rapid nondestructive characterization of a wide range of materials.

For soils, the basic application range consists of the visible (350 to 700 nm), near-infrared (NIR) (700 to 2500 nm), and mid-infrared range (MIR) (2500 to 25,000 nm). While NIRS has developed into a major tool for analytical determinations over the past two decades, the main use of mid-infrared has been for research or qualitative analysis involving spectral interpretation.[72] A number of

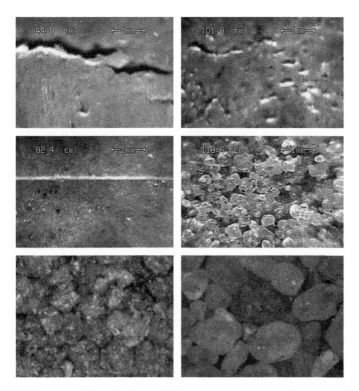

FIGURE 5.5 Images collected with a soil imaging penetrometer. (Courtesy of Soil and Topography Information, LLC, Madison, WI.) (See color version on the accompanying CD.)

scientists[72,73] found that MIR diffuse reflectance spectroscopy can be performed with accuracy equal to or greater than that achieved using NIR spectroscopy. For example, soil organic matter is related to reflectance in the 2.5 to 25 μm range, but their overtones (at one half, one third, one fourth, etc. of the wavelength of the fundamental feature) occur in the near-infrared 0.7 to 1.0 μm and shortwave 1.0 to 2.5 μm regions.[74]

Spectral reflectance signature libraries of numerous material samples and composites have been cataloged.[74–76] From these libraries, unknown samples can be interpreted for soil properties. First, the spectral signatures of soil samples are scanned with a spectroradiometer in the field or laboratory. Soil properties are then measured, designated to sample the variation in the spectral library, and calibrated to soil reflectance. Chemometric models to predict soil properties from soil spectra can be built using classification and regression trees (CARTs), multivariate adaptive regression splines (MARSs), partial least square regression (PLSR), and other statistical methods. The resultant functions are then employed to predict the soil properties for new samples that belong to the same population as the library soils.[74] The models are evaluated using statistical tools (e.g., cross-validation, validation, coefficient of determination). Therefore, the success of using spectral libraries to characterize soil properties depends primarily on the ability to build robust models between measured soil properties and soil reflectance spectra.

Research has demonstrated the ability of reflectance spectroscopy to provide nondestructive, rapid prediction of soil physical, chemical, and biological properties (e.g., calcium, magnesium, iron, manganese, and potassium),[77] soil texture,[78] total carbon, total nitrogen, and pH.[72] Soil structure has influence on the reflectance behavior of soil, which is probably why Daniel et al.,[70] Udelhoven et al.,[77] Couillard et al.,[78] and Sudduth and Hummel[81] found laboratory spectrometry performed better than field spectrometry.

Despite the tremendous scope of reflectance spectroscopy to characterize soils, there is need for research in the use of spectroscopy for soil characterization and mapping. Considering the problems of atmospheric interferences, shade and shadow, etc., associated with remote sensing from space-based platforms,[76] soil spectroscopy offers a reliable alternative for accurate on-the-go mapping of soil properties. Since most satellite and airborne sensors are limited to the topsoil, spectroscopy offers rapid analysis of multiple soil properties from one soil sample. Rapid assessment of soils using spectral models will dramatically cut sample processing time and cost. Reduced sample processing costs may facilitate sampling soils with much higher density and frequency. Research in data acquisition (e.g., instrument sensitivity) and data analysis (statistical chemometric modeling) remains an area of continuous research.

5.2.3.3 Remote Sensors

Remote sensing is a relatively cheap and rapid method of acquiring up-to-date information over large geographical areas. Although satellite and airborne images contain tremendous information in various spectral bands, data transformation techniques such as tasseled cap[80] and vegetation indices[76] (e.g., normalized difference vegetation index (NDVI)) reduce the number of bands and provide a more direct association between signal response and physical processes on the ground, highlighting users' interest. Passive remote sensing employs sensors that measure radiation naturally reflected or emitted from the ground, atmosphere, and clouds in the visible, near-infrared, and short-wave infrared range. In contrast, active remote sensing techniques employ an artificial source of radiation as a probe. The resulting signal that scatters back to the sensor characterizes the atmosphere or Earth.

Various satellite systems operating within the optical range generate data on Earth resources at different resolutions.[76] Landsat Enhanced Thematic Mapper (ETM 7+) scenes have a grid resolution of 30 m. Hyperspectral sensors have multiple bands (e.g., hundreds of bands) with spectral resolutions less than 20 nm. Hyperspectral images, such as NASA's Airborne Visible/Infrared Imaging Spectrometer (AVIRIS), with 224 contiguous spectral channels and approximately 20-m resolution, and IKONOS multispectral images, with 1-m panchromatic and 4-m multispectral resolution, provide a spatial resolution we have not seen before.

These spatial resolutions contrast readily available crisp soil survey maps derived from few auger borings per hundreds of hectares. For example, currently SSURGO soil maps derived at a map scale of 1:24,000 provide the most detailed soil data in the U.S. In Florida, the average soil map unit size in SSURGO is 605,176 m^2, which contrasts the resolution of remote sensing imagery.

Exhaustive pixel-based thematic maps can be derived from satellite imagery that are valuable to support soil mapping efforts (e.g., the selection of representative sites for field investigations), to provide auxiliary data to predict soil properties, assess environmental quality, and develop recommendations for land resource management. Georeferenced soil properties can be integrated with remotely sensed images and upscaled to regional scale. Remote sensing combined with GIS offers a powerful toolset to support soil mapping. Current limitations of remote sensors for mapping of soil properties have been (1) the lack of penetration depth, which is often limited to the top centimeters of the soil, and (2) vegetation cover, which obscures the mapping of soil properties. Radar and gamma radiometry have overcome these limitations. Radar sensors are noted for their ability to penetrate clouds, fog, and rain, as well as an ability to provide nighttime reflected imaging by virtue of their own active illumination. The synthetic aperture radar (SAR) aboard the European remote sensing satellites ERS-1 and ERS-2 is an imaging technology in which radiation is emitted in a beam from a moving sensor, and the backscattered component returned to the sensor from the ground is measured. These images have been used to map soil moisture and climate characteristics at landscape scale.

Remote sensing applications are manifold and include soil organic matter mapping,[80] land cover mapping,[82,83] land classification,[84] delineation of salt-affected areas,[85] change detection of land use,[86] and many more. Kasischke et al.[87] used ERS SAR imagery for monitoring surface hydrologic conditions in wetlands of southern Florida. The results showed wide variation in ERS backscatter in individual sites when they were flooded and nonflooded. Surface soil moisture was retrieved using microwave radiometry by Pardé et al.[88] using the L-band. To date, there is no space-borne sensor measuring the microwave emission of the soil surface at this frequency, although several new programs are scheduled. Hyperspectral images of AVIRIS and Hyperspectral Mapper (HyMap) were used to map expansive clay soils in Colorado, focusing on smectites, smectites/illites, and kaolinites.[89] Mapping of vegetation, geology, and soils using AVIRIS imagery was conducted by Drake et al.[90] in a semiarid shrubland/rangeland soil-landscape using short-wavelength infrared (2 to 2.5 µm).

5.3 GEOGRAPHIC AND SOIL INFORMATION SYSTEMS

5.3.1 GEOGRAPHIC INFORMATION SYSTEMS

Geographic information systems, which help to manipulate, analyze, and present information that is tied to a spatial location, have revolutionized the way we manage soil and other environmental datasets. Modern GIS emerged in the 1960s. The history of GIS is unique in that it was developed nearly concurrently by separate research teams at different locations with different backgrounds. One of the earliest accounts of a computerized GIS is the Canada Geographic Information System (CGIS) developed in the early 1960s. Meanwhile, the Harvard Laboratory for Computer Graphics and Spatial Analysis (HLCGSA) created an automated mapping application, SYMAP, which served as the training ground for many of the scientists that developed and created the precursors to the popular GIS packages used today. For the U.S. Census study in 1966, new methods were developed incorporating topology, i.e., spatial relationships between connecting or adjacent vector features. While the 1960s served as the decade of GIS development, the 1970s were years of lateral diffusion.[91] More universities and government agencies became interested in the technology, expanding the user base of GIS. Since the early 1970s, the Environmental Systems Research Institute, Inc. (ESRI) became one of the dominating players providing GIS software, GIS datasets, and GIS services to a worldwide audience. While hardware became cheaper, faster, and more powerful, software evolved, providing sophisticated spatial operations to users. For example, GIS enabled users to integrate soil-forming factor layers and relate them to soil data within a spatially explicit framework. Geographic information systems are versatile to provide representations in the form of crisp and raster-based soil maps. They are scalable, expandable, and provide ease to update soil and other datasets. Soil scientists were able to implement the conceptual soil-landscape models developed several decades earlier within a GIS environment. Spatial operations include overlying, extracting, and generalizing spatial data to name only a few. Complex statistical and geostatistical methods were embedded within the GIS environment providing ease of use.[92] For example, advanced geostatistical techniques within a GIS environment were used by Grunwald et al.[93] to create soil quality maps that characterize the spatial distribution of soil phosphorus. Environmental variables derived from auxiliary sources (e.g., remote sensing, GPR, EM, etc.) can be combined with primary soil data (e.g., soil texture) in a GIS to derive secondary/functional soil data (e.g., soil quality index). Spatially explicit relationships between topographic, geologic, vegetation, and soil properties can be quantified using GIS. Soil GIS scientists have been creative in developing predictive soil-landscape models with the support of GIS. Most important, GIS forced soil scientists to quantify previously descriptive soil datasets. To summarize, GIS has revolutionized how we manage, analyze, and present spatial data, including soil-landscape data. It has served as an integrator of soil and other environmental datasets, providing the platform for quantitative soil-landscape modeling. Geographic information systems are used by state and federal agencies, universities, consulting companies, and others, becoming as commonplace as a pencil to a traditional soil surveyor.

To ensure the sharing of spatial data between different user groups, agencies, applications, and platforms, a variety of spatial data standards were developed. The Spatial Data Transfer Standard (SDTS) was developed by the Federal Geographic Data Committee (FGDC). In the Open GIS Consortium (OCG), numerous private companies, government agencies, and academic institutions work side by side to develop publicly available geoprocessing specifications. For example, the Open GIS Consortium developed the Geographic Markup Language (GML), which is an open standard to enable transfer of spatial data between different vendors. The Geographic Markup Language is the geographic counterpart of the eXtensible Markup Language (XML) used for metadata documentation. GML encodes geographic information, including both the geometry and properties of geographic features. The FGDC of the U.S. has defined a Content Standard for Digital Geospatial Metadata (CSDGM) providing standardized guidelines for the documentation of spatial datasets. The CSDGM was adopted by the USDA NRCS to disseminate soil data via SIS. Besides these formalized spatial standards, a double standard based on ESRI's vector and raster data formats exists (e.g., ESRI's shapefile data format).

In recent years, the adoption of GIS technology by soil scientists resulted in dramatic changes. Geographic information systems developed from mainframe GIS to desktop GIS and WebGIS. WebGIS refers to the use of the World Wide Web (WWW, in short Web) as a primary means to integrate, disseminate, and communicate geographic information.[94,95] Spatial applications provided on the Web include (1) data services that enable users to retrieve data from Web-based geodatabases and (2) map services that provide users with online display capabilities of thematic maps. Distributed geographic information (DGI) refers to the use of Internet technologies to distribute geographic information in a variety of forms, including maps, images, datasets, spatial analysis operations, and reports.[96] DGI is the most encompassing framework, including both WebGIS and mobile GIS (e.g., pocket PCs with wireless connections). Peng and Tsou[97] suggest expanding GIS data services by including the dissemination of spatial science knowledge and GIS output, which is commonly called GIScience. Examples of state-of-the-art WebGIS applications are given by Mathiyalagan et al.,[98] who developed a WebGIS and geodatabase for Florida's wetlands, providing data and map services related to soil and vegetation data (Figure 5.6). The Florida Geographic Data Library (FGDL) (http://www.fgdl.org/) provides a repository of spatial data for Florida using the data service concept.

Only a few GIS-based representations of soil-landscapes are three-dimensional, and even less are four-dimensional, considering changes of soil-landscape properties through time. Roshannejad and Kainz[99] developed a logical data model for a space–time information system. The model considers x (easting), y (northing), z (depth), t (time), and a (attributes). Time is considered the fourth dimension, relegating the attribute values to the fifth dimension (Figure 5.7). Data evolution can be described by a sequence of records $\{(s, t, e)\}$, where s is a spatial coordinate, t is a time stamp, and e is an ecosystem function representing behavior (e.g., illuviation, mineralization, nitrate leaching). To maintain data consistency, spatiotemporal indexing can be used. Few prototypes of space–time information systems have been presented. Abraham and Roddick[100] presented a detailed review of spatiotemporal database concepts. Koeppel and Ahlmer[101] distinguished between attribute-oriented spatiotemporal systems that track

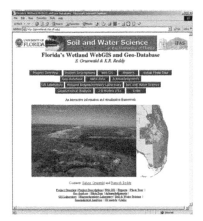

FIGURE 5.6 WebGIS and geodatabase for Florida's wetlands, including map and data services (http://GISWetlands.ifas.ufl.edu). (See color version on the accompanying CD.)

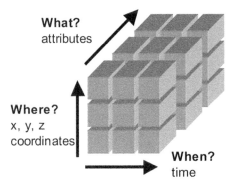

FIGURE 5.7 Logical data model. (After Roshannejad, A.A. and Kainz, W., *ACSM/ASPRS Annu. Convention Exposition Tech. Pap.*, 4, 119–126, 1995.)

changes in spatial entities and topology-oriented spatiotemporal systems that track changes in positional information about features and their spatial relationships. Peuquet and Duan[102] suggested an Event-Based Spatio-Temporal Data Model (ESTDM) focusing on events that are represented along a temporal vector in chain-like fashion. Yuan[103] suggested a three-domain model representing semantics, space, and time separately and providing links between them to describe geographic processes and phenomena. Ramasundaram et al.[104] extended two-dimensional GIS operations to the third (space) and fourth (time) dimensions. They developed three-dimensional soil-landscape models and interactive space–time hydrologic simulations for a flatwood site in Florida that are disseminated via the WWW. Results from a space–time GIS application using nitrate–nitrogen data for a site in northern Florida are shown in Figure 5.8. The model was created using multidimensional ordinary kriging of nitrate–nitrogen

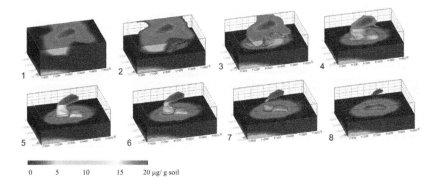

0   5   10   15   20 µg/ g soil

FIGURE 5.8 Nitrate–nitrogen plume at different periods ($n = 1$ to 8) implemented within a GIS (EVS-PRO). The space–time model shows change in attribute values and geometry of the plume. (See color version on the accompanying CD.)

at eight different time periods using EVS-PRO (CTech Development Corp., Huntington Beach, CA). The Spatio-Temporal Environment Mapper (STEM), a GIS system that handles time and depth of an entity in addition to mapping the entity horizontally, was presented by Morris et al.[105] They argued to treat time and depth (z) as a dimension rather than an attribute, which is a prerequisite to effective multidimensional visualization and analysis. This opposes the common view of modeling three-dimensional solid objects (e.g., representation of soils as three-dimensional objects). Rather, the x (easting) and y (northing) coordinates are extended by time and a depth dimension. Though concepts and prototype space–time information systems exist, more research is necessary to develop a universal tool that can manage, analyze, and visualize soil-landscape data in space and through time.

5.3.2 Soil Information Systems

To disseminate soil data flat files, hard-copy and digital maps, geodatabases, GIS, or the WWW are used in numerous variations. Specialized geodatabases for soil data have been developed at different spatial scales, for different geographic regions and data formats. Commonly, they are referred to as soil information systems.

At the global scale the Food and Agriculture Organization of the United Nations (FAO)–United Nations Educational, Scientific and Cultural Organization (UNESCO) *Soil Map of the World* was the first attempt to cooperatively develop a standardized soil map covering all continents. A uniform legend was developed, with the main objective to obtain an inventory of the world soil resources based on integration of existing soil classification systems. The first world soil map at a scale of 1:5 million was published in 1974. Based on soil development status, material, and major geographical zones, 24 major soil groups (MSGs) and 106 soil units were distinguished. The definitions and nomenclature of the diagnostic horizons and properties were adopted from *U.S. Soil Taxonomy*,[106] but the definitions have been summarized and sometimes simplified to serve the purpose of the legend.[107] Revised versions of the FAO-UNESCO legend of the *Soil Map of the World* were issued in 1988 that distinguished 28 MSGs and 153 soil units.[108] At the global level the FAO-UNESCO *Soil Map of the World* is still the only worldwide, consistent, harmonized soil inventory that is readily available in digital format and provides a set of estimated soil properties for each mapping unit.[109] The development of the Soil Terrain (SOTER) program started in 1986 and focused on standardized mapping of areas with distinctive, often repetitive patterns of landform, morphology, slope, parent material, and soils at a 1:1 million scale. Each SOTER unit is linked through a GIS with a database containing attributes of landform, terrain, soils, climate, vegetation, and land use.[109] The World Reference Base for Soil Resources (WRB) project was initiated by the International Soil Science Society (ISSS), FAO, and the International Soil and Reference Information Center (ISRIC) to provide scientific depth and background to the 1988 *Soil Map of the World, Revised Legend*.[110] The WRB is a first step toward standardization of our global soil resources.[111] All global soil maps are based on the double-crisp paradigm lumping over geographic and taxonomic space, which is due to the lack of high-resolution soil data in many nations, the density of soil observations, and the predominance of soil classification systems around the globe.

In the U.S., three complementing soil information systems (SISs) were developed and are maintained by USDA NRCS: (1) Soil Survey Geographic Database (SSURGO, 1:15,840 to 1:31,680; typical map scale of 1:24,000; the conversion of SSURGO into a new database format called Soil Data Mart is in progress), (2) State Soil Geographic Database (STATSGO, 1:250,000), and (3) National Soil Geographic Database (NATSGO, 1:7,500,000). The SSURGO provides the most detailed digital soil data using a relational database management system (RDBMS). Data mining of hard-copy soil survey maps to produce vector GIS soil data is still ongoing across the U.S. Each map unit is assigned a unique identifier, known as the map unit identification number (MUID), which links to soil attributes stored in separate data tables. The SSURGO was intended for use by landowners, farmers, and planners at the county level. However, it lacks the capabilities for site-specific application (e.g., septic tank installation, site-specific management). The STATSGO was developed generalizing SSURGO data and is useful at the regional scale. The NATSGO was designed for national and multistate resource appraisal, planning, and monitoring. All U.S. SIS are based on the double-crisp paradigm. Many more double-crisp SIS were adopted in different nations.[112]

Soil information systems that store georeferenced soil property data are invaluable for future soil-landscape modeling projects. These data can be readily integrated with other spatial environmental layers, providing a valuable resource for various land resource applications. The U.S. National Soil Characterization Database provides morphological, chemical, physical, and biological soil properties for thousands of pedons. The Hydraulic Properties of European Soils (HYPRES) is a central database having a flexible relational structure consisting of soil pedological and hydraulic properties from different institutions in

Europe.[113] The Australian Soil Resources Information System contains over 160,000 soil profile descriptions complemented by laboratory data. The International Soil and Reference Information Center has been instrumental in developing numerous SIS, including the world inventory of soil emission potentials (WISE) international and global soil profile dataset, the International Geosphere Biosphere Programme–Data and Information System (IGBP-DIS) soil dataset for pedotransfer function development, and the ISRIC SIS, which assembles monoliths accompanied by soil profile descriptions, environmental data, soil reports, and a slide collection. These and other SIS provide a valuable resource to build soil-landscape models.

5.4   SIGNIFICANCE FOR FUTURE SOIL-LANDSCAPE MODELING

Previous shortcomings in soil-landscape modeling have been related to soil mapping paradigms, the labor-intensive collection of soil data, and database management. Soil survey maps have been produced extensively using the double-crisp soil mapping paradigm. Such soil polygon maps and associated datasets provide representative soil property values for map units without providing statistics about the within-unit variability and statistics such as the variance, coefficient of variation, range, and minimum and maximum values. Deficiencies of map unit boundary placement and location-specific mapping of map components have been acknowledged by soil surveyors.[114] In contrast, pixel-based soil mapping is much more data intensive and requires knowledge about quantitative spatial modeling techniques to produce accurate soil-landscape models. Our technical progress is impressive, encompassing how we collect, manage, and analyze soil and environmental datasets to characterize soil-landscapes. Yet there is still an urgent need to improve the training of the next generation of scientists and soil mappers. As pointed out by the current chair of commission of 1.5 Pedometrics of the International Union of Soil Science, Gerard Heuvelink, we must introduce pedometrics in the soil science curricula of higher education. Much more emphasis has to be placed on holistic soil-landscape modeling, integrating soil with other environmental datasets. This requires an education that is not limited to soil taxonomy and soil genesis, but includes GIS, remote sensing, quantitative methods, and more. We have to learn to grasp beyond tacit knowledge of soil surveyors to produce state-of-the-art soil-landscape models. Available geographic information technology combined with mathematical and (geo)statistical methods will produce the next generation of quantitative soil-landscape models.

Numerous studies have shown that it is possible to reconcile both soil mapping paradigms using a hybrid approach. Since the demand for pixel-based soil property maps is evident, there is no doubt that shifts from the double crisp to georeferenced soil property mapping will naturally occur. Geographic information systems have no limitation in handling hundreds of thousands of soil property data, which can be retrieved and combined with other environmental datasets on demand. Rapid soil mapping devices and techniques are finding more widespread use to generate denser soil datasets. Such dense datasets provide the input for advanced geostatistical methods to create quantitative soil-landscape models. Combinations of dense and sparse soil and environmental datasets, collected with different sensors and techniques, and model- and design-based mapping strategies are pivotal to create accurate, high-resolution soil-landscape models. Such models can be readily shared using state-of-the-art Web-based geographic information technology. We are optimistic that eventually we can satisfy the demand for quantitative georeferenced soil-landscape property data using emerging geographic information technologies.

ACKNOWLEDGMENTS

This research was supported by the Florida Agricultural Experiment Station and approved for publication as Journal Series R-10025.

REFERENCES

1. RM Lark. Variation in soil conditions and crop performance. In *Precision Agriculture-'97*, Proc Eur Conf Precision Agric., Warwick University Conference Center, JV Stafford, Ed. BIOS Scientific Publishers Ltd., Oxford, 1997, pp. 127–135.
2. J Bouma, J Stoorvogel, BJ van Alphen, HWG Booltink. Pedology, precision agriculture and the changing paradigms of agricultural research. *Soil Sci Soc Am J* 63: 1763–1768, 1999.
3. AB McBratney, MJ Pringle. Estimating average and proportional variograms of soil properties and their potential use in precision agriculture. *Precision Agric* 1: 125–152, 1999.
4. Natural Resources Conservation Service (NRCS). *Data Rich and Information Poor*, Report to the Chief of the NRCS by the Blue Ribbon Panel on Natural Resource Inventory and Performance Management. USDA, Washington, DC, 1995.
5. MR Hoosbeek, RG Amundson, RB Byrant. Pedological modeling. In *Handbook of Soil Science A*, ME Sumner, Ed. CRC Press, New York, 2000, pp. 321–352.
6. PA Burrough, RA McDonnell. *Principles of Geographical Information Systems*. Oxford University Press, New York, 1998.
7. D Peuquet. Presentations of geographic space: towards a conceptual synthesis. *Ann Assoc Am Geogr* 78: 375–394, 1988.
8. M Goodchild, G Sun, S Yang. Development and test of an error model for categorical data. *Int J Geogr Inf Syst* 6: 87–104, 1992.
9. SW Buol, FD Hole, TJ McCracken, RJ Southard. *Soil Genesis and Classification*. Iowa State University Press, Ames, 1997.
10. JC Bell. Assessment, Description and Delineation of Soil Spatial Variability at Hillslope to Landscape Scales. Paper presented at Proceedings of Pedometrics, Madison, WI, 1997.
11. S Nortcliff. Soil variability and reconnaissance soil mapping: a statistical study in Norfolk. *J Soil Sci* 29: 403–418, 1978.

12. PA Burrough. Sampling design for quantifying map unit composition. In *Spatial Variabilities of Soils and Landforms*, SSSA Special Publication 28, MJ Mausbach, LP Winding, Eds. SSSA, Madison, WI, 1991, pp. 89–125.

13. GBM Heuvelink, JA Huisman. Choosing between abrupt and gradual spatial variation? In *Quantifying Spatial Uncertainty in Natural Resources: Theory and Applications for GIS and Remote Sensing*, HT Mowrer, RG Congalton, Eds. Ann Arbor Press, Chelsea, MI, 2000, pp. 111–117.

14. PA Burrough, PFM van Gaans, R Hootsmans. Continuous classification in soil survey: spatial correlation, confusion and boundaries. *Geoderma* 77: 115–135, 1997.

15. SW Bie, PHT Beckett. Comparison of four independent surveys by air-photo interpretation, Paphos area (Cyprus). *Photogrammetria* 29: 189–202, 1973.

16. EH Isaaks, RM Srivastava. *An Introduction to Applied Geostatistics*. Oxford University Press, New York, 1989.

17. P Goovaerts. *Geostatistics for Natural Resources Evaluation*. Oxford University Press, New York, 1997.

18. R Webster, MA Oliver. *Geostatistics for Environmental Scientists*. John Wiley & Sons, New York, 2001.

19. JP Chilès, P Delfiner. *Geostatistics: Modeling Spatial Uncertainty*. John Wiley & Sons, New York, 1999.

20. P Goovaerts. Geostatistical modeling of uncertainty in soil science. *Geoderma* 103: 3–26, 2001.

21. AB McBratney, IOA Odeh, TFA Bishop, MS Dunbar, TM Shatar. An overview of pedometric techniques for use in soil survey. *Geoderma* 97: 293–327, 2000.

22. LA Zadeh. Fuzzy sets. *Inf Control* 8: 338–353, 1965.

23. PA Burrough. Fuzzy mathematical methods for soil survey and land evaluation. *J Soil Sci* 40: 477–792, 1989.

24. PA Burrough, RA MacMillan, WPA van Deursen. Fuzzy classification methods for determining land suitability from soil profile observations and topography. *J Soil Sci* 43: 193–210, 1992.

25. AB McBratney, JJ De Gruijter. A continuum approach to soil classification using fuzzy k-means with extragrades. *J Soil Sci* 43: 159–176, 1992.

26. AB McBratney, IOA Odeh. Application of fuzzy sets in soil science: fuzzy logic, fuzzy measurement and fuzzy decisions. *Geoderma* 77: 85–113, 1997.

27. IOA Odeh, AB McBratney, DJ Chittleborough. Soil pattern recognition with fuzzy-c-means: application to classification and soil-landform interrelationships. *Soil Sci Soc Am J* 56: 505–516, 1992.

28. S Grunwald, K McSweeney, DJ Rooney, B Lowery. Soil layer models created with profile cone penetrometer data. *Geoderma* 103: 181–201, 2001.

29. A-Z Zhu. A similarity model for representing soil spatial information. *Geoderma* 77: 217–242, 1997.

30. H Bandemer, S Gottwald. *Fuzzy Sets, Fuzzy Logic and Fuzzy Methods with Applications*. John Wiley, Chichester, U.K., 1995.

31. AK Bregt, J Bouma, M Jellinek. Comparison of thematic maps derived from a soil map and from kriging of point data. *Geoderma* 39: 281–291, 1987.

32. D Leenhardt, M Voltz, M Bornand, R Webster. Evaluating soil maps for prediction of soil water properties. *Eur J Soil Sci* 45: 293–301, 1994.

33. AS Rogowski. Quantifying soil variability in GIS applications. II. Spatial distribution of soil properties. *Int J Geogr Inf Syst* 10: 445–475, 1996.

34. S Shekhar, S Chawla. *Spatial Databases*. Prentice Hall, Upper Saddle River, NJ, 2003.

35. PA Burrough. The technologic paradox in soil survey: new methods and techniques of data capture and handling. *Enschede ITC J* 3: 15–22, 1993.

36. R Dudal. How good is our soil classification? In *Soil Classification: A Global Desk Reference*, H Eswaran, T Rice, R Ahrens, BA Stewart, Eds. CRC Press, New York, 2003, pp. 11–18.

37. J Thurston, TK Poiker, JP Moore. *Integrated Geospatial Technologies*. John Wiley & Sons, New York, 2003.

38. P Bolstad. *GIS Fundamentals*. Eider Press, White Bear Lake, MN, 2000.

39. LB Conyers, D Goodman. *Ground-Penetrating Radar: An Introduction to Archaelogists*. AltaMara Press, Walnut Creek, CA, 1997.

40. A Tabbagh, M Dabas, A Hesse, C Panissod. Soil resistivity: a non-invasive tool to map soil structure horizonation. *Geoderma* 97: 393–404, 2000.

41. CK Johnson, JW Doran, HR Duke, BJ Wienhold, KM Eskridge, JF Shanahan. Field-scale electrical conductivity mapping for delineating soil condition. *Soil Sci Soc Am J* 65: 1829–1837, 2001.

42. GW Schellentrager, JA Doolittle, TE Calhoun, CA Wettstein. Using ground-penetrating radar to update soil survey information. *Soil Sci Soc Am J* 52: 746–752, 1988.

43. JD Rhoades, NA Manteghi, PJ Shouse, WJ Alves. Estimating soil salinity from saturated soil paste electrical conductivity. *Soil Sci Soc Am J* 53: 428–433, 1989.

44. J Doolittle, E Ealy, G Secrist, D Rector, M Crouch. Reconnaissance soil mapping of a small watershed using electromagnetic induction and global positioning system techniques. *Soil Surv Horizons* 36: 86–94, 1995.

45. NJ Hartsock, TG Mueller, GW Thomas, RJ Barnhisel, KL Wells, SA Shearer. Soil electrical conductivity variability. In *Proceedings: Fifth International Conference on Precision Agriculture*, ASA Miscellaneous Publication, PC Robert, RH Rust, WE Larson, Eds. ASA, Madison, WI, 2000.

46. E De Jong, AK Suddeth, NR Kitchen, SJ Indorante. Measurement of apparent electrical conductivity of soils by an electromagnetic induction probe to aid in salinity surveys. *Soil Sci Soc Am J* 43: 810–812, 1979.

47. CLS Morgan, JM Norman, CC Molling, K McSweeney, B Lowery. Evaluating soil data from several sources using a landscape model. In *Scaling Methods in Soil Physics*, Y Pachepsky, DE Radcliffe, HM Selim, Eds. CRC Press, New York, 2003, pp. 243–260.

48. RE Yoder, RS Freeland, JT Ammons, LL Leonard. Mapping agricultural fields with GPR and EMI to prevent offsite movement of agrochemicals. *J Appl Geophys* 47: 251–259, 2000.

49. KA Sudduth, NR Kitchen, ST Drummond. Soil conductivity sensing on clay pan soils: comparison of electromagnetic induction and direct methods. In *Proceedings of the Fourth International Conference on Precision Agriculture*, St. Paul, MN, 1998, pp. 979–990.

50. JA Doolittle, ME Collins. A comparison of EM induction and GPR methods in areas of karst. *Geoderma* 85: 83–102, 1998.

51. R Rashad, D Kawamura, H Nemoto, T Miyata, K Nakagawa. Ground penetrating radar investigations across the Uemachi fault, Osaka, Japan. *J Appl Geophys* 53: 63–75, 2003.

52. JL Davis, AP Annan. Applications of ground penetrating radar to mining, ground-water, and geotechnical projects: selected case histories. In *Ground Penetrating Radar*, JA Pilon, Ed. Waterloo Center for Groundwater Research, Waterloo, Canada, 1992, pp. 1115–1133.

53. TJ Gish, WP Dulaney, KJS Kung, CST Daughtry, JA Doolittle, PT Miller. Evaluating use of ground-penetrating radar for identifying subsurface flow pathways. *Soil Sci Soc Am J* 66: 1620–1629, 2002.

54. V Basile, MT Carrozzon, S Negri, L Nuzzo, T Quarta, AV Villani. A ground-penetrating radar survey for archaeological investigations in an urban area (Lecce, Italy). *J Appl Geophys* 44: 15–32, 2000.

55. S Grunwald, B Lowery, DJ Rooney, K McSweeney. Profile cone penetrometer data used to distinguish between soil materials. *Soil Tillage Res* 62: 27–40, 2001.

56. American Society for Testing and Materials (ASTM). ASTM Standard D5778-95: Standard test method for performing electric friction cone and piezocone penetration testing of soils. In *Annual Book of ASTM Standards*. ASTM, West Conshocken, PA, 1995, p. 576.

57. American Society of Agricultural Engineers (ASAE). ASAE Standard S313.1: Soil cone penetrometer. In *Agricultural Engineers Yearbook of Standards*. ASAE, St. Joseph, MI, 2000, pp. 832–833.

58. DJ Rooney, B Lowery. A profile cone penetrometer for mapping soil horizons. *Soil Sci Soc Am J* 64: 2136–2139, 2000.

59. AG Kasim, MY Chu, CN Jensen. Field correlation of cone and standard penetration tests. *J Geotech Eng* 112: 368–372, 1986.

60. PU Kurup, GZ Voyiadjis, MT Tumay. Calibration chamber studies of peizocone test in cohesive soils. *J Geotech Eng* 120: 81–107, 1994.

61. AJ Puppala, YB Acar, MT Tumay. Cone penetration in very weakly cemented sand. *J Geotech Eng* 121: 589–600, 1995.

62. PD Ayers, JV Perumpral. Moisture and density effect on cone index. *Trans ASAE* 25: 1169–1172, 1982.

63. AR Quiroga, DE Buschiasso, N Peinemann. Soil compaction is related to management practices in the semi-arid Argentine pampas. *Soil Tillage Res* 52: 21–28, 1999.

64. B Lowery, RT Schuler. Duration and effects of compaction on soil and plant growth in Wisconsin. *Soil Tillage Res* 29: 205–210, 1994.

65. S Grunwald, DJ Rooney, K McSweeney, B Lowery. Development of pedotransfer functions for a profile cone penetrometer. *Geoderma* 100: 25–47, 2001.

66. Earth Information Technologies Corporation (EarthIT Corp.). http://www.earthit.com. 2004.

67. DG Schulze, JL Nagel, GE Van Scoyoc, TL Henderson, MF Baumgardner, DE Stott. The significance of organic matter in determining soil color. In *Soil Color*, Soil Science Society of America Special Publication 31, JM Bigham, EJ Ciolkosz, Eds. SSSA, Madison, WI, 1993, pp. 71–90.

68. E VanRanst, M Verloo, A Demeyer, JM Pauwels. *Manual for the Soil Chemistry and Fertility Laboratory.* University of Ghent, Ghent, Belgium, 1999.

69. A Dent, A Young. *Soil Survey and Land Evaluation.* George Allen & Unwin Publication, Boston, 1981.

70. K Daniel, NK Tripathi, K Honda, E Apisit. Analysis of Spectral Reflectance and Absorption Patterns of Soil Organic Matter. Paper presented at 22nd Asian Conference on Remote Sensing, Singapore, 2001.

71. HW Siesler, Y Ozaki, S Kawata, HM Heise. *Near-Infrared Spectroscopy: Principles, Instruments, Applications*. Wiley-VCH, Weinheim, Germany, 2002.

72. JB Reeves, GW McCarty, VB Reever. Mid-infrared diffuse reflectance spectroscopy for the quantitative analysis of agricultural soils. *J Agric Food Chem* 49: 766–772, 2001.

73. LJ Janik, RH Merry, JO Skjemstad. Can mid infrared diffuse reflectance analysis replace soil extractions. *Aust J Exp Agric* 38: 681–696, 1998.

74. KD Shephard, MG Walsh. Development of reflectance spectral libraries for characterization of soil properties. *Soil Sci Soc Am J* 66: 988–998, 2002.

75. RN Clark. Spectroscopy of rocks and minerals, and principles of spectroscopy. In *Remote Sensing for the Earth Sciences: Manual of Remote Sensing*, Vol 3, N Rencz, Ed. John Wiley & Sons, New York, 1999, pp. 3–52.

76. TM Lillesand, WW Kiefer. *Remote Sensing and Image Interpretation*. John Wiley & Sons, New York, 2000.

77. T Udelhoven, C Emmerling, T Jarmer. Quantitative analysis of soil chemical properties with diffuse reflectance spectrometry and partial least-square regression: a feasibility study. *Plant Soil* 251: 319–329, 2003.

78. A Couillard, AJ Turgeon, MO Westerhaus, JS Shenk. Determination of soil separates with near infrared reflectance spectroscopy. *J Near Infrared Spectrosc* 4: 201–212, 1997.

79. L Kooistra, J Wanders, GF Epema, RSEW Leuven, R Wehrens, LMC Buydens. The potential of field spectroscopy for the assessment of sediment properties in river floodplains. *Anal Chem Acta* 484: 189–200, 2003.

80. RJ Kauth, GS Thomas. The Tasselled Cap: A Graphic Description of Spectral-Temporal Development of Agricultural Crops as Seen by Landsat. Paper presented at Proceedings of the Second International Symposium on Machine Processing of Remotely Sensed Data, Purdue University, West Lafayette, IN, 1976.

81. KA Sudduth, JW Hummel. Near infrared spectrophotometry for soil property sensing. In *Proceedings of SPIE Conference on Optics in Agriculture and Forestry No. 1836*, Bellingham, WA, 1992, pp. 14–25.

82. DR Oetter, WB Cohen, M Berterretche, TK Maiersperger, RE Kennedy. Land cover mapping in an agricultural setting using multi-seasonal thematic data. *Remote Sensing Environ* 76: 139–155, 2000.

83. MA Friedl, CE Brodley. Decision tree classification of land cover from remotely sensed data. *Remote Sensing Environ* 61: 399–409, 1997.

84. SA Sader, D Ahl, WS Lio. Accuracy of Landsat-TM and GIS rule-based methods for forest wetland classification in Maine. *Remote Sensing Environ* 53: 133–144, 1995.

85. GI Metternicht, JA Zinck. Remote sensing of soil salinity: potentials and constraints. *Remote Sensing Environ* 85: 1–20, 2003.

86. MK Ridd, J Liu. A comparison of four algorithms for change detection in an urban environment. *Remote Sensing Environ* 63: 95–100, 1998.

87. ES Kasischke, KB Smith, LL Bourgeau-Chavez, EA Romanowicz, S Brunzell, CJ Richardson. Effects of seasonal hydrologic patterns in south Florida wetlands on radar backscatter measured from ERS-2 SAR imagery. *Remote Sensing Environ J* 88: 423–441, 2003.

88. M Pardé, J-P Wigneron, A Chanzy, P Waldteufel, Y Kerr, S Huet. Retrieving surface soil moisture over a wheat field: comparison of different methods. *Remote Sensing Environ J* 87: 334–344, 2003.

89. S Chabrillat, AFH Goetz, L Krosley, HW Olsen. Use of hyperspectral images in the identification and mapping of expansive clay soils and the role of spatial resolution. *Remote Sensing Environ J* 82: 431–445, 2002.

90. NA Drake, S Mackin, JJ Settle. Mapping vegetation, soils, and geology in semiarid shrublands using spectral matching and mixture modeling of SWIR AVIRIS imagery. *Remote Sensing Environ J* 68: 12–25, 1999.

91. RF Tomlinson. The impact of the transition from analogue to digital cartographic representation. *Am Cartogr* 15: 249–261, 1988.

92. PA Burrough. GIS and geostatistics: essential partners for spatial analysis. *Environ Ecol Stat* 8: 361–377, 2001.

93. S Grunwald, KR Reddy, S Newman, WF DeBusk. Spatial variability, distribution, and uncertainty assessment of soil phosphorus in a south FL wetland. *Environmetrics J*, 15: 811–825, 2004.

94. ZR Peng. An assessment framework of the development strategies of Internet GIS. *Environ Planning B Planning Design* 26: 117–132, 1999.

95. S Grunwald, KR Reddy, V Mathiyalagan, SA Bloom. Florida's Wetland WebGIS. Paper presented at ESRI User Conference, San Diego, CA, July 7–11, 2003.

96. B Plewe. *GIS Online: Information Retrieval, Mapping, and the Internet.* OnWord Press, Santa Fe, NM, 1997.

97. Z-R Peng, M-H Tsou. *Internet GIS: Distributed Geographic Information Services for the Internet and Wireless Networks.* John Wiley & Sons, New York, 2003.

98. V Mathiyalagan, S Grunwald, KR Reddy, SA Bloom. A WebGIS and geodatabase for Florida's wetlands. *Comput Electron Agric.* 47: 69–75, 2005.

99. AA Roshannejad, W Kainz. Handling identities in spatio-temporal databases. *ACSM/ASPRS Annu Convention Exposition Tech Papers* 4: 119–126, 1995.

100. T Abraham, JF Roddick. Survey of spatio-temporal databases. *Geoinformatica* 3: 61–99, 1999.

101. IJ Koeppel, SD Ahlmer. Integrating the Dimension of Time into AM/FM Systems. Paper presented at Proceedings of the AM/FM XVI International Annual Conference, Aurora, CO, 1993.

102. DJ Peuquet, N Duan. An Event-Based Spatio-Temporal Data Model (ESTDM) for temporal analysis of geographical data. *Int J Geogr Inf Syst* 9: 7–24, 1995.

103. M Yuan 1997. Modeling semantics, spatial and temporal information in a GIS. In *Progress in Trans-Atlantic Geographic Information Research*, M Craglia, H Couleclis, Eds. Taylor & Francis, New York, 1997, pp. 334–347.

104. V Ramasundaram, S Grunwald, A Mangeot, NB Comerford, CM Bliss. Development of a virtual field laboratory. *J. Comput Educ.* 45: 21–34, 2005.

105. K Morris, D Hill, A Moore. Mapping the environment through three-dimensional space and time. *Comput Environ Urban Syst* 24: 435–450, 2000.

106. Soil Survey Staff. *Soil Taxonomy*, Agricultural Handbook 436. USDA Soil Conservation Service, Washington DC, 1999.

107. OC Spaargaren. Other systems of soil classification. In *Handbook of Soil Science*, ME Sumner, Ed. CRC Press, New York, 2000, pp. 137–174.

108. FAO. *FAO-UNESCO Soil Map of the World, Revised Legend*, World Soil Resources Report 84. FAO, Rome, 1988.

109. FOF Nachtergaele. The future of the FAO legend and the FAO/UNESCO *Soil Map of the World*. In *Soil Classification: A Global Desk Reference*, H Eswaran, T Rice, R Ahrens, BA Stewart, Eds. CRC Press, New York, 2003, pp. 147–156.

110. FAO, ISRIC, ISSS. *World Reference Base for Soil Resource*, World Soil Resources Report 84. FAO, Rome, 1988.

111. AR Mermut, H Eswaran. Some major developments in soil science since the mid-1960s. *Geoderma* 100: 403–426, 2001.

112. H Eswaran, T Rice, R Ahrens, BA Stewart. *Soil Classification: A Global Desk Reference*. CRC Press, New York, 2003.

113. JHM Wösten, A Lilly, A Names, CL Bas. Development and use of a database of hydraulic properties of European soils. *Geoderma* 90: 169–185, 1999.

114. C Ditzler. *Soil Survey and Soil Variability: A Literature Review and Recommendations*. Report presented to the deputy chief of the Natural Resource Conservation Service, Maury Mausbach, December 3, 2003.

6 Topographic Mapping

Scot E. Smith

CONTENTS

ABSTRACT

Topographic mapping has been important to mankind from the start and was practiced since ancient Egyptian times. While the world is nearly entirely mapped topographically, large areas are barely covered in detail, and in many regions topography undergoes rapid topographic change and must be remapped regularly. The technology of topographic mapping is also undergoing rapid change with the advent of digital data, satellite imagery, and other rapid and accurate mapping methods. This chapter describes (1) topographic maps, (2) means by which data are collected for their production (i.e., plane table surveying, land surveying, photogrammetry, lidar (light detection and ranging), radar, and satellite imagery), and (3) how these data are transformed into a topographic map. Several authors have described the relationship between soil and landscape modeling and topography.[1,2] Without an understanding of topography, soils mapping is one-dimensional and limited in scope. Landscape modeling can only be accomplished with a thorough understanding of topography. Increasingly, visualization tools utilize topography not merely as an interesting backdrop, but as an actual layer and quantifiable element for analysis. This is becoming more true as the means to produce increasingly accurate topographic maps improve and become cheaper. This chapter attempts to provide an understanding of the state of the art of topographic mapping so that people interested in soil and landscape modeling can incorporate it to its fullest advantage.

6.1 WHAT ARE TOPOGRAPHIC MAPS?

Maps exist in a variety of forms. They may be on paper, Mylar, or in a digital form. They consist of simple half-tone line drawings or multidimensional and multicolored, multilayered geographic information systems. The popularity and utility of topographic maps has remained high through time. Topographic maps usually contain map feature information such as political boundaries, transportation routes, water, and forested areas, but the feature that distinguishes topographic maps from others is contour lines that portray elevation of the land. Topographic maps render the three-dimensional aspect of terrain on a two-dimensional surface.[15]

Topographic maps usually indicate three-dimensional shape through contour lines, which are lines drawn on a map connecting points of equal elevation. Along a contour line, elevation is neither gained nor lost. For example, if you walk on a beach along the line where the water meets the shore, the water surface marks an elevation known *as sea level*, and as you follow the shoreline, your elevation remains constant. This is, by definition, a *contour line*. If you walk upslope away from the sea, your elevation rises above sea level and you leave the contour line. If you walk into the sea, you also leave the contour line and go below sea level.

Topographic maps are usually composed of a series of contour lines separated by a prescribed contour interval. In the margin of a map is a declaration of the contour interval, such as "Contour Interval: 5 feet," but the interval can also be determined by looking at *index contours*, where the elevation is directly written on the line. Many U.S. Geological Survey (USGS) 7.5-feet topographic maps have a contour interval of 50 feet, although in some low-lying and flat regions, such as Florida, the interval may be 10 feet. In especially low-lying areas, where it is critical to precisely and accurately know the topography for drainage purposes, the contour interval can be smaller. This is the case for parts of southern Florida, for example, where topographic maps with a contour interval of 0.5 feet exist.

Topographic maps are used to determine elevation, but they are also used to visualize topography, which is useful, among other things, for recreation, planning, transportation routing, defense, and aviation. The important thing is to understand the pattern of the contour lines and not simply the elevation they represent. One of the most basic topographic observations is the gradient or slope of the ground surface. Steep gradients occur in areas where there is a significant change in elevation over a short distance, whereas gentle gradients occur where there is little change over the same distance.

Topographic maps do not necessarily have to contain contour lines. They are topographic maps in the sense that they show the "lay of the land," and so are useful in visualizing the landscape. Examples include:

- **Slope maps** that show the rate of change of elevation or the steepness of an area
- **Aspect maps** that show the direction (north, south, east, west) to which slope is oriented
- **Curvature maps** that show the curvature of the land as a measure of the rate of change of slope

6.1.1 National Topographic Mapping Efforts

Most topographic mapping exists on a national basis, although the USGS produces topographic maps for the entire world. The mandate of the USGS is to be the nation's domestic mapping agency. For the most part, international mapping is not part of its mission, except in small study areas and on the continent of Antarctica. The British Ordinance Survey also produces maps of areas outside the U.K., but at a relatively low level of detail.

There is a wide disparity of topographic mapping efforts in the world. Some countries have had an ongoing, uninterrupted mapping program for centuries and have mapped the topography of nearly every square meter of their territory in detail. Other regions, such as central Africa, have not been well mapped until now. Some countries were at one time well mapped topographically, but have allowed their maps to become out of date.

In addition to this disparity, countries have different means of producing topographic mapping and distributing those maps. In many cases, military agencies are responsible for this task. This organizational structure has made public access to the data difficult or, in some cases, impossible since the military's first priority is national defense. That responsibility may preclude distribution of domestic maps of any kind. In the U.S., responsibility for mapping by national agencies is clearly divided. The primary civilian agency responsible for domestic mapping is the USGS, whereas the National Geospatial-Intelligence Agency (NGA) (http://www.geoplace.com/gw/2003/0312/0312cnf.asp) is responsible for military and international mapping efforts. The National Geodetic Survey is a part of the National Oceanographic and Atmospheric Administration and conducts topographic and bathymetric mapping of the world's shorelines. Therefore, it is relatively easy to obtain topographic maps of the U.S. This ease of access has contributed to the country's economic and physical development. In places where topographic maps are not easily available, land development schemes are less likely to occur due to lack of basic land characteristic information.

Much of the problem will probably be alleviated in the near future as public satellite-based topographic mapping capability improves. Public satellite imagery is under the United Nation's sanctioned "open skies" policy and, as such, can be used by anyone and is not subject to national defense agencies arbitrarily rendering the information unavailable to the public. As technologies such as lidar (light detection and ranging) are improved, detailed topographic mapping of any place on Earth will be possible, regardless of national boundaries or military secrecy.

As matters stand now, however, most public topographic mapping material is held and distributed by national (military or civilian) mapping programs. Countries can be classified into one of three categories with respect to topographic mapping: (1) topographically well mapped and available to the public, (2) topographically well mapped, but not generally available to the public, or (3) not well mapped. It is beyond the scope of this chapter to describe each and every national topographic mapping program in the world, but examples of some national topographic mapping efforts follow:

Australia: Australia has one of the most comprehensive sets of topographic maps of any collection in the world. The maps are accurate, detailed, and widely available. The GeoScience Australia (www.geoaustralia.com) organization publishes topographic maps for the country. Maps of 1:50,000 scale are available for some areas, as well as digital data.

Brazil: Detailed topographic map coverage of Brazil is not comprehensive, but it is improving. The mountainous and coastal regions are well mapped, but the vast pampa and Amazon regions are less well covered. The Geological Survey of Brazil (www.cprm.gov.br/ingles) is the source of all cartographic products for the country. Most maps are still available only in a paper format, but collaboration with the USGS has resulted in digitization of some regions.

Canada: Canadians pioneered many of the concepts of topographic mapping from aerial photography and coined the phrase geographic information systems. The government's national mapping program has produced an excellent topographic series. Due to the country's large size and relatively small population, work is still being carried out for remote regions, but all major areas have been mapped and are kept up to date. The Centre for Topographic Information (maps.nrcan.gc.ca) is the repository and publisher of all Canadian topographic information. The entire country is mapped at scales of 1:50,000 and 1:250,000, and topographic maps are available in great detail for many areas.

China: Topographic maps of China are generally not publically available, and little is known outside China regarding the extent of coverage. The Science Foundation for Surveying and Mapping (www.casm.ac.cn/) is the source for all cartographic products for China.

Germany: Germany has been mapped topographically continuously for decades. Not all regions are mapped to the same level of detail, but the entire country (including the former East Germany) has topographic maps at various scales. The Bundesamt für Kartographie und Geodäsie produces and provides cartographic data for Germany, including 1:100,000 and larger-scale topographic maps.

India: India was mapped by the British Ordinance Survey, and continuous efforts to improve the maps have been made ever since. Not all of the country has topographic coverage, but critical areas, such as floodplains, have been mapped in detail by the Geological Survey of India, which is the source of all map products for the country.

Scandinavia: The government agencies responsible for topographic mapping of Scandinavia are the Finnish Hydrographic Office, National Land Survey of Iceland, Norweign Mapping Authority, and Swedish Mapping Agency. All have both paper and digital products of topography for the entire region, and they are updated regularly. The National Survey and Cadastre is the official producer of paper and digital cartographic data for Denmark. Paper maps are available for the country at scales as large as 1:24,000.

U.K.: The U.K. has, perhaps, the most complete set of topographic maps of any nation on Earth. They are detailed, accurate, and regularly updated. Most areas are available in a digital format as well as paper. The Ordnance Survey handles maps in the U.K. There are several series in the Pathfinder/Explorer series ranging from 1:25,000. The digital datasets include raster map images and digital height data.

U.S.: The central agency responsible for topographic mapping in the U.S. is the U.S. Geological Survey (USGS). USGS has produced a series of large-scale topographic maps, which includes approximately 53,000 map sheets for the conterminous U.S. and is the only uniform map series that covers this area at such a large scale. Most USGS topographic maps are produced at a scale of 1:24,000, while some are produced at a scale of 1:25,000. In addition to the 1:24,000 scale maps, complete topographic coverage of the U.S. is available at scales of 1:100,000 and 1:250,000. All these maps are available in both printed-paper form and scanned digital data form. The digital files are referred to as digital raster graphics (DRGs).

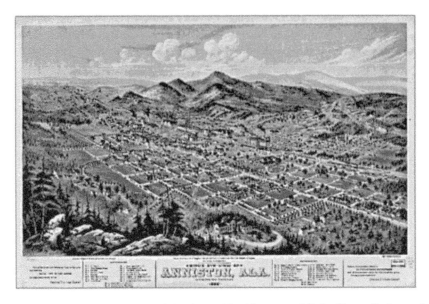

FIGURE 6.1 Perspective view of Anniston, AL. (Courtesy of E.S. Glover Shober and Carqueville Lithography Company, 1888.)

6.2 PRODUCTION OF TOPOGRAPHIC MAPS

6.2.1 GATHERING DATA: GROUND SURVEYS AND AERIAL PHOTOGRAPHY

The earliest topographic maps were probably produced by Egyptians over 4000 years ago. Flood protection from the Nile River necessitated an understanding and mapping of topography, and so Egyptians produced maps showing the flow of water. From these maps dikes and other flood protection devices could be designed and built. These maps do not survive today, but an early map showing contour lines, drawn by the Greek geographer Herodotus in the fifth century B.C., showing the Nile flowing toward the Mediterranean Sea, can be considered a form of topographic map.

Henceforth, *perspective maps* were produced from a high elevation (i.e., from a hilltop), where the cartographer or artist simply drew what he could see from a high vantage point from the bottom of the hill to the horizon. Sometimes an attempt was made to incorporate perspective concepts, such as correcting for convergence, but most of the time the maps were drawn to maintain an oblique perspective. An example of a perspective map is shown in Figure 6.1.

6.2.1.1 Plane Table Surveying

Most of the original topographic maps produced during the 1700s and part of the 1800s were made using a cumbersome technique called *plane table surveying*.

Plane table surveying took great skill and was backbreaking work, but produced reasonably accurate topographic maps for the times. Carrying a plane table, which is essentially a portable drawing board on a tripod with a sighting device, the cartographer/surveyor would hike to the area's best vantage point and plot on the map those features that could be seen and measured in the field by triangulation. Plane table surveying remained the dominant mapping technique until the 1800s, when it gave way to ground surveying.

6.2.1.2 Ground Surveying

Ground surveying was used for topographic map production during the 19th century and continues today when high accuracy is required. Early surveyors used instruments such as transits, levels, stadia, and chains. Modern surveying employs global positioning satellites and electronic measuring devices, but the concepts remain the same. The objective is to produce an accurate map of natural and cultural features of interest by precisely measuring the locations of features and changes in elevation.

One common process is called a *topographic survey*.[7] For example, transportation construction requires that accurate estimates of cut-and-fill volume be determined. One way to calculate this volume is to conduct a ground topographic survey. Fieldwork for a topographic survey consists of two processes: (1) establishing a network of horizontal and vertical control points of known location and (2) determining the horizontal and vertical locations of features near each instrument station.

Topographic control consists of two elements: horizontal and vertical. The horizontal element locates the horizontally fixed position of specified control points, and vertical control determines the elevations of benchmarks. This provides the framework from which topographic details are determined.

Traversing, triangulation, or both are used to locate horizontal control points. There are two levels of control: *primary*, where a small number of points are located with a high degree of accuracy, and *secondary*, where less accurately located control points are established within the network.

Benchmarks serve as beginning and closing points for determining the location of the control points. A series of permanent benchmarks exist in what is referred to as a *geodetic network* in most countries of the world. Topographic maps are tied into these marks, which in turn are to tied into other geodetic networks.

Vertical control is determined by differential leveling. When the primary vertical control is required, the following four standard *degrees of precision* are used:

- $0.05 \text{ foot } \sqrt{\text{distance in miles}}$ is used in relatively flat terrain.

- $0.1 \text{ foot } \sqrt{\text{distance in miles}}$ is used for a contour interval of 2 feet.

- $0.3 \text{ foot } \sqrt{\text{distance in miles}}$ is used for a contour interval of 5 feet.

- $0.5 \text{ foot } \sqrt{\text{distance in miles}}$ is used for a contour interval of 10 feet.

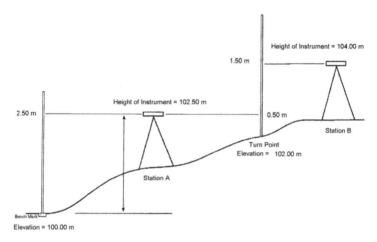

FIGURE 6.2 Differential leveling schematic. (From USGS, 2001.)

Once vertical control has been set, it is necessary to locate horizontal and vertical features near the control points. These are features that will eventually appear on the map.

How a ground survey is conducted depends on the intended use of the final topographic map. For high-accuracy maps, angles should be measured with a theodolite, and horizontal distances should be measured with an electronic distance measurement device. Highly accurate and precise elevations are determined with a differential leveling device, as shown in Figure 6.2.

In summary, ground surveying is suitable for topographic mapping when the area to be mapped is relatively small and the accuracy required is high. For many large-scale applications, however, ground surveying is impractical, and hence aerial photogrammetry is used.

6.2.2 TOPOGRAPHIC RELIEF REPRESENTATION

Topographic relief can be represented on a map in a number of different ways, such as *contours* that define equal elevation differences, as shown in Figure 6.3, or a *digital elevation model* (DEM) grid, as shown in Figure 6.4.[12]

Digital models of maps are made in one of two formats: vector or raster. Maps in a vector format represent spatial relationships through points, lines, and polygons. Contour lines are represented in the vector format by inputing each curve as a line. The line is modeled by joining adjacent points with straight line segments. A plotter then draws the curve by moving in straight line segments from one point to an adjacent point.

In raster format, maps are represented by grid cells, which are formed by squares superimposed on the area to be mapped. Topographic information, such as elevation, is stored in the grid cell. Therefore, the highest resolution depends on the size of the cell, which can be a problem for detailed maps, such as those

FIGURE 6.3 Topographic contours. (From USGS, 2001.)

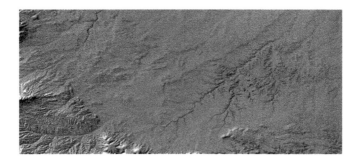

FIGURE 6.4 DEM grid representation of topographic relief. (From USGS, 2001.) (See color version on accompanying CD.)

designed for transportaion or drainange. An example of a topographic map derived from raster data is shown in Figure 6.5.

6.2.2.1 Digital Elevation and Digital Terrain Models

A digital elevation model (DEM) is a digital file consisting of terrain elevations for ground positions at regularly spaced horizontal intervals. DEMs are used for the generation of three-dimensional graphics displaying terrain slope, aspect (direction of slope), and terrain profiles between selected points. DEMs have been used in combination with digital raster graphics (DRGs), digital line graphs (DLGs), and digital orthophoto quadrangles (DOQs) to both enhance visual information for data extraction and revision purposes and create dramatic hybrid digital images. A digital terrain model is data model that attempts to provide a three-dimensional representation of a continuous surface. They are often used to represent relief.

The USGS produces five different digital elevation products. Although all are the same with respect to the manner in which the data are structured, each varies in sampling interval, geographic reference system, areas of coverage, and accu-

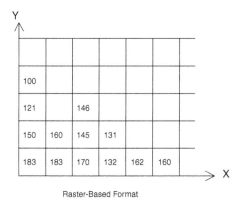

Raster-Based Format

Numbers in the cells represent digital numbers

FIGURE 6.5 Representation of topographic relief in raster format.

racy. The primary difference between them is the spacing interval of the data. The five current USGS DEM products are:

- 7.5-minute DEM, 30 × 30 m data spacing
- 1° DEM, 3 × 3 arc-second data spacing
- 2 arc-second DEM, 2 × 2 arc-second data spacing
- 15-minute Alaska DEM, 2 × 3 arc-second data spacing
- 7.5-minute Alaska DEM, 1 × 2 arc-second data spacing

The USGS collects digital elevation data using the following five approaches: (1) manual profiling from photogrammetric stereo models, (2) stereo model digitizing of contours, (3) digitizing of topographic map contour plates, (4) performing autocorrelation via automated photogrammetric systems, and (5) converting hypsographic and hydrographic tagged vector files. Of these five, vector hypsographic and hydrographic data produce the most accurate model.[15]

The manner in which elevations are coded affects the accuracy of the DEM. Coding on a single byte (8 bits) makes it possible to record 256 elevation values. The best solution would be to store elevations as four bytes so that all elevations from 0 to 1000 m could be described to the nearest millimeter.

6.2.2.2 Triangulated Irregular Networks

Triangulated irregular networks (TINs) are built up from grids. The idea is to eliminate elevation points that do not contribute necessary information. This is done by storing vertices of triangles with the maximum spacing possible with respect to relief and aspect. Figure 6.6 is an example of a topographic map derived from a TIN.

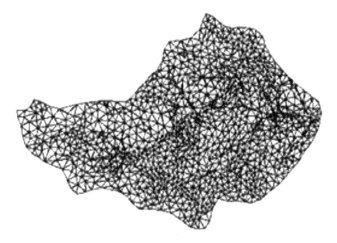

FIGURE 6.6 Example of topographic map derived from a TIN.

6.2.3 PHOTOGRAMMETRY

Photogrammetry is the technique of measuring two- or three-dimensional objects from photography.[8] The imaging device may be on the ground, in an airplane, or on a spacecraft. Targets may be as large as a field of wheat or as small as a cancer cell in radiometeric photogrammetry. For most topographic mapping applications, an airplane is used.

Aerial photogrammetry resulted from two technological events coming together: (1) stable aircraft and (2) large-film-format cameras. It began during the early 1920s from research and development for the war effort (World War I) and resulted when the technologies matured and merged.[9]

Most of the photographs used for topographic mapping today use traditional film media, although the popularity of aerial digital cameras is rapidly increasing. Usually, the film imagery is electronically stored on disk after being scanned.[10]

Topographic maps made from aerial photography take advantage of the principle of *stereoscopic vision.* Stereoscopic vision refers to visualizing of topographic relief by using images taken at different perspectives. Stereoscopic vision is based on the idea that the brain registers an image in three dimensions when viewing two images of the same place on the ground taken from different perspectives.[11] Viewing images in simulated three dimensions gives image interpreters an impression of topographic relief.

For topographic mapping, stereoscopic vision is a tool that uses the separate perspective taken by two cameras to recreate three-dimensional space. Parallax (the difference in perspective) can be quantified and measured, and from it vertical dimensions can be determined. Four steps are needed to map topography using stereo aerial photography:

- Acquire photograph with a minimum 50% end lap along the flight line.
- Locate control points on the ground (locations are well known and are also visible in the photographs).
- Calculate and recreate in a stereo plotting device the geometry of the conditions of the airplane (tilt, row, and yaw) at the moment the images were taken. This is called *exterior orientation.*
- Map elevations using a plotting device.

6.2.3.1 Geometry of Photogrammetric Topographic Mapping

As previously mentioned, photogrammetry is the science of measuring spatial relationships from a photograph. This includes linear measurements, area, volume, angles, and height. When photographic images are taken, every light ray that passes through the camera lens and reaches the film during exposure is interpreted as a single point. In order to measure objects, these rays must be reconstructed. Therefore, the internal geometry of the camera (i.e., focal length, image center, lens distortion, etc.) must be known. This process is called *interior orientation.*

The conditions when the photograph was taken (*exterior orientation*) can be recreated through modeling the orientation of the imaging device with control points. Since the location of a point on the ground is identified in each image in the stereo pair, its position in three-dimensional space can be determined by the intersection or convergence of straight lines or rays of light.

Since the overlapping portion of photographs taken parallel to each other can be viewed in three dimensions, parallax and heights can be determined, as shown in Figure 6.7.

There are three methods by which topography is derived from stereo pairs of aerial images: (1) analog photogrammetry, (2) analytical photogrammetry, and (3) digital photogrammetry. Prior to any of these methods, however, the images must be oriented with respect to exterior, interior, relative, and absolute orientation.

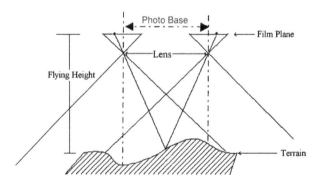

FIGURE 6.7 Model of stereo photographs along a flight line.

6.2.3.2 Exterior, Interior, Relative, and Absolute Model Orientation

Image orientation is the process of recreating the geometry of what happened when the image was taken and placing everything into a vertical perspective. Orientation parameters are determined by reconstructing the geometry of the image when it was taken and using control points both on the ground and in the image.

More and more *direct georeferencing* is used for exterior orientation. It is a process that establishes the on-flight measurements for the exterior orientation of each image by the global positioning systems (GPS) and inertial navigation systems (INS). Using these instruments, the exterior orientation (the orientation of the camera with respect to its position and altitude) can be determined.

Interior orientation is the process by which the relationship between the position of the film and photo coordinate system is determined. A two-dimensional transformation converts pixel coordinates to photo coordinates. Converting the image pixel coordinate system to the image space coordinate system is done mathematically by combining information about the image center and lens distortion and attenuation due to the atmosphere.

As described above in the stereoscopic vision section, three-dimensional measurement of objects with height is possible from stereo pairs of images. The *relative orientation* is needed to form a stereo model for these measurements. Therefore, for relative orientation, the relative positions of rays of light with respect to each other have to be determined.

Absolute orientation is used to transform an arbitrarily oriented model of images into a *real-world coordinate system*. The coordinate system of the image model is brought into the space of the control point coordinate system. At the end of this operation, coordinates anywhere in the image can be determined in a real-world coordinate system, such as the Universal Transverse Mercator (UTM) system or state plane coordinate system.

6.2.3.3 Aerial Triangulation

Aerial triangulation is used to (1) reduce the number of control points needed, (2) achieve higher accuracy, (3) ensure good edge matching between images, and (4) create additional control points through control point extension.[3] Ground control points are used to establish a geometric relationship between a set of tie points, with unknown ground coordinates measured on at least two images. These results are used to determine overall accuracy.

The geometric relationship between adjacent and other nearby images, the side lap between adjacent flight lines, must be known for aerial triangulation and is calculated through a bundle block adjustment (BBA). A BBA is an iterative process that solves image orientations and determines locations of the perspective centers simultaneously as one large image block, rather than as individual images.

The model of the stereo images is viewed through a stereoscope, which allows someone to view photographs simultaneously through a magnification device. The intersection of light rays can be measured by the stereoscope point by point using a measuring dot. When viewing the stereo model, the two points fuse into a three-dimensional point (floating dot), which can be moved and raised until the desired point of the three-dimensional object is found. The movements of the point are mechanically transmitted to a plotting device, and a map with contour lines is then drawn.

6.2.3.4 Analytical Photogrammetry

Although the first analytical plotter was introduced in the late 1950s, it was not until the 1970s, when fast and inexpensive computers became available, that analytical plotters became common. The basic concept is similar to analog instruments, but in analytical photogrammetry a computer determines the spatial relationship between image coordinates and real-world coordinates.

The images must still be oriented in analytical photogrammetry. After recreating the inner orientation, both images are also relatively oriented and then coorections are applied for lens distortion and atmospheric attenuation. A three-dimensional model is created and an absolute orientation is done so that features can be measured in three dimensions.

A difference between analog plotters and analytical plotters is that the latter draws into a computer. Analytical plotters use a computer program to calculate real-world coordinates, and three-dimensional drawings are created that are stored in the computer. Another difference between the analog and analytical systems is cost. Analytical stereo plotters cost several times more than analog plotters, and so in some developing countries, where labor costs are relatively low, analog plotters may still make economic sense.[20]

6.2.3.5 Analog Aerial Photogrammetry

Analog aerial photogrammetry is the use of film-based aerial photography to measure objects on the ground. It is highly labor intensive, but results in a high degree of accuracy for both two- and three-dimensional measurements. The concepts are simple and the instruments are widely available today. A drawback to the instruments is that they are cumbersome and do not remain calibrated, and thus are expensive to maintain.

The first attempts to make maps from photography started soon after photography was discovered in the mid-1850s. Cameras were mounted from a high perspective such as a hilltop, and an oblique image was taken. Attempts were made to produce a planimetric perspective map from the oblique image by retracing the lines on the photograph. In terms of topographic mapping, however, it was unsuccessful.

Innovations from the World Wars and their mapping requirements for speed and accuracy resulted in precise photogrammetry as we know it. A second impetus for improvement of photogrammetry came as a result of the planning, design, and construction of the interstate highway system in the U.S. following World War II. Many photogrammetric engineers who worked for the U.S. Army Topographic Laboratory worked after the war for the U.S. Department of Transportation and steadily developed the technology of analog aerial photogrammetry.

Stable aircraft, instruments, and plastic-film (vs. glass plates) photography all made aerial photography feasible. Cameras and planes were relatively crude compared with today's equipment, however, and so the metric quality of the photographs was low. Another drawback was the fact that much of the film used was nitrate based. Nitrate film was chemically unstable and prone to emulsion deterioration over time. It was replaced by more stable emulsion and film backing after World War II. This is mentioned because a great deal of aerial photography was taken with nitrate-base film and still exists in archives. For historical studies of topography or land use and land cover, this film can be used, but care must be taken in its handling and its interpretation.

Photogrammetry has two major advantages over ground-based surveying and mapping. It reduces fieldwork and can map landscapes independent of their terrain characteristics. Photogrammetry, however, is numerically intensive and so requires a large number of calculations. Therefore, it was inappropriate for large mapping projects until the advent of fast and affordable computers in the 1970s.

6.2.3.6 Digital Photogrammetry

Photogrammetry based on analog or photographs dominated aerial photogrammetry until the end of the 20th century and will probably continue to do so for at least another decade. Digital cameras were introduced in the late 1990s and may eventually replace film cameras for topographic mapping. The primary reason that digital photogrammetry has not yet replaced analog is due to the relatively small size of the arrays (charged couple devices) available for large-format aerial cameras. Since digital cameras have relatively small arrays, the area sensed on the ground is also small. Therefore, they require a higher number of flight lines to cover the same area as can be done with a film camera, making them impractical for large areas.

Digital photogrammetry, therefore, is usually performed on analog aerial photographs that have been scanned or digitized. Nearly all topographic maps produced today use scanned analog aerial photography as a base.[4]

Scanning is usually conducted on a high-precision drum scanner with an accuracy of up to 25 μm. Flatbed scanners can be used, but they have more optical distortions. These distortions can be modeled out through geometric calibration, but this is an imperfect procedure. Scanners result in loss of detail.

For example, aerial photos typically have a resolution of a 40 line pairs per millimeter (lp/mm). The relation between a line pair and pixel size is 1 line pair = 2 pixels. Therefore, 40 lp/mm corresponds to 80 pixels/mm, or a pixel size of 12 μm. Expressed in dots per inch (dpi), a 12-μm pixel size corresponds to 2000 dpi, or 8-μm pixel size = 3000 dpi.

Contemporary scanners have a gray value resolution of 4096 gray values (12 bit). In order to conserve memory, images can be reduced to a gray value resolution of 256 gray levels (8 bit). A typical black-and-white 23 × 23 cm aerial image with 12-μm pixel size and 8-bit resolution requires approximately 370 Mb of storage space. A color image requires over 1.0 Gb because color requires three bands of 370 Mb each. Storage requirements for digital imagery are an important consideration in any large mapping effort, although this is less of an issue as computer memory becomes more efficient.

6.2.3.7   Digital Photogrammetric Workstations

A digital photogrammetric workstation (DPWS) represents the state of the art in topographic mapping instrumentation.[13] It consists of four major components: (1) stereo viewing devices, (2) a three-dimensional mouse, (3) a PC, and (4) software.

Software packages come in four basic types: (1) database software for vector, raster, and attribute data; (2) image-handling, compression, processing, and display software; (3) photogrammetric applications software for image orientation and generation of digital terrain models (DTMs); and (4) user interface software.

Digital stereo workstations are based on exactly the same concept as analytical stereo plotters. The difference is that images are moved on a computer monitor rather than photographs on a plate. Contemporary plotters have the floating mark fixed and the images are scrolled continuously. For stereoscopic viewing, each eye sees the adjacent images separately in time or space. The simplest method of image separation is the use of an anaglyph (red-and-green image with green and red eye filters). This technique is limited to black-and-white images.

For color, a technique called crystal eye, or polarized light, is commonly used. For this technology, images are shown, one followed by the other, on a cycle of 50 Hz. Crystal shutters in the form of goggles are synchronized by infrared light, with one lens opaque and the other transparent. The left eye can see only the left image and the right eye can see only the right image at one time. The brain remembers the previous image and forms a three-dimensional image, and the viewer perceives three dimensions.

In the polarization technique, a polarization filter is in front of the viewing screen, which changes the polarization orientation from horizontal to vertical orientations back and forth on a screen. The image is viewed through polarization filters, with the left lens horizontally polarized and the right lens vertically polarized.

The principal advantage of DPWS over analytical plotters is automation.[5] Examples of automated procedures include generation of digital terrain models, extraction of tie points for relative orientation, and generation of orthophotos. Automatic image matching is also a potential advantage of DPWS, but it only works well in flat, featureless terrain. For many applications, however, there is vitually no difference between results derived from a DPWS and those obtained from analytical plotters.[6]

6.2.3.8 Aerial Digital Cameras

Stereo plotters were designed to use standard 23 × 23 cm size film. Most aerial photography is still taken with this format, but the film is usually digitized. Scanned imagery taken by film camera dominates the type of imagery used in topographic mapping since the ideal digital replacement for film cameras would be based on a very large area array sensor, which has not yet been built. There would need to be a sufficient number of pixels to be equal to or better than precision-scanned aerial film. A 20-µm pixel size produces 11.5 × 11.5 k pixels, so this would be the desired array to compare with a film camera. The largest array digital camera manufactured in 2004 was only 9 × 9 k.[5]

There are two manufacturers of digital cameras used in aerial topographic mapping today: Leica Geosystems and Z/I Imaging (a joint venture between Carl Zeiss Corporation and Intergraph Corporation). Leica makes the ADS-40, which has multiple linear arrays analogous to multispectral scanners onboard satellites such as Landsat. From a flying altitude of 3000 m, a swath of 3.75 km is covered with a 15-cm ground pixel size.

The Leica ADS-40 Airborne Digital Sensor is capable of delivering photogrammetric accuracy and coverage as well as multispectral data. The ADS-40 differs from the older and very common RC30 film camera in many ways. It is digital rather than analog, with three panchromatic charged couple device lines capturing panchromatic information in views forward, nadir, and backward from the aircraft. Photogrammetric restitution is supplemented by four multispectral lines, resulting in the simultaneous capture of seven bands of information. Raw data are rectified using position and attitude data supplied by a position and orientation system. These features virtually eliminate the need for image orientation.

The Leica ADS-40 has the following advantages over a conventional analog aerial photographic camera:

- Three sensors (black and white, color, and false color)
- RGB coregistration through trichroid filter
- DTMs are automatically generated from three-line stereo sensor data
- Reduced ground control requirements
- No film processing or scanning

Z/I Imaging manufactures the digital modular camera (DMC). The DMC uses a modular design. It comprises eight synchronously operating charged couple device (CCD) matrix-based camera modules. Four parallel cameras can generate multispectral imagery for color composites. Four panchromatic images from converging cameras are mosaicked digitally to form a single image.

DMC imagery is based on the central perspective approach, which allows the camera to achieve, in theory, better than 2-inch ground resolution size. Therefore, it could produce the same resolution as a film-based aerial camera system. The DMC has an electromechanical shutter placed in the center of the lens. The advantage of this design is that it results in a nearly distortion-free image since all image points are exposed through the same optical path at the same moment in time.

At full resolution (12 bits) and color mode, the DMC can capture and store more than 2000 images, which is more than three rolls of the 500-foot film taken with typical camera system. Up to four multispectral channels can be mounted in the DMC, allowing for the collection of images, for instance, in the red, green, blue, and a separate infrared channel for taking simultaneous true and false color images. A wide-angle relatively fast (aperature of f/4) lens is combined with a 3 × 2 k charged couple device chip in the camera. The ground coverage of the multi-spectral channels and the four high-resolution panchromatic channels are identical.[17]

6.2.3.9 Light Detection and Ranging (Lidar)

The conventional methods of topographic data collection, as described previously, are land surveying and aerial photography. These techniques have been time tested and result in topographic maps for which standards have been well established. Lidar is a relatively new technology that is gaining acceptance in commercial mapping as a tool for topographic measurement and mapping. The basic concepts of lidar are similar to other active remote sensing systems, such as radar. Lidar transmits laser pulses while scanning a swath of terrain centered on and colinear with the flight path. The beam's travel time from the aircraft to the ground is measured and the time intervals are converted to distance or range.

The position of aircraft is determined by the global positioning system (GPS), and rotational positions of the laser pulse direction are combined with aircraft attitude values determined with an inertial navigation system (INS). They are also combined with range measurements to obtain range vectors from the aircraft to ground points. When these vectors are combined with the aircraft location, they give the coordinates of ground points (x, y, z).

Lidar has several potential advantages over both ground-based topographic mapping and photogrammetry for topographic mapping. They include (1) a relatively fast data capture rate (90 km²/hour), (2) minimum human handling of data, (3) ability to measure subtle changes in terrain due to the fact that it generates up to 100,000 pulses per second, and (4) each pulse being individually georeferenced using the onboard, and so only one GPS ground station is required for improving the accuracy differentially.

Drawbacks to lidar are (1) accuracy on the of order of 10 to 15 cm (vertical) and 50 to 100 cm horizontal, which is unacceptable for some topographic mapping applications, (2) high capital cost for equipment ($1,000,000+) and high leasing rates (in the range of $3000 to $10,000/hour), and (3) critical image features such as break lines being misinterpreted. Therefore, the imagery requires thorough interpretation and editing.

Points to remember when using lidar for a topographic mapping project are:

- Lidar instrument manufacturers' published values for accuracy usually focus on vertical accuracy, but horizontal accuracy is also very important.

FIGURE 6.8 Orthophotograph/lidar drape of Devil's Millhopper sinkhole in Alachua County, Florida: View 1. (Courtesy of 3001 Spatial Data Corporation.) (See color version on accompanying CD.)

- The accuracy of lidar-derived height varies according to the terrain. Steep slope can be impossible to measure accurately with lidar.
- Very high reflectors in the lidar's field of view send it away from the geometric center of the collecting optics, which results in an erroneous range measurement.
- Targets such as the painted centerline on a road can be misinterpreted as a break line. Other artifacts that cause problems with the imagery are vegetation, which must be "removed" if a bare-earth model is desired for the final product.

Lidar now plays a complementary role to the traditional photogrammetric processes in topographic mapping. For example, it can improve the performance of automated point measurement for the triangulation process or of automated DEM generation by image matching.[5]

An example of applying lidar imagery to soils mapping is shown in Figure 6.8 and Figure 6.9. Figure 6.8 is a 1-m orthophotograph draped over a lidar-derived surface model of the Devil's Millhopper sinkhole in Alachua County, Florida. Figure 6.9 shows the same data, but at a different view angle.

Figure 6.10 is a lidar-derived shaded relief surface with a 30-m DEM drape of the Amite River Basin in Louisiana. Data were thinned to every hundredth point prior to creating the surface. Non-bare-earth features, such as trees and buildings, have been removed (edited) from the southern half of the area.

Figure 6.11 is a 5-m DEM derived from lidar data of Profit Island in the Mississippi River Basin. It shows the drainage pattern indicative of water and wind-laid silt and so is useful for soil mapping of sediment.

Figure 6.12 is a profile showing the interaction of a laser pulse and a forested region. The bottom part of the image is a profile view of lidar data collected over a forested area along the Louisiana–Mississippi border. Shades of red represent

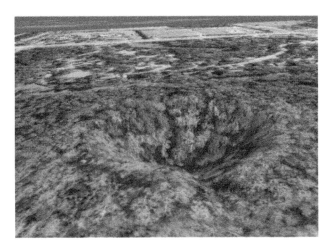

FIGURE 6.9 Orthophoto/lidar image of Devil's Millhopper sinkhole in Alachua County Florida: View 2. (Courtesy of 3001 Spatial Data Corporation.) (See color version on accompanying CD.)

FIGURE 6.10 Shaded relief map with 30-m drape of Amite River Basin in Louisiana. (Courtesy of 3001 Spatial Data Corporation.) (See color version on accompanying CD.)

the upper tree canopy, and shades of yellow represent mid-story growth. Green respresents the ground surface.

Figure 6.13 is a 2-m posting surface of variance draped on top of a shaded relief model created from lidar data collected in March 1999 at an altitude of 8000 feet. Areas covered with vegetation have a heavily textured appearance,

FIGURE 6.11 Five-m DEM derived from lidar data of Profit Island in the Mississippi River Basin. (Courtesy of 3001 Spatial Data Corporation.) (See color version on accompanying CD.)

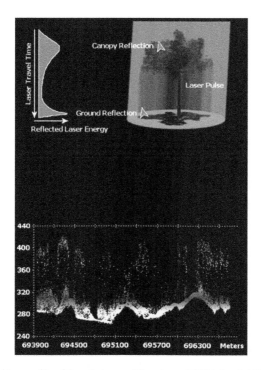

FIGURE 6.12 Lidar profile of forested area. (Courtesy of 3001 Spatial Data Corporation.) (See color version on accompanying CD.)

while open farmland appears smooth. Variance ranges from low (cool tones) to high (hot tones). High variance is an indication of high vegetation density or tall vegetation. Though variance analysis approximation of stand density can be made, the data can also provide insight on canopy structure, approximate age, and harvest potential.

FIGURE 6.13 Combined lidar and aerial photograph. (Courtesy of 3001 Spatial Data Corporation.) (See color version on accompanying CD.)

6.2.3.10 Radio Detection and Ranging (Radar)

Radar measures the strength and time of return of microwave signals emitted by an antenna and reflected off an object. The radar antenna alternately transmits and receives pulses at wavelengths between 1 cm and 1 m. For imaging radar, approximately 1500 pulses per second are transmitted with specific pulse duration. The pulse typically has bandwidths in the range of 10 to 200 MHz.[18]

Energy from the transmitted radar pulse is reflected back to the antenna from the ground and converted to a digital number. The number is recorded and displayed as an image. The pulse length determines resolution in the range direction of the radar (range resolution). Shorter pulses result in higher range resolution. The radar moves along a flight path, building an image as the area illuminated by the radar moves along the surface in a swath.[14] The radar's beam width determines its azimuth resolution with high beam widths resulting in lower azimuth resolution.

Radargrammetry refers to measurement of objects using radar imagery. Highly accurate height measurements of the terrain can be made from radar images. Nonimaging radars such as altimeters can measure elevations to within centimeters. Making a topographic map for most purposes, however, requires a higher degree of spatial resolution than is presently available from current radar systems. Radar images are also comparatively expensive. In 2005, the radar satellite Terra SAR-X, with a 1-m pixel size, is scheduled for launch. If the imagery works as planned, it will prove that radar can be used for topographic mapping. Also, the cost of the imagery will be relatively low due to the fact that it is on a satellite (vis-à-vis aircraft) platform.

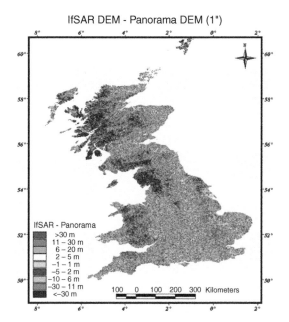

FIGURE 6.14 SRTM topographic image. (From the Landmark Project, University of Manchester, U.K.) (See color version on accompanying CD.)

6.2.3.11   The Space Shuttle Radar Topographic Mapping Experiment

The Shuttle Radar Topography Mission (SRTM) was flown in the year 2000.[19] The key SRTM technology was interferometric synthetic aperature radar (IfSAR), which compares two radar images taken at slightly different locations to obtain elevation or surface change information. The SRTM used single-pass interferometry, meaning that the two images were acquired at the same time — one from the radar antennas in the shuttle's payload bay and the other from the radar antennas at the end of a mast extending from the shuttle. Combining the two images produced a single three-dimensional image, shown in Figure 6.14.

The mission was a partnership between NASA and the Department of Defense's National Geospatial-Intelligence Agency (NGA). SRTM collected radar data over nearly 80% of Earth's land surface. Analysts are in the process of using the SRTM data to generate three-dimensional topographic DEMs. These data are being systematically processed on a continent-by-continent basis, with North America first. As each continent is completed, the data are delivered to the NGA, where they are edited, verified, and brought into conformance with National Map Accuracy Standards. These finished data will then be returned to NASA for distribution to the public through the USGS.

Each site covers a number of 1° latitude by 1° longitude "cells," and each processed dataset consists of unedited digital elevation maps, images, and ancilliary data. As these datasets are completed, they will be made available to the scientific community and the general public.

6.3 MAP PRODUCTION

6.3.1 MAP COMPILATION

The following describes the process used by the USGS in its production of topographic maps. While many of the steps have been or are in the process of being automated, it is still a highly labor intensive procedure.

Map features and contour lines are traced automatically, semiautomatically, or by hand as they appear in the stereo model. As the stereo plotter operator moves a reference mark, the tracing is transmitted to a tracing table that produces the map manuscript.

Figure 6.15 shows (1) a complete portion of a USGS 7.5-minute topographic quadrangle sheet, (2) the forested area separate in green, (3) the topographic contour layer in brown, and (4) features that have been interpreted via aerial photography, but not yet field verified in purple. There are three other separate layers in USGS topographic maps: the transportation layer (red), urban areas (pink), and water (blue).

FIGURE 6.15 Map separates. (From USGS, 2001.) (See color version on accompanying CD.)

6.3.2 Map Scribing, Editing, and Printing

Prior to digital technology, map production was labor intensive and involved the following steps. After the map manuscript was compiled, a map-size film negative of the compiled manuscript was made. This negative was then photographically reproduced on thin plastic sheets, to which a soft translucent coating had been applied. Using a light table, a scriber used engraving instruments to etch the map's lines and symbols by removing the soft coating from the hard plastic guide. All features to be printed in the same color on the map, such as blue for water, were etched onto separate sheets.

Type sets for the words on the map were selected according to standards that would ensure consistency of type sizes and styles for all maps in that series. Type was positioned on clear plastic sheets that were overlaid on the scribed separations. Photographic negatives were then made of the type for printing.

The final step before printing was preparation of a color proof. This was accomplished by making multiple exposures of the type negatives and scribed sheets. A press plate was made for each map color by exposing the appropriate scribed sheets and type negatives. Repeated runs of the map paper through the lithographic printing press accomplished printing with one for each color, or one run through a press capable of printing several colors in sequence (USGS, 2001).

Most topographic maps in circulation today were made using these techniques, but computer technology has changed everything. For example, most map compilation and revision is performed from digital images. Color separations are plotted from digital data rather than manually scribed separations, and even the type for words on the map is positioned and plotted from digital data. These new procedures are being introduced to mapping agencies around the world and have made map production and updating a much faster, easier, and more accurate process.

6.3.3 Map Accuracy Standards

Every mapping agency has a set of vertical and horizontal mapping accuracy standards published somewhere on the map. It is critical that the user of a map understand the accuracy limitations of each map and apply data from the map accordingly.

In order to meet U.S. National Map Accuracy Standards, the following specifications must be met for USGS 1:24,000 scale 7.5-minute quadrangle topographic maps:

- Horizontal accuracy: Positions of 90% of all points tested must be accurate within 1/50th of an inch (0.05 cm) on the map. At 1:24,000 scale, 1/50th of an inch is 40 feet.
- Vertical accuracy: The elevation of 90% of all points tested must be correct within half of the contour interval. On a map with a contour interval of 10 feet, the map must correctly show 90% of all points tested within 5 feet of the actual elevation.

All maps produced by the USGS at 1:250,000 scale and larger are prepared by methods designed to meet these accuracy standards and carry the statement "This map complies with National Map Accuracy Standards." Exceptions to this practice involve areas covered by dense woodland or obscured by fog or clouds; in those areas, aerial photographs cannot provide the detail needed for accurate mapping. The USGS samples a sufficient number of its maps to ensure it is producing maps that meet accuracy standards.

One disturbing activity is the marketing of maps derived from a small scale as large-scale maps. For example, USGS 1:100,000 scale maps have been marketed as 1:24,000 scale maps. The map producer, in this case, simply rescaled the 1:100,000 scale map to a 1:24,000 scale one.

The user, in this case, might think that he is getting National Map Accuracy Standards at a scale of 1:24,000, whereas he is actually getting 1:100,000 scale accuracy. The accuracy standard only applies to the original scale and does not improve if the map is simply enlarged.

6.3.4 DIGITIZING MAP DATA

Map digitization is an ongoing process at most mapping agencies in the world. It resembles the original map scribing process in that it requires that each feature on each map separate be located, classified, and traced. Typically, maps have 10 or more layers, such as roads, topographic contours, boundaries, surface cover, and manmade features, which require digitization. Maps can be digitized by hand, in which each map's lines are manually traced with a cursor or automatically traced with scanners. Obviously, scanning is preferred to hand digitization for time's sake. However, scanners often miss features and have other problems that necessitate human intervention.

After digitizing, several editing operations must be performed. For example, attribute codes must be added to identify what each digitized line or symbol represents. A variety of other tasks must be performed to ensure that information is complete and correct, including matching features with adjoining files, matching features relative to each other within the file, and controlling the accuracy of attribute coding and positions.

6.4 SUMMARY AND CONCLUSION

Modern topographic mapping started with analog photogrammetry, continued with analytical photogrammetry, and now is becoming digital. Digital systems (cameras, plotters, and scanners) have several advantages, including a high degree of automation, greater accuracy, and overall efficiency.

Film cameras have probably nearly reached the end of their logical development with respect to optimal optics, image motion compensators, and gyro-stabilized platforms. On the other hand, digital cameras are in their formulative stages and will certainly improve in terms of resolution and speed.

The advent of high-resolution satellite systems such as IKONOS and Quickbird will challenge airborne systems. IKONOS has a 1-m instantaneous field of view (the same footprint as USGS orthoquads cells), and Quickbird has a 0.67-m cell size. However, space-based mapping systems are still relatively expensive due to high launch costs and are not currently cost competitive with aerial systems. Also, topographic map rendering requires a stereoscopic perspective, which only one civilian satellite program (SPOT) offers today. SPOT's vertical accuracy is on the order of 50 m, a tolerance not useful for many topographic mapping applications.

Other remote sensing technologies, such as lidar and radar, will complement photogrammetric approaches to topographic mapping in the near future. These technologies will improve with time and may even eventually replace traditional aerial photography for topographic mapping as we know it.

REFERENCES

JOURNAL ARTICLES

1. A McBratney, I Odeh, T Bishop, M Dunbar, T Shatar. An overview of pedometrics techniques for use in soil survey. *Geoderma* 97: 293–327, 2000.
2. G Heuvelink, R Webster. Modeling soil variation: past, present, and future. *Geoderma* 100: 269–301, 2001.
3. R Kaczynski, J Ziobro. Digital aerial triangulation for DTM and orthophoto generation. *Int Arch Photogrammetry Remote Sensing* 32 (Part B4), 281–283, 1998.
4. J Fallow, K Murray. Digital photogrammetry: options and opportunities. *Int Arch Photogrammetry Remote Sensing* 29 (Part B2), 397–403, 1992.
5. B Wald. The Synergies between LIDAR and Photogrammetry Lie Primarily in Areas of Softwater and Support. Interview in *GIS Development Magazine*, August 2000.
6. C Dekeyne. Applications of Digital Photogrammetric Workstations: Digital Plotting, OEEPE Official Publication 33. OEEPE, 1996, pp. 283–290.

BOOKS

7. R Davis, F Foote, J Anderson, E Mikhal. *Surveying: Theory and Practice*, 6th ed. McGraw-Hill, New York, 1981, p. 896.
8. P Wolf, B Dewitt. *Photogrammetry*, 3rd ed. John Wiley & Sons, New York, 2000, p. 432.
9. T Avery, G Berlin. *Fundamentals of Remote Sensing and Airphoto Interpretation*, 5th ed. Prentice Hall, New York, 1992.
10. E Mikhail, J Bethel, J McGlone. *Introduction to Modern Photogrammetry*. John Wiley & Sons, New York, 2001.
11. A Rencz, R Ryerson, Eds. *Manual of Remote Sensing*, Vol. 3, *Remote Sensing for Earth Science*. John Wiley & Sons, New York, 1999, p. 728.
12. D Maune, Ed. *Digital Elevation Model Technologies and Applications*. American Society of Photogrammetry and Remote Sensing, Bethesda, MD, 2001, p. 540.

13. T Schenk. *Digital Photogrammetry*. TerraScience Publishers, Laurelville, OH, 1999, p. 428.
14. F Henderson, A Lewis, Eds. *Principles and Applications of Imaging Radar*. John Wiley & Sons, New York, 1998, p. 896.

ELECTRONIC PUBLICATIONS

15. USGS. http://erg.usgs.gov/isb/pubs/booklets/topo/topo.html. 2001.
16. Leica Corporation. http://www.leica-geosystems.com/products/. 2003.
17. Zeiss Imaging Corporation. http://ww2.ziimaging.com/. 2003.
18. T Freeman. What Is Imaging Radar? http://southport.jpl.nasa.gov/desc/imagingradarv3.htm. 2004.
19. NASA. http://www-radar.jpl.nasa.gov/srtm/. 2001.

MISCELLANEOUS

20. H Amoyaw. Optimizing the Workflow in a Hybrid Production System of Analytical and Digital Production of Geodata from Aerial Photographs. M.S. thesis, The Institute for GeoInfo Systems and Earth Observations, Netherlands, 2002.

Section III

Pedometrics

7 Digital Soil-Terrain Modeling: The Predictive Potential and Uncertainty

Thomas F.A. Bishop and Budiman Minasny

CONTENTS

ABSTRACT

Research in the past 20 years has demonstrated that digital terrain models are a useful secondary information source for the prediction of soil properties and classes. This chapter begins with a brief introduction to digital terrain modeling; in particular, the types of terrain attributes that can be calculated from a digital elevation model (DEM) are described. The next section reviews soil-terrain modeling, with an emphasis on the variety of prediction methods that have been used. A summary of published soil-terrain studies is given. The second half of the chapter presents a case study aimed at illustrating the impact that the source DEM spatial resolution and uncertainty have on soil-terrain prediction models. The study site is a 74-ha field in Australia. The datasets include a 5-m DEM created from a carrier-phase global positioning systems (GPS) survey, a 25-m

DEM created from digitized contour lines, and 111 measurements of soil clay content. Cokriging was used to map clay content, with slope as the secondary information source. The correlation between clay and slope was −0.53 for the 5-m DEM and −0.02 for the 25-m, DEM which illustrates the potential impact that resolution has on soil-terrain modeling. Monte Carlo simulation with a modified version of Latin hypercube sampling (LHS) was used to perform an uncertainty analysis of the clay-slope modeling process. Maps of the mean and standard deviation of clay content across 20 realizations were created. The values in the standard deviation of clay content maps were generally small (<2%) and, in most parts of the field, less than the analytical accuracy of the hydrometer method, which was used to measure soil clay content in the laboratory. The implication of the uncertainty analysis is that in this case, the DEM is accurate enough for the calculation of slope and subsequent modeling of the clay–slope relationship for the cokriging prediction model. As the spatial resolution is coarsened and the uncertainty increases, this may not be the case. Therefore, it is recommended that uncertainty analysis should become a routine part of any soil-terrain modeling process.

7.1 BASIC CONCEPTS: DIGITAL TERRAIN MODELING

Previous chapters have discussed sources of elevation data, but a brief review of concepts in digital terrain modeling is given below. Interested readers should refer to the literature for more detail; some excellent examples include Moore et al.[1] and Wilson and Gallant.[2]

A digital elevation model (DEM) represents the spatial distribution of elevation across a landscape. The term *digital* has been used since the 1970s, when digital cartography replaced conventional paper cartography. Today, geographic information systems (GISs) are commonly used to store, display, and manipulate elevation (and other spatial) data. Analogous to a DEM, a digital terrain model (DTM) is an ordered array of numbers that represent the spatial distribution of terrain attributes across a landscape. A DEM is the basis for calculation of surface attributes, which include slope, aspect, and curvature. This has also been called parametization of the surface model or a numerical description of the continuous landscape surface.[3,4]

Terrain attributes can be parameterized from a DEM and are traditionally divided into primary and secondary (or compound) attributes. Primary terrain attributes are calculated directly from the DEM, whereas secondary attributes are calculated from two or more primary terrain attributes.[1] Primary terrain attributes can be further divided into those that are derived locally (using local neighborhood points) and those that are derived regionally following prescribed rules.[5] The local and regional approaches can be further divided into scale specific and scale invariant.

The basic and most commonly used *primary terrain attributes* include *surface derivatives* or *local attributes*, such as slope, aspect, and curvature. Slope is defined as the gradient or rate of change of elevation[6] and is generally measured either in percent rise or in degrees. Aspect is defined as the azimuth of slope and is generally measured in degrees.[1] Slope may be thought of as the first derivative of the elevation surface *down* the slope, perpendicular to the contours, and aspect as the first derivative of the elevation surface *across* the slope, parallel to the contours.[6] Curvature is the second derivative of the elevation surface in a particular direction; it can be thought of as the first derivative of slope in the case of profile curvature, or as the first derivative of aspect in the case of plan curvature.[6] Evans[7] presented an excellent summary of the interpretation of primary terrain attributes in relation to their geomorphological meaning.

With respect to *regional* attributes, upslope contributing area (also called drainage or catchment area) is one of the most important. It is defined as the area above a certain pixel (or length of contour) that contributes flow across that pixel (or contour interval). To calculate upslope contributing area, a method for calculating flow direction is first needed. The simplest method for specifying flow directions is to assign flow from each grid cell to one of its eight neighbors, either in the cardinal directions or diagonally, in the direction with steepest downward slope. This widely used method, designated D8 (for the eight flow directions), was introduced by O'Callaghan and Mark.[8] The D8 approach has disadvantages arising from the discretization of flow into only one of eight possible directions, separated by 45°. This limitation has motivated the development of other methods, including multiple flow direction methods, random direction methods, and stream tube methods.[9,10] The relative merits of flow direction algorithms are discussed by Tarboton.[10] As an alternative, Dobos et al.[11] proposed potential drainage density (PDD), designed to highlight relative terrain differences even on a relatively level land surface.

As mentioned previously, *secondary terrain attributes* are calculated from the combination of two or more primary terrain attributes. The purpose of secondary terrain attributes is to model the spatial variation of hydrological, geomorphological, and ecological processes across the landscape.[12] In most cases, sophistication is sacrificed to enable the representation of spatial variability across a landscape.[1]

The most widely used secondary terrain attribute in soil science and hydrology is the topographic wetness index (TWI), or the compound topographic index (CTI):

$$\omega = \ln\left(\frac{A_s}{\tan\beta}\right) \tag{7.1}$$

where
      ω = TWI or CTI
      A_s = specific catchment area
      β = slope

TWI originated from studies in hydrological modeling.[13] Large TWI values indicate an increased likelihood of saturated conditions; the larger values are usually found in the lower parts of watersheds and convergent hollow areas associated with soils with small hydraulic conductivity or areas of small slope.[14]

Other secondary attributes that can be derived for specific catchment areas are stream power index (SPI) and sediment transport index (STI).[1,15] In addition, Wilson and Gallant[2] provided routines for the calculation of erosion index, solar irradiation, and dynamic wetness index.

7.2 SOIL-TERRAIN MODELING

Topography (or relief) is one of the five factors of soil formation as described by Jenny[16]:

$$S = f(C, O, R, P, T) \tag{7.2}$$

where
S = soil
C = climate
O = organisms
R = relief
P = parent material
T = time

Ever since Jenny first presented this equation, soil scientists have attempted to build quantitative predictive models of soil formation based on topography ($S = f(R)$) with ever-increasing complexity in terms of prediction methods (f) and terrain variables (R). Aandahl[17] related the distribution of soil nitrogen to slope length in what is possibly the first published attempt to quantitatively model soil-terrain relationships. Another early example was that of Walker et al.,[18] where slope, curvature, aspect, and distance from the local summit in combination with multiple linear regression were used to predict soil morphological properties such as A-horizon depth and depth to mottling and carbonates.

Until the 1980s, a major limitation to soil-terrain modeling was the concurrent availability of elevation data, computing power to create two-dimensional maps of terrain attributes and algorithms to calculate contextual area (e.g., upslope area), dispersal area, and secondary terrain attributes (e.g., topographic wetness index). Chapter 6 provides a detailed review of the history of topographic mapping and availability of elevation data. Today, the lack of elevation data is generally not a problem for anything but fine-resolution soil mapping, though the quality of elevation data may vary greatly.

In the 1980s, elevation data became more readily available, geographic information systems (GIS) rose in prominence, and algorithms improved for calculating terrain attributes. This has resulted in GIS-based studies that have related digital terrain attributes to soil observations, thus resulting in the term *digital soil-terrain modeling*. Soil observation points are intersected with layers of terrain attributes, a model is fitted to predict soil variables at the observation points, and then the model is used to predict soil variables for all other locations on the raster.

Rather than using primary terrain attributes (e.g., slope, curvature), many of studies have used secondary terrain attributes, which indirectly represent soil and hydrological processes. Such attributes have often been found to be more useful than primary terrain attributes for soil prediction purposes. Examples include:

- Upslope contributing area (used by McBratney et al.[19] to predict clay content)
- Topographic wetness index (used by Gessler et al.[20] to predict A-horizon depth)
- Stream power index (used by Moore et al.[15] to predict extractable P)
- Drainage proximity index (used by Bell et al.[21] to predict A-horizon depth)
- Accumulated flow index (used by Bell et al.[21] to predict depth to carbonates)

Furthermore, the prediction methods have increased in complexity as researchers have shifted away from multiple linear regression (MLR). Examples include modern regression techniques such as generalized linear models (GLMs),[20] generalized additive models (GAMs),[22] and regression and classification trees (RT/CTs).[23] These techniques are described extensively in Hastie et al.,[24] and a summary of the most widely used statistical techniques is given in Table 7.1 and Table 7.2.

Multivariate geostatistical methods such as kriging with external drift[25] and cokriging[26] have also been used in combination with terrain information. In addition, hybrid methods have been used where MLR,[27] GLMs,[22] GAMs,[22] and RTs[19] were employed to model the deterministic component of soil variation when regression kriging was performed.

McBratney et al.[28] proposed the inclusion of soil (S) and spatial information (N) as predictors in addition to Jenny's five soil-forming factors. They called the resultant function the SCORPAN spatial soil prediction function. Spatial information may be simply represented as eastings or northings, or a linear or nonlinear (nonaffine) transformation of the original spatial coordinates. For example, relative position in the landscape has been found to be related to soil properties, and also to be useful for characterizing landform. Gessler et al.[20] found elevation above local stream, distance to local stream, and distance to local drainageway to be good predictors of soil attributes. Moran and Bui[29] similarly found distance downhill to channels and distance downhill from hilltops to be good predictors of soil classes for coarse resolution mapping (250-m pixel size). It is arguable whether such predictor variables are N or R SCORPAN factors. S factors may be crude proximal sensors, e.g., electromagnetic induction instruments,[30] which are increasingly used for field-extent soil mapping or soil class maps for regional soil mapping.

TABLE 7.1
Statistical Methods for Soil-Terrain Modeling

Multiple linear regression
Multiple regression analysis typically assumes a linear relationship between several independent
variables or predictors and a dependent or predicted variable. Multiple regression analysis fits a
straight line (or plane in an n-dimensional space, where n is the number of independent variables)
to the data.

Discriminant analysis
A procedure for the determination of the group to which an individual belongs based on the
characteristics of the individual. Discriminant analysis investigates the differences among
multivariate classes, to determine which attributes discriminate between the classes, and to determine
the most parsimonious way to distinguish among classes.

k-means clustering
Clustering is a method for grouping multivariate data into clusters where k-means clustering is a
method for nonhierarchical clustering of multivariate data. Data are grouped into clusters, each
having its means or centroid. The membership of an individual to each class is determined by the
relative distance of its attributes to the centroid of that class. Hard k-means only allows an individual
to lie in one mutually exclusive class, while fuzzy k-means allows for an individual to lie as bridges
between classes.

Generalized linear model (GLM)
This term describes a class of models that arises from a natural generalization of ordinary linear
models. Values for the transformed dependent variable values are predicted from (or are linked to)
a linear combination of predictor variables; the transformation is referred to as the link function.
Also, different distributions can be assumed for the dependent variable values.

Generalized additive model (GAM)
These are models that use smoothing techniques to identify and represent possible nonlinear
relationships between the predictor and predicted variables. GAM is a generalization of GLM where
the linear function of the predictor is replaced by an unspecified (nonparametric) function (e.g.,
splines).

Artificial neural network
These describe a mathematical structure modeled after the functioning of the human nervous system.
The essential feature is a network of simple processing elements joined together by weights.

Classification and regression tree
A classification tree is a rule for predicting the class of an individual from the value of its predictor
variables, while a regression tree predicts continuous data. Rather than fitting a model to the data,
a tree structure is generated by dividing the sample recursively into a number of groups, each division
being chosen so as to maximize some measure difference in the predicted variable in the resulting
two groups. The resulting structure provides easy interpretation, as variables most important for
prediction can be identified quickly.

Compiled from Everitt, B.S., *The Cambridge Dictionary of Statistics*, Cambridge University Press,
Cambridge, U.K., 2002; Upton, G. and Cook, I., *A Dictionary of Statistics*, Oxford University Press,
Oxford, 2002; and *StatSoft Statistics Glossary*, available online at http://www.statsoftinc.com/text-
book/glosfra.html.

TABLE 7.2
Comparison of Different Statistical Prediction Models

Feature	Linear Models	GLM	GAM	Classification and Regression Tree	Neural Net
Ease of use	☺	😐	😐	☺	☹
Parsimony	☺	😐	☹	☺	☹
Interpretability	☺	😐	☹	☺	☹
Nonlinearity	☹	☹	☺	☺	☺
Prediction of qualitative data (e.g., soil classes)	☹	☺	☺	☺	☺
Handling of mixed data type (both qualitative and quantitative)	😐	☺	☺	☺	☹
Computational efficiency (for large datasets)	☺	☺	😐	☺	☹
Predictive power	☹	😐	😐	😐	☺

Note: ☺ = good; 😐 = fair; ☹ = poor.

Adapted from Table 10.1 in Hastie, T. et al., *The Elements of Statistical Learning: Data Mining, Inference and Prediction*, Springer-Verlag, New York, 2001.

Table 7.3 presents examples from the literature of the use of terrain attributes to predict soil classes or soil variables for mapping purposes, arranged in chronological order. From the table, we can see that terrain attributes have been used to predict soil classes and soil attributes. Terrain attributes predict soil taxonomic and drainage classes quite well. Continuous soil attributes can also be predicted with reasonable accuracy; soil horizon depth/thickness, clay content, organic matter content, cation exchange capacity (CEC), and phosphorus have all been successfully predicted. Recently, it has also been suggested that terrain attributes can be used to predict not only basic soil properties, but also other, more expensive soil physical properties, i.e., soil-water retention.[31–33]

Quantitative soil spatial prediction models have not been restricted to terrain as the predictor variable; other variables in Jenny's state-factor equation have also been exploited (Equation 7.2). McBratney et al.[28] surveyed published soil mapping studies and found that a DEM was the most common source of secondary information; furthermore, in 80% of the studies a terrain attribute was used in the final soil prediction model. From this it can be concluded that digital terrain information is the most useful (or at least the most readily available) secondary information source for digital soil mapping. For interested readers, McBratney et al.[28] thoroughly reviewed digital soil mapping and the emerging paradigm shift based on quantitative prediction methods and geospatial technologies.

The usefulness of digital terrain attributes for soil mapping is largely dependent on the landscape (does topography have a major impact on soil variation?)

TABLE 7.3
Examples of the Use of Terrain Features to Predict Soil Classes or Soil Variables for Mapping (Arranged in Chronological Order)

Ref.	Predicted Soil Variables	Terrain Attributes as Predictors	DEM Resolution (m)	Other Soil-Forming Factors Used as Predictors	Fitting Model	Study Area (km²)
36	Soil drainage classes	Slope, curvature			Linear regression	
18	Soil thickness, subsoil mottling, depth to mottling	Elevation, slope gradient, slope length direction, slope length curvature, slope width curvature, distance from local summit	10		Linear regression	0.007
37	Soil classes	Elevation, slope, aspect	500	Geology, lithology	Principal component analysis	624
38	Thickness of A-horizon, depth to $CaCO_3$	Elevation, slope, curvature	10, 50		Discriminant analysis, linear regression	
26, 39	Soil morphological, physical and chemical properties	Gradient, profile and plan curvatures, upslope distance, upslope area	10		Ordination technique, linear regression, kriging	0.26
40, 41	Soil units	Slope, wetness index, terrain position	10	Aerial photograph	Bayesian and expert system	1
21, 42	Soil drainage classes	Slope, slope–curvature ratio, elevation above local stream, gradient to local stream, distance to local stream, distance to local drainageway	30	Geology	Discriminant analysis	0.26

43	Clay content, CEC, electrical conductivity (EC), pH, bulk density, water content at –10 and –1500 kPa	100	Slope, relief, slope position, landform	Air photo	Ordination technique, GLM	500
15	A-horizon thickness, organic matter content, pH, extractable P, silt and sand contents	15	Slope and wetness indices		Linear regression	0.05
44	Soil series, A-horizon thickness	30	Elevation, aspect, curvature	Landsat TM, geology map	Fuzzy logic	36
45	Organic matter, clay content	1000	Distance from upstream	Climate data	Linear regression	5000
20	A-horizon depth, solum depth, presence of E-horizon	20	Plan curvature, TWI		Linear regression	
31	Available water capacity	100	TWI		Linear regression	2600
46	Soil drainage classes	10	Elevation, aspect	Airborne Visible/Infrared Imaging Spectrometer (AVIRIS) normalized difference vegetation index (NDVI)	Classification tree	24

TABLE 7.3 (Continued)
Examples of the Use of Terrain Features to Predict Soil Classes or Soil Variables for Mapping (Arranged in Chronological Order)

Ref.	Predicted Soil Variables	Terrain Attributes as Predictors	DEM Resolution (m)	Other Soil-Forming Factors Used as Predictors	Fitting Model	Study Area (km^2)
47	Soil classes	Elevation from river, slope, curvature, distance from the nearest drainage axis and river bank	50	Geology	Classification tree	35
48	Profile darkness index (hydromorphic features)	Slope, profile curvature, elevation above local depression	10	Geology	Linear regression	
49	Topsoil clay	Elevation, slope, plan curvature, stream power index, wetness index	5		Fuzzy k-means	0.2
50	B- and C-horizon depth	Slope, aspect, mean, horizontal, and vertical curvatures, specific catchment area, TWI, SPI	11, 8, 5, 4		Linear regression	63.7
51	Soil classes	Aspect, slope, curvature	250	Geology	Decision tree, Bayesian model	1300
52	Presence of a noncalcareous clay-loam	Slope and aspect	20	Flow path, solar radiation, wind intensity	GLM	0.16
53	Water content at two depths, soil mineral N in the subsoil	Slope, profile curvature, plan curvature	5		Fuzzy k-means	0.06

23	Soil profile depth, total phosphorus, total carbon	Elevation, slope, aspect, specific catchment area, TWI, flow direction, dispersal area, stream power index, erosion index	25	Geology map, climate data, gamma radiometrics	Regression tree, GLM	500
54	Soil depth	Elevation, depth, slope, upslope catchment area	50		Discriminant analysis	1.5
55	Soil types	Altitude, slope, aspect, profile and plan curvature, distance to the thalweg	50		Discriminant analysis	60
56	Hydromorphic index	Elevation above stream bank, slope, specific catchment area, TWI	10		Linear regression, kriging	0.02
11, 57	Soil types	PDD, slope, elevation	1000	Advanced very high resolution radiometer (AVHRR)	Linear regression	9200
58	Soil organic C	Flow direction, flow accumulation, slope, profile and plan curvature, TWI	2, 4, 6, 8, 10		Linear regression	
19	Clay content, CEC	Elevation, slope, curvature, TWI	2, 200, 500	Crop yield data, ECa	GLM, GAM, regression tree, neural networks, kriging	0.42, 1100, 45,600
59	Soil depth, C, P, available water capacity	Elevation, slope, aspect, curvature, contributing area, dispersal area, stream power index, erosion index	10, 25	Landsat TM, climate data, gamma-radiometrics	Regression tree, linear regression	2.7, 484
60	CEC	Elevation, slope, curvature, TWI	5	Aerial photograph, Landsat TM, crop yield, EM induction	GAM, regression tree, regression kriging	0.74
61	Thickness of soil horizon	Elevation, slope, aspect, curvature, surface curvature, upslope contributing area, wetness index	10		Linear regression	0.9

TABLE 7.3 (Continued)
Examples of the Use of Terrain Features to Predict Soil Classes or Soil Variables for Mapping (Arranged in Chronological Order)

Ref.	Predicted Soil Variables	Terrain Attributes as Predictors	DEM Resolution (m)	Other Soil-Forming Factors Used as Predictors	Fitting Model	Study Area (km²)
32	Soil water retention	Slope, aspect, curvature, TWI, stream power index, slope-aspect index	30	Slope, profile and tangential curvature	Linear regression with correlated residual	0.2
62	Soil drainage classes		25	Landsat TM	Logistic regression	589
29	Soil classes	Upslope contributing area, distance downhill to channels, distance downhill from hilltops	250	Landsat TM, lithology map	Boosted classification tree	1300
63	Silt content, CEC, total exchangeable bases, Mn	Slope, aspect, curvature, upslope area, wetness index	10	Vegetation map, soil map	Artificial neural network (ANN), regression tree, GLM	0.03
33	Water retention curve	Elevation, slope, aspect, curvature, distance from middle stream, flow path length, specific contributing area, TWI, solar radiation	25		Linear regression	32
64	Soil drainage classes	Elevation, slope	10	Landsat TM, IKONOS, aerial photograph	Supervised classification	0.06

and the quality of the digital elevation model. Quality can be expressed in terms of the spatial resolution (is the DEM too coarse to represent topographic variability?) or in terms of uncertainty in the DEM. Therefore, the next two sections will consider separately (1) the impact of DEM spatial resolution and (2) the propagation of DEM uncertainty into soil prediction models.

7.3 THE IMPACT OF SPATIAL RESOLUTION ON DIGITAL SOIL-TERRAIN MODELS

7.3.1 INTRODUCTION

Scale and resolution of DEM significantly affect the terrain attributes. For example, slope will generally decrease with increasing scale or coarser resolution. Thus, the correlation between soil and terrain depends on the scale or resolution of interest. For small scale (D1–3 surveys with resolution of <100 m, as described by McBratney et al.[28]), local terrain attributes (slope, aspect, curvature) are found to be good predictors of soil variability. A fine-resolution DEM provides a more accurate representation of terrain's shape and can be justified by mechanistic soil formation models.[65,66] The influence of elevation, slope, and curvature is illustrated by a mechanistic soil-landscape model[66] in Figure 7.1, where elevation is the driving force behind soil erosion processes, and the transport of soil material in the landscape is a function of slope. Aspect plays an important role in soil formation, as it creates microclimatic and vegetation differences.[67] For resolutions of >100 m, the local terrain attributes sometimes are no longer relevant, especially in physiographical complex areas[57]; e.g., slope at a coarse resolution is associated

$$\frac{\partial Z}{\partial t} = -kZ$$

Rate of landscape lowering is proportional
to elevation.
Higher regions are more severely eroded.

$$\frac{\partial Z}{\partial t} = D\frac{\partial^2 Z}{\partial x^2}$$

Rate of landscape lowering is proportional
to profile curvature.
Lateral transport proportional to slope.

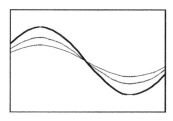

FIGURE 7.1 Models of landscape evolution based on elevation and curvature. (Modified after Pollack, H.N. *Four Corners Geological Society Guidebook*, 1969.)

with broad-scale relief. In this case, the position in the landscape appears to be a more important predictor for soil attributes.[29,45] Nevertheless, the success of the local topographic variables as soil predictors depends on the terrain.

In terms of spatial resolution, it is important to have a DEM with sufficient detail that can characterize the topographic variability that impacts on soil formation and variation (Figure 7.1). Thompson et al.[69] compared the predictive performance of terrain attributes for soil mapping of two DEMs at resolutions of 10 and 30 m. While the 30-m terrain attributes were actually generalizations of the 10-m terrain attributes, the predictive performance of each for A-horizon depth was similar. In this case, both DEMs represented those surface features that controlled soil formation and variation. A similar example is presented next.

7.3.2 EXAMPLE

One dataset will be used for illustrative purposes in this chapter. The study site is a 74-ha field, East Creek, on a farm located 25 km east of Moree in northern New South Wales, Australia (Figure 7.2). In April 1999, 113 soil cores were taken within the field to a depth of 90 cm. For this study, the clay content of the 30- to 90-cm soil layer was used for comparing prediction models. The hydrometer method was used to measure clay content.[70]

In December 1999, elevation was surveyed using two Ashtech GG-24 carrier-phase GPS units. One GPS unit was used as a base station, and the other unit was mounted on a four-wheeled all-terrain vehicle, which was driven across the field mapping elevation at a logging rate of one measurement per second. The raw elevation information was postprocessed using proprietary software, PNAV,[71] which outputs the measured elevation with an associated root mean square error (RMSE).

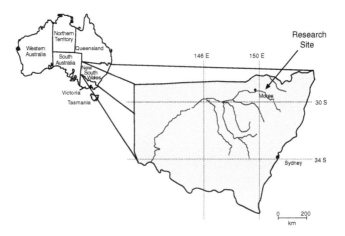

FIGURE 7.2 Location of study site.

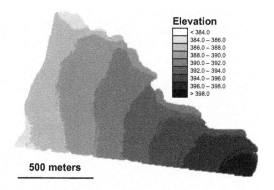

FIGURE 7.3 DEM of East Creek (5 m raster).

The point elevation data were rasterized to a 5-m digital elevation model using the TOPOGRID tool in Arc Info,[72] which is an earlier version of ANU-DEM.[73] The RMSE data were interpolated onto the same grid as the DEM, using kriging with local semivariograms.[74] In addition, the spot heights and contour lines of a 1:50,000 topographic map sheet were digitized and TOPOGRID was used to create a 25-m DEM of the same field. Figure 7.3 and Figure 7.4 present the 5- and 25-m DEM; both represent the general trend of elevation rising from west to east ($\rho_{pearson} = 0.81$ between each), but the 25-m DEM does not adequately represent the sharp drop in elevation along the northern boundary of the field where a seasonal creek is located. Slope was calculated for each DEM using the method described by Evans[75] using code written in S-PLUS.[76] The slope map based on the 5-m DEM (Figure 7.5) clearly shows the steeper slopes along the northern edge of the field. The linear features running approximately north–south are human-made contour banks aimed at controlling erosion (Figure 7.5). Other local primary terrain attributes exhibited poor relationships with clay

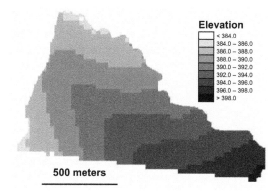

FIGURE 7.4 DEM of East Creek (25 m raster).

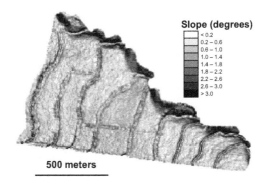

FIGURE 7.5 Slope of East Creek (5 m raster).

TABLE 7.4
Partial Correlation Matrix

	Clay	DEM, 5 m	DEM, 25 m	Slope, 5 m	Slope, 25 m
Clay	1.00				
DEM, 5 m	0.18	1.00			
DEM, 25 m	0.20	0.88	1.00		
Slope, 5 m	−0.53	0.02	0.03	1.00	
Slope, 25 m	−0.05	−0.09	−0.23	−0.01	1.00

and were not included for the rest of this study. Regional primary and secondary terrain attributes were not included as the elevation of the entire catchment was not measured.

Table 7.4 presents the partial correlation matrix between clay and the terrain attributes. There is a weak relationship between clay and elevation in this field for both DEMs. The clay-slope relationship is strong for the 5-m DEM, with the principal reason being that fine topsoil materials have been eroded along the northern boundary of the field where the slope is greatest, bringing the coarser subsoil closer to the surface in this portion of the field. The change in texture is directly related to the slope (Figure 7.5). The clay-slope relationship for the 25-m DEM is nonexistent as the slope at that resolution is not representative of the field. For example, the 25-m slope map (Figure 7.6) shows a large linear depression running diagonally through the field. This is an artefact and illustrates the importance of examining the DEM quality before use. One method for doing this is to calculate terrain attributes, as any errors are accentuated and easily spotted in the first and second derivatives of an elevation surface.[77]

To examine the impact of DEM resolution on the soil-terrain prediction models, cokriging was performed for three soil-terrain combinations: 25-m clay-slope, 5-m clay-slope, and 5-m clay-DEM. In addition, ordinary kriging (OK)

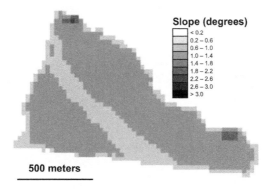

FIGURE 7.6 Slope of East Creek (25 m raster).

TABLE 7.5
Quality of Prediction Models

Prediction Model	RMSE (%)	ME (%)	Variance Explained (%)
OK	7.5	−0.2	49.1
CK, 5-m slope	7.3	0.0	56.1
CK, 25-m slope	7.8	−0.4	49.5
CK, 25-m DEM	7.5	−0.3	43.2

was performed to test whether incorporating secondary information, i.e., terrain attributes, into the prediction models improved the prediction quality. Full details concerning the geostatistical methods can be found in Chapter 9. The geostatistical analyses were performed using ISATIS,[78] and cross-validation was used as the validation method. The results are shown in Table 7.5.

The CK 5-m slope model is marginally the best, but neither of the 25-m prediction models surpassed the prediction quality of OK. This example illustrates a few points: DEMs are only useful if they are related to soil variability, and in this example, terrain was a dominant soil-forming factor along the northern edge of the field. Other SCORPAN factors are related to the clay variability for the rest of the field. Second, the DEM resolution has to be sufficient to characterize the terrain surface. In this study area, the 25-m DEM is too coarse, especially for representing a drop in elevation along the northern boundary, which turned out to be the main contributor to the clay-terrain–slope relationship. Within 5 years, the elevation of much of the Earth's surface will have been mapped at a spatial resolution of 10 to 30 m via remote sensing missions (Chapter 6). In flat, undulating landscapes, as shown in this example, it is questionable whether such elevation will have sufficient resolution to be useful for fine-resolution soil map-

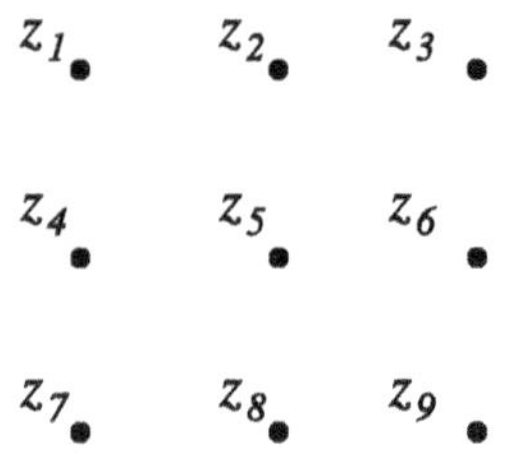

$$z_1 \quad\quad z_2 \quad\quad z_3$$

$$z_4 \quad\quad z_5 \quad\quad z_6$$

$$z_7 \quad\quad z_8 \quad\quad z_9$$

FIGURE 7.7 Analysis window for calculation of primary terrain attributes (w = grid spacing or raster size).

ping. Only resolution was considered in this example; the next section deals with DEM attribute uncertainty.

7.4 UNCERTAINTY ANALYSIS OF DIGITAL SOIL-TERRAIN MODELS

7.4.1 INTRODUCTION

Uncertainty in spatial information and its effect on spatial modeling has become an increasingly important research issue during the last decade.[79,80] The data models implemented within GIS contain various errors, and frequently, inexperienced users are able to perform complex analyses without adequately considering issues of data quality. Uncertainty analysis provides the means to quantitatively examine the impact of input (ε_i) and model error (ε_m) on the error (ε) in the outputs of a modeling process:

$$\varepsilon = \varepsilon_m + \varepsilon_i \tag{7.3}$$

When calculating slope from a DEM, a quadratic trend surface is usually fitted to the local neighbors (Figure 7.7), such as[75]

$$z = \frac{rx^2}{2} + \frac{ty^2}{2} + sxy + px + qy + u \tag{7.4}$$

Slope (G) is calculated using finite differences:

$$G = \sqrt{p^2 + q^2} \tag{7.5}$$

where

$$p = \frac{z_3 + z_6 + z_9 - z_1 - z_4 - z_7}{6w} \tag{7.6}$$

and

$$q = \frac{z_1 + z_2 + z_3 - z_7 - z_8 - z_9}{6w} \qquad (7.7)$$

The model's uncertainty is mainly a function of how well the quadratic function fits the real surface and the approximation of the finite differences to estimate p and q (Equations 7.4 to 7.7). The input uncertainty constitutes the uncertainty in the DEM. At present, most research in the uncertainty analysis of digital terrain modeling has focused on DEM input uncertainty rather than model error. While studies have attempted to compare methods for calculating terrain attributes, the major obstacle is obtaining estimates of true values: for example, how do we measure the true upslope area of a point? Therefore, the uncertainty within a DTM is dependent on the algorithm used to calculate a particular terrain attribute and the uncertainty of the DEM.

In soil-terrain modeling and other soil mapping research using SCORPAN factors, most of the emphasis has been on improvements in prediction that can be made by improvements in the quality of f and the associated SCORPAN factors (Equation 7.2). Little or no research has considered the propagation of errors based on uncertainty in the SCORPAN factors. This is increasingly important when secondary information sources, each with an associated uncertainty, are now commonly used for soil mapping. The following example illustrates the propagation of DEM uncertainty via the soil mapping process using the best prediction model from the previous example, the cokriging 5-m slope model for mapping clay content.

7.4.2 EXAMPLE

A program called Digeman was used to estimate the propagation of uncertainty in the DEM during the calculation of slope.[81] As mentioned previously, a map of RMSE of the elevation estimate derived from the raw GPS elevation data was available for each raster cell (Figure 7.9). For the estimation of input uncertainty, it was assumed that ME = 0 and that the uncertainty had a Gaussian distribution; therefore, the RMSE was equal to the standard deviation of the uncertainty. Monte-Carlo simulation with a modified version of Latin hypercube sampling (LHS) was used to create spatially correlated uncertainty fields (Figure 7.8).[82]

Twenty realizations of the slope surface were generated; the standard deviation of the slope is shown in Figure 7.10. Table 7.6 and Table 7.7 present summary statistics of the slope uncertainty and important bivariate relationships.

Distributions of the RMSE, slope, and standard deviations (SDs) of the slope were strongly positively skewed. Therefore, the median and interquartile range (IQR) were reported rather than the mean and SD. The median and IQR of the slope uncertainty indicate that the DEM is of sufficient quality for the calculation of slope. Previous research has been mixed when considering the relationship between slope magnitude and corresponding errors in slope. Studies have found

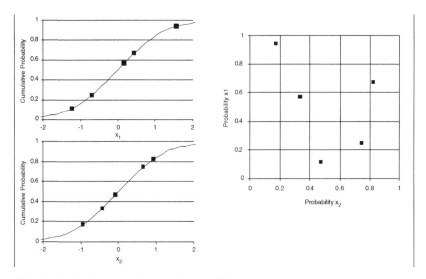

FIGURE 7.8 Latin hypercube sampling (LHS).

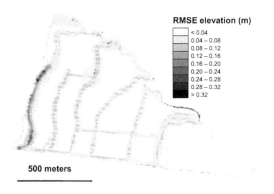

FIGURE 7.9 RMSE of East Creek DEM (5 m raster).

that larger errors occur on steeper slopes,[85,86] whereas Carter[87] found that absolute values of slope RMSE were similar for all slopes, but in relative terms (in terms of the magnitude of slope), the steeper slopes had the least errors. Alternatively, Davis and Dozier[88] found that slope errors were greatest where there was a rapid change in elevation. In this study, visual trends indicate that as slope increases, so does the uncertainty (Figure 7.5 and Figure 7.10). Further evidence of this is the moderate correlations (Table 7.7). The reality is that the DEM RMSE had the greatest effect on the standard deviation of slope ($\rho_{pearson} = 0.64$). While stating the obvious, this implies that when creating DEM, maximum effort must be made to reduce uncertainty in the final DEM.

The spatial variation in RMSE is generally due to variations in satellite visibility or geometry and to problems with line of sight between the GPS rover

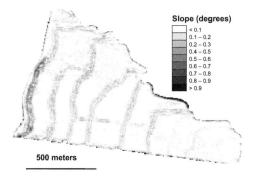

FIGURE 7.10 Standard deviation of the slope (5 m raster).

TABLE 7.6
Summary Statistics of Slope Uncertainty Analysis

Attribute	Median	IQR	Minimum	Maximum
DEM RMSE	0.04	0.03	0.01	0.35
Slope	0.75	0.71	0.00	9.14
SD slope	0.06	0.10	0.00	3.02

TABLE 7.7
Partial Correlation Matrix of Slope Uncertainty Statistics

	Slope[a]	DEM RMSE	SD Slope
Slope			
RMSE	0.26/0.27		
SD slope	0.24/0.43	0.64/0.61	

[a] Numbers to the left of each slash are Pearson's correlation coefficients and numbers to the right are Spearman's correlation coefficients.

and the GPS base station. The base station GPS receiver was placed approximately 400 m north of East Creek when the elevation was surveyed. This explains the linear transect of large RMSE (dark gray colors) on the western side of the field. This transect actually corresponds to a contour, behind which line of sight with the base station was poor. The larger RMSE values along each contour bank are probably due to a combination of two factors, the elevation changing sharply over a short distance and too fast a vehicular speed, both resulting in the GPS receiver updating its position too slowly and also inaccurately. The same reasons

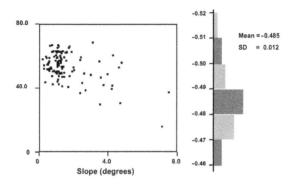

FIGURE 7.11 Stability of the clay–slope relationship.

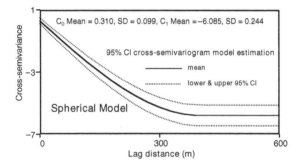

FIGURE 7.12 Stability of clay–slope cross-semivariogram.

explain the larger RMSE values along the northern boundary of the field, where the elevation drops sharply toward a creek. In addition, poor line of sight with the base station may explain the larger RMSE in this area. Other interesting features include the linear patterns where the RMSE value is greater than that for the surrounding area. This probably corresponds to sampling transects where either satellite geometry or rover base station–satellite communication was poor. Improvements to future surveys and subsequent terrain modeling results should (1) ensure good line of sight with the base station over the entire survey area and (2) reduce vehicular speed in areas where the elevation changes rapidly.

For each of the 20 slope realizations, cokriging of clay was performed in ISATIS[78]; Figure 7.11 to Figure 7.14 present the results. Figure 7.11 shows the plot of mean slope at each soil sampling location vs. the clay content, in addition to the histogram of Pearson's correlation coefficients for clay vs. slope for each of the 20 realizations. Figure 7.12 is the mean clay-slope cross-semivariogram model, with associated 95% confidence intervals based on the 20 realizations. These figures indicate that both the statistical and spatial correlation between clay and slope are quite stable between realizations.

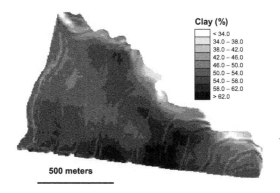

FIGURE 7.13 Mean clay content (5 m raster).

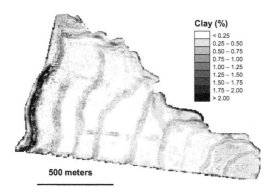

FIGURE 7.14 Standard deviation clay content (5 m raster).

Figure 7.13 and Figure 7.14 present the mean clay content map and standard deviation of clay across all realizations. Figure 7.13 indicates that the smaller clay contents along the northern edge of the field are due to erosion caused by larger slopes. The dark gray colors, particularly along the southern boundary, are associated with the heavy-textured Grey Vertosols.[89] In terms of uncertainty, the majority of the field has an uncertainty that is less than the analytical accuracy of the actual method used to measure clay content, ~1% (Figure 7.14).[70]

The implication of the uncertainty map of clay content is that in this case, the DEM is accurate enough for the calculation of slope and subsequent modeling of the clay-slope relationship for the cokriging prediction model. It would be expected that if the 25-m DEM were used to cokrige clay content, the uncertainty values would be larger, as the DEM uncertainty would be greater. The same approach could not be used for the 25-m prediction models because no corresponding estimates of the uncertainty were available.

7.5 CONCLUSIONS

In this chapter digital terrain modeling has been reviewed with particular reference to soil-terrain modeling. Examples have been presented to illustrate two important concepts in soil-terrain modeling: the impacts of both DEM spatial resolution and uncertainty on the quality of soil-terrain models. As the use of terrain (and other SCORPAN factors) becomes more commonplace in soil prediction models, more research is needed to examine the propagation of errors in the modeling process. A major impediment to research in this area is the paucity of quantitative information on the spatial distribution of error in GIS data layers used as inputs into the soil prediction models. It is no longer suitable to provide only prediction estimates, for it is equally important to estimate the uncertainty of those predictions. To conclude, a quote from the philosopher Vroomfondel seems appropriate:

We demand rigidly defined areas of doubt and uncertainty!

— **Douglas Adams,** *Hitchhiker's Guide to the Galaxy*

ACKNOWLEDGMENTS

The authors acknowledge Prof. A.B. McBratney at the University of Sydney as this chapter developed from research carried out under his supervision. Thomas Bishop acknowledges S. Grunwald for the portion of this work conducted at the Soil and Water Science Department, University of Florida.

REFERENCES

1. ID Moore, RB Grayson, AR Ladson. Digital terrain modelling: a review of hydrological, geomorphological, and biological applications. *Hydrol Process* 3: 3–30, 1991.
2. JP Wilson, JC Gallant, Eds. *Terrain Analysis: Principles and Applications.* John Wiley & Sons, New York, 2000.
3. RJ Pike. The geometric signature: quantifying landslide terrain types from digital elevation models. *Math Geol* 20: 491–511, 1988.
4. J Wood. The Geomorphological Characterisation of Digital Elevation Models. Ph.D. dissertation, University of Leicester, U.K., 1996. (Available online at http://www.soi.city.ac.uk/~jwo/phd.)
5. PA Shary, LS Sharayab, AV Mitusov. Fundamental quantitative methods of land surface analysis. *Geoderma* 107: 1–43, 2002.
6. JC Gallant, JP Wilson. Primary topographic attributes. In *Terrain Analysis: Principles and Applications*, JP Wilson, JC Gallant, Eds. John Wiley & Sons, New York, 2000, pp. 51–85.
7. IS Evans. What do terrain statistics really mean? In *Land Monitoring, Modelling and Analysis*, SN Lane, KS Richards, H Chandler, Eds. John Wiley, Chichester, U.K., 1998, pp. 119–138.

8. JF O'Callagahan, DM Mark. The extraction of drainage networks from digital elevation data. *Comput Vision Graphics Image Process* 28: 323–344, 1984.

9. MC Costa-Cabral, SJ Burges. Digital elevation model networks (DEMON): a model of flow over hillslopes for computation of contributing and dispersal areas. *Water Resour Res* 30: 1681–1692, 1994.

10. DG Tarboton. A new method for the determination of flow directions and contributing areas in grid digital elevation models. *Water Resour Res* 33: 309–319, 1997.

11. E Dobos, L Montanarella, T Negre, E Micheli. A regional scale soil mapping approach using integrated AVHRR and DEM data. *Int J Appl Earth Obs Geoinf* 3: 30–41, 2001.

12. JP Wilson, JC Gallant. Secondary topographic attributes. In *Terrain Analysis: Principles and Applications*, JP Wilson, JC Gallant, Eds. John Wiley & Sons, New York, 2000, pp. 87–131.

13. K Beven, MJ Kirkby. A physically based, variable contributing area model of basin hydrology. *Hydrol Sci Bull* 24: 1–10, 1979.

14. K Beven, EF Wood. Catchment geomorphology and the dynamics of runoff contributing areas. *J Hydrol* 65: 139–158, 1983.

15. ID Moore, PE Gessler, GA Nielsen, GA Peterson. Soil attribute prediction using terrain analysis. *Soil Sci Soc Am J* 57: 443–452, 1993.

16. H Jenny. *Factors of Soil Formation: A System of Quantitative Pedology.* McGraw-Hill, New York, 1941.

17. AR Aandahl. The characterization of slope positions and their influence on the total N content of a few virgin soils in Western Iowa. *Soil Sci Soc Am Proc* 13: 449–454, 1948.

18. PH Walker, GF Hall, R Protz. Relation between landform parameters and soil properties. *Soil Sci Soc Am Proc* 32: 101–104, 1968.

19. AB McBratney, IOA Odeh, TFA Bishop, MS Dunbar, TM Shatar. An overview of pedometric techniques for use in soil survey. *Geoderma* 97: 293–327, 2000.

20. PE Gessler, ID Moore, NJ McKenzie, PJ Ryan. Soil-landscape modelling and spatial prediction of soil attributes. *Int J Geogr Inf Syst* 9: 421–432, 1995.

21. JC Bell, CA Butler, JA Thompson. Soil-terrain modelling for site-specific agricultural management. In *Site-Specific Management for Agricultural Systems*, PC Robert, RH Rust, WE Larson, Eds. ASA-CSSA-SSSA, Madison, WI, 1994, pp. 209–228.

22. IOA Odeh, AB McBratney, BK Slater. Predicting soil properties from ancillary information: non-spatial models compared with geostatistical and combined methods. In *Geostatistics Wollongong '96*, Vol. 2, EY Baafi, NA Schofield, Eds. Kluwer Academic Publishers, Netherlands, 1997, pp. 1008–1019.

23. NJ McKenzie, PJ Ryan. Spatial prediction of soil properties using environmental correlation. *Geoderma* 89: 67–94, 1999.

24. T Hastie, R Tibshirani, J Friedman. *The Elements of Statistical Learning: Data Mining, Inference and Prediction.* Springer-Verlag, New York, 2001.

25. H Bourennane, D King, P Chery, A Bruand. Improving the kriging of a soil variable using slope gradient as external drift. *Eur J Soil Sci* 47: 473–483, 1996.

26. IOA Odeh, PE Gessler, NJ McKenzie, AB McBratney. Using attributes derived from digital elevation models for spatial prediction of soil properties. In *Resource Technology '94*, Melbourne, Australia, 1994, pp. 451–463.

27. IOA Odeh, AB McBratney, DJ Chittleborough. Further results on prediction of soil properties from terrain attributes: heterotopic cokriging and regression-kriging. *Geoderma* 67: 215–225, 1995.

28. AB McBratney, ML Mendonça Santos, B Minasny. On digital soil mapping. *Geoderma* 117: 3–52, 2003.

29. CJ Moran, EN Bui. Spatial data mining for enhanced soil map modelling. *Int J Geogr Inf Sci* 16: 533–549, 2002.

30. J Triantafilis, AI Huckel, IOA Odeh. Comparison of statistical prediction methods for estimating field-scale clay content using different combinations of ancillary variables. *Soil Sci* 166: 415–427, 2001.

31. D Zheng, ER Hunt, SW Running. Comparison of available soil water capacity estimated from topography and soil series information. *Landscape Ecol* 11: 3–14, 1996.

32. YA Pachepsky, DJ Timlin, WJ Rawls. Soil water retention as related to topographic variables. *Soil Sci Soc Am J* 65: 1787–1795, 2001.

33. N Romano, M Palladino. Prediction of soil water retention using soil physical data and terrain attributes. *J Hydrol* 265: 56–75, 2002.

34. BS Everitt. *The Cambridge Dictionary of Statistics*. Cambridge University Press, Cambridge, U.K., 2002.

35. G Upton, I Cook. *A Dictionary of Statistics*. Oxford University Press, Oxford, 2002.

36. FR Troeh. Landform parameters correlated to soil drainage. *Soil Sci Soc Am Proc* 28: 808–812, 1964.

37. JP Legros, P Bonneric. Modelisation informatique de la repartition des sols dans le Parc Naturel Régional du Pilat. *Annales de l'Université de Savoie*, Tome 4, *Sciences Naturelles*, 63–68, 1979.

38. DJ Pennock, BJ Zebarth, E De Jong. Landform classification and soil distribution in hummocky terrain, Saskatchewan, Canada. *Geoderma* 40: 297–315, 1987.

39. IOA Odeh, DJ Chittleborough, AB McBratney. Elucidation of soil-landform interrelationships by canonical ordination analysis. *Geoderma* 49: 1–32, 1991.

40. AK Skidmore, PJ Ryan, W Dawes, D Short, E O'Loughlin. Use of an expert system to map forest soils from a geographical information system. *Int J Geogr Inf Sci* 5: 431–445, 1991.

41. AK Skidmore, F Watford, P Luckananurug, PJ Ryan. An operational GIS expert system for mapping forest soils. *Photogrammetric Eng Remote Sensing* 62: 501–511, 1996.

42. JC Bell, RL Cunningham, MW Havens. Calibration and validation of a soil-landscape model for predicting soil drainage class. *Soil Sci Soc Am J* 56: 1860–1866, 1992.

43. NJ McKenzie, MP Austin. A quantitative Australian approach to medium and small scale surveys based on soil stratigraphy and environmental correlation. *Geoderma* 57: 329–355, 1993.

44. AX Zhu, LE Band. A knowledge-based approach to data integration for soil mapping. *Can J Remote Sensing* 20: 408–418, 1994.

45. D Arrouays, I Vion, JL Kicin. Spatial analysis and modeling of topsoil carbon storage in temperate forest humic loamy soils of France. *Soil Sci* 159: 191–198, 1995.

46. AT Cialella, R Dubayah, W Lawrence, E Levine. Predicting soil drainage class using remotely sensed and digital elevation data. *Photogrammetric Eng Remote Sensing* 63: 171–178, 1997.
47. P Lagacherie, S Holmes. Addressing geographical data errors in a classification tree soil unit prediction. *Int J Geogr Inf Sci* 11: 183–198, 1997.
48. JA Thompson, JC Bell, CA Butler. Quantitative soil-landscape modeling for estimating the areal extent of hydromorphic soils. *Soil Sci Soc Am J* 61: 971–980, 1997.
49. S De Bruin, A Stein. Soil-landscape modeling using fuzzy c-means clustering of attribute data derived from a digital elevation model (DEM). *Geoderma* 83: 17–33, 1998.
50. IV Florinsky, HA Arlashina. Quantitative topographic analysis of gilgai soil morphology. *Geoderma* 82: 359–380, 1998.
51. EN Bui, A Loughhead, R Corner. Extracting soil-landscape rules from previous soil surveys. *Aust J Soil Res* 37: 495–508, 1999.
52. D King, H Bourennane, M Isambert, JJ Macaire. Relationship of the presence of a non-calcareous clay–loam horizon to DEM attributes in a gently sloping area. *Geoderma* 89: 95–111, 1999.
53. RM Lark. Soil-landform relationships at within-field scales: an investigation using continuous classification. *Geoderma* 92: 141–165, 1999.
54. W Sinowski, K Auerswald. Using relief parameters in a discriminant analysis to stratify geological areas with different spatial variability of soil properties. *Geoderma* 89: 113–128, 1999.
55. AL Thomas, D King, E Dambrine, A Couturier, A Roque. Predicting soil classes with parameters derived from relief geologic materials in a sandstone region of the Vosges mountains (northeastern France). *Geoderma* 90: 291–305, 1999.
56. V Chaplot, C Walter, P Curmi. Improving soil hydromorphy prediction according to DEM resolution and available pedological data. *Geoderma* 97: 405–422, 2000.
57. E Dobos, E Micheli, MF Baumgardner, L Biehl, T Helt. Use of combined digital elevation model and satellite radiometric data for regional soil mapping. *Geoderma* 97: 367–391, 2000.
58. PE Gessler, OA Chadwick, F Chamron, K Holmes, L Althouse. Modeling soil-landscape and ecosystem properties using terrain attributes. *Soil Sci Soc Am J* 64: 2046–2056, 2000.
59. PJ Ryan, NJ McKenzie, D O'Connell, AN Loughhead, PM Leppert, D Jacquier, L Ashton. Integrating forest soils information across scales: spatial prediction of soil properties under Australian forests. *For Ecol Manage* 138: 139–157, 2000.
60. TFA Bishop, AB McBratney. A comparison of prediction methods for the creation of field-extent soil property maps. *Geoderma* 103: 149–160, 2001.
61. SJ Park, K McSweeney, B Lowery. Identification of the spatial distribution of soils using a process-based terrain characterization. *Geoderma* 103: 249–272, 2001.
62. P Campling, A Gobin, J Feyen. Logistic modeling to spatially predict the probability of soil drainage classes. *Soil Sci Soc Am J* 66: 1390–1401, 2002.
63. SJ Park, LG Vlek. Prediction of three-dimensional soil spatial variability: a comparison of three environmental correlation techniques. *Geoderma* 109: 117–140, 2002.

64. W Peng, DB Wheeler, JC Bell, MG Krusemark. Delineating patterns of soil drainage class on bare soils using remote sensing analyses. *Geoderma* 115: 261–280, 2003.

65. RJ Huggett. Soil landscape systems: a model of soil genesis. *Geoderma* 13: 1–22, 1975.

66. B Minasny, AB McBratney. A rudimentary mechanistic model for soil production and landscape development. II. A two-dimensional model. *Geoderma* 103: 161–179, 2001.

67. PW Birkeland. *Soils and Geomorphology*. Oxford University Press, New York, 1999.

68. HN Pollack. A numerical model of Grand Canyon. In *Four Corners Geological Society Guidebook*. Geology and Natural History of the Fifth Field Conference, Powell Centennial River Expedition, 1969, pp. 61–62.

69. JA Thompson, JC Bell, CA Butler. Digital elevation model resolution: effects on terrain attribute calculation and quantitative soil-landscape modeling. *Geoderma* 100: 67–89, 2001.

70. PR Day. Particle fractionation and particle-size analysis. In *Methods of Soil Analysis*, Part 1, CA Black, Ed. American Society of Agronomy, Madison, WI, 1965, pp. 545–567.

71. Ashtech. *Ashtech™ Precise Differential GPS Navigation (PNAV) Trajectory Software*, Version 2.1. Ashtech, Sunnyvale, CA, 1994.

72. ESRI. *Arc Info 7.2.1*. ESRI, Redlands, CA, 1998.

73. MF Hutchinson. A new procedure for gridding elevation and stream line data with automatic removal of spurious pits. *J Hydrol* 106: 211–232, 1989.

74. B Minasny, AB McBratney, BM Whelan. *VESPER*, Version 1.0 (1999). Australian Centre for Precision Agriculture, McMillan Building A05, The University of Sydney, NSW 2006. (Available online at http://www.usyd.edu.au/su/agric/acpa.)

75. IS Evans. An integrated system of terrain analysis for slope mapping. *Z Geomorph* 36: 274–295, 1980.

76. StatSci. *S-PLUS 2000*. Statistical Sciences, Seattle, 2000.

77. MF Hutchinson, JC Gallant. Digital elevation models and representation of terrain shape. In *Terrain Analysis: Principles and Applications*, JP Wilson, JC Gallant, Eds. John Wiley & Sons, New York, 2000, pp. 29–50.

78. Geovariances. *ISATIS*. Geovariances, Avon Cedex, France, 2002.

79. GM Foody, PM Atkinson, Eds. *Uncertainty in Remote Sensing and GIS*. John Wiley & Sons, Chichester, U.K., 2002.

80. J Zhang, M Goodchild, Eds. *Uncertainty in Geographical Information*. Taylor & Francis, London, 2002.

81. B Minasny, TFA Bishop. Uncertainty Analysis of Digital Terrain Modelling. Paper presented at Pedometrics 2003, International Conference of the IUSS Working Group on Pedometrics, Reading University, Reading, U.K., September 11–12, 2003.

82. RL Iman, WJ Conover. A distribution-free approach to inducing rank correlation among input variables. *Commun Stat* B11: 311–334, 1982.

83. ML Stein. Large sample properties of simulations using Latin hypercube sampling. *Technometrics* 29: 143–151, 1987.

84. EJ Pebesma, GBM Heuvelink. Latin hypercube sampling of Gaussian random fields. *Technometrics* 41: 303–312, 1999.

85. PV Bolstad, T Stowe. An evaluation of DEM accuracy: elevation, slope, and aspect. *Photogrammetric Eng Remote Sensing* 60: 1327–1332, 1994.

86. KC Sasowsky, GW Petersen, BM Evans. Accuracy of SPOT digital elevation model and derivatives: utility for Alaska's North Slope. *Photogrammetric Eng Remote Sensing* 58: 815–824, 1992.

87. JR Carter. The effect of data precision on the calculation of slope and aspect using gridded data. *Cartographica* 29: 22–34, 1992.

88. FW Davis, J Dozier. Information analysis of a spatial database for ecological land classification. *Photogrammetric Eng Remote Sensing* 56: 605–613, 1990.

89. RF Isbell. *The Australian Soil Classification.* CSIRO, Melbourne, Australia, 1996.

8 Fuzzy Logic Models

A.-Xing Zhu

CONTENTS

ABSTRACT

This chapter provides an introduction to the application of fuzzy set theory to soil science with an emphasis on how it helps scientists to better represent soil as a continuum in both the spatial and attribute domains. After the basic introduction to the fundamental concepts of fuzzy set theory and the notion of soil as a continuum, this chapter focuses on the discussion of how fuzzy logic (fuzzy set theory) can help to address the two basic limitations currently faced by practitioners in representing soils: generalization of soils in the spatial domain and generalization of soil in the attribute domain. Through the SoLIM (Soil Land Inference Model) example, this chapter illustrates how fuzzy set theory can be combined with the fundamental concepts and theories in geographic information systems (GIS) and artificial intelligence to map the spatial distribution of soils as a continuum in both the spatial and attribute domains. The chapter concludes with remarks on the current challenges and possible future research directions.

8.1 INTRODUCTION

Lofti Zadeh introduced fuzzy logic in his 1965 paper in *Information Control*.[1] The conception of fuzzy logic brought many changes in the way information is represented, processed, and presented. Many even went as far as saying the arrival of fuzzy set caused paradigm shifts in many fields. The widespread use of fuzzy logic concepts and their applications in many fields are not accidental. Fuzzy logic provides a very viable alternative to Boolean logic for many problems. The fundamental difference between fuzzy logic and Boolean logic is that the former deals with similarity between objects and the latter addresses occurrence of objects.

In recent years many researchers have explored the use of fuzzy logic concepts within soil science and have found it to be a powerful tool for soil classification,[2,3] soil information retrieval and soil interpretation of design and ratings, and soil resource inventory.[4] This chapter provides the basics of fuzzy set theory and highlights the application of fuzzy logic within the field of soil science. In the next section, the basic concepts of fuzzy logic are introduced, followed by the explanation of soil as a continuum. This will shed some light on why fuzzy logic is an appropriate and powerful asset for soil classification, soil interpretation, and related works. Based on discussions provided in these two sections, Section 8.4 presents soil information representation under fuzzy logic. Section 8.5 provides an overview of recent applications of fuzzy logic in soil survey. Section 8.6 highlights the challenges and possible future research directions.

8.2 BASICS OF FUZZY SET THEORY

8.2.1 SET AND CRISP SET

A set is a group of objects that share a common list of properties or attributes. For example, an Olympic team for a given nation is a set because all of its members meet the set selection criteria through internal competition, and they are the best athletes from that nation. We often use "she or he is on the Olympic team" to refer to individual members of this particular group. By doing so, we acknowledge the unique properties or status of this group. Another example would be the concept of "tall people," although it is a bit hard to define how tall is tall. Nevertheless, it is a set (group) of people considered to be tall, however it is defined. A soil type (or class) is also a set. For example, "Miami silt loam" is a set containing all soil individuals that share the same soil properties defined to represent Miami silt loam, although we know that not every individual soil profile classified as Miami silt loam exactly matches the typical soil properties/profile of the class perfectly.

A crisp (Boolean) set is a type of set whose members must share the properties defined for the set perfectly (exactly). In other words, in order to be a member of a given set, the object (individual) needs to meet the criteria of the set fully or 100%. Once assigned to the set, an individual will have full membership in that set. Just like being a member of a club, once you pay the dues (one qualification or criteria), you will have full access (membership) to the facilities of the club. For example, the "national Olympic team" set is one that can be described using the concept of crisp set because a person is either qualified or not qualified for the Olympic team. There is no halfway about this. Once on the team for a given Olympic sport, the person can compete at the Olympics.

However, there are set concepts that cannot be described appropriately using the concept of crisp set. For example, the concept of "tall people" is difficult to be described using the crisp set concept because the definition of tall is vague. Would a 180-cm tall person be considered tall? If so, what about a person of 179 cm? Furthermore, what about a person that is 200 cm tall? Would this person be considered the same as that of 180 cm? Often we use a threshold (a cutoff value) to define a set (crisp set), but we ignore the differences among the individuals that meet the cutoff value and exaggerate the differences between members of the set and those that do not meet the cutoff value (nonmembers).

8.2.2 Fuzzy Set

Fuzzy logic is an infinite-valued logic, which is different from the classic two-valued (yes or no) logic (Boolean logic or crisp logic). Membership in a set under fuzzy logic is not characterized by yes (1) or no (0), but is more adequately considered in terms of degrees (often referred to as grade of membership). In other words, a fuzzy set is a set whose elements share the properties defined for the set at certain degrees, which can range from 0.0 to 1.0, with 0.0 meaning no membership in the set and 1.0 full membership. Thus, a fuzzy set is characterized by a set of membership values, each of them is defined as a real number in the interval [0, 1]. A formal definition of fuzzy set is given as follows.

8.2.2.1 Definition 1: Fuzzy Set[5]

If X is a collection of objects denoted generically by x, then a fuzzy set $\tilde{A}$ in X is a set of ordered pairs:

$$A = \{x, \mu_{\tilde{A}}(x)\} \qquad x \in X \tag{8.1}$$

where x is an object that belongs to the set of objects X, $\mu_{\tilde{A}}(x)$ is the degree of membership, and $\mu_{\tilde{A}}()$ is the membership function of x in $\tilde{A}$, which maps X to the membership space M.

To illustrate the concept of a fuzzy set, let us use "tall people" as an example. The fuzzy set of tall people is defined by a function mapping individual's heights into membership space. Let us assume that there are six individuals whose heights and fuzzy membership in "tall people" are listed in Table 8.1 and the membership is calculated using the function given in Equations 8.2 through 8.4:

$$\mu_{Tall}(x) = 0 \quad \text{for } x \leq 165 \text{ cm} \tag{8.2}$$

$$\mu_{Tall}(x) = \frac{x - 165}{20} \quad \text{for } 165 \leq x \leq 185 \text{ cm} \tag{8.3}$$

$$\mu_{Tall}(x) = 1 \quad \text{for } x > 185 \text{ cm} \tag{8.4}$$

The fuzzy set of tall people for the set of objects (individuals) can be written as { (165, 0.0), (170, 0.25), (175, 0.5), (180, 0.75), (185, 1.0), (190, 1.0)}. The important aspect of a fuzzy set is its membership value, which tells us the level or degree of belonging. This degree of belonging provides us information about the certainty (or uncertainty) associated with assigning the object to the set (a class). For different objects, the degrees of belonging can be different and their respective uncertainty values in the given set (class) are known, which is impossible under crisp set (logic) (see row 4 of Table 8.1). Information on certainty (or uncertainty) is very important in risk assessment and decision making.

There are times when we need to identify members of a fuzzy set whose membership values exceed a certain level. This is accomplished by the idea of

---

TABLE 8.1
Heights of People, the Fuzzy Set of "Tall People,"
and Its Boolean Counterparts

Object	1	2	3	4	5	6
Height (cm)	165	170	175	180	185	190
Membership	0	0.25	0.5	0.75	1	1
Boolean set	0	0	0	1	1	1

Note: Considered tall if $x > 180$ cm under the Boolean counterparts.

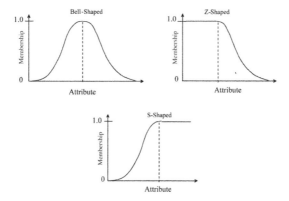

FIGURE 8.1 Three basic forms of membership functions.

an α-cut of a fuzzy set. An α-cut set is a set of individuals whose membership values exceed a predefined threshold, α. For example, the α-cut set of tall people for the subjects listed above is $\{(175, 0.5), (180, 0.75), (185, 1.0), (190, 1.0)\}$ when α is 0.3. The α-cut is often used to provide a finer control on the classification of objects.

The most important part of a fuzzy set is its membership function, because that determines the membership value of an object in the given fuzzy set. In other words, a fuzzy set is defined once its membership function is defined. There are three basic forms of membership functions that are often used in fuzzy mathematics (Figure 8.1): the bell-shaped, Z-shaped, and S-shaped curves. The bell-shaped curve describes that there is an optimal attribute value or range over which membership in the set is at unity (1.0), and as the attribute of the object deviates from this value or range, the membership value decreases. For example, the concept of moderately thick A-horizon can be captured using this membership function. The Z-shaped curve describes the scenario that there is a threshold value for the attribute of an object, smaller than which the membership is at unity (1.0) and greater than which the membership decreases. Thin A-horizon can be expressed using this function (the thinner the A-horizon, the higher the membership). The S-shaped curves define the relationships opposite to those characterized by the Z-shaped curves. The concept of thick A-horizon can be depicted using the S-shaped membership function (the thicker the A-horizon, the higher the membership).

Membership functions are domain specific as well as set specific. For example, the membership function for tall people will be different from that for short people. The functions used for defining height concepts will be different from those defining other concepts. Even for the same concept, there may be different functions used, depending on the application domain or the perception of these concepts. For example, the concept for tall people in the ordinary sense, which might be captured well using Equations 8.2 through 8.4 (dashed line in Figure 8.2), would be different from that in the context of basketball players, under

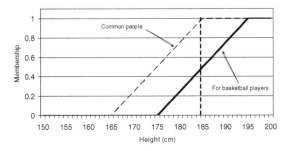

FIGURE 8.2 Two different membership functions for tall people: one for common people and the other for basketball players. For common people, a person with a height of 185 cm will have full membership (1.0) as tall, but for a basketball player, membership in tall is only about 0.5.

which a person with a height of 185 cm is not considered as tall at all. A better membership function for the latter version of tall people could be that portrayed by the solid line in Figure 8.2. Section 8.4.2.2 provides different ways of defining membership functions in soil science.

8.2.3 SIMPLE FUZZY SET OPERATIONS

Because membership function is a crucial component of a fuzzy set, fuzzy set operations are defined via their membership functions. There are many ways to define the fuzzy set operations,[6] and it is not within the scope of this paper to discuss them all. Here we discuss the basic fuzzy set operations.

8.2.3.1 Definition 2: Intersection

The membership function $\mu_{\tilde{N}}(x)$ of the intersection (logical "and") set of fuzzy sets $\tilde{A}$ and $\tilde{O}$ is defined by

$$\mu_{\tilde{N}}(x) = \min\{\mu_{\cdot}(x), \mu_{\tilde{O}}(x)\}, \qquad x \in X \tag{8.5}$$

8.2.3.2 Definition 3: Union

The membership function $\mu_{\tilde{M}}(x)$ of the union (logical "or") set of fuzzy sets $\tilde{A}$ and $\tilde{O}$ is defined by

$$\mu_{\tilde{M}}(x) = \max\{\mu_{\cdot}(x), \mu_{\tilde{O}}(x)\}, \qquad x \in X \tag{8.6}$$

8.2.3.3 Definition 4: Complement

The membership function $\mu_{\tilde{C}}(x)$ of the complement (logical "not") set of fuzzy set $\tilde{A}$ is defined by

$$\mu_{\bar{C}}(x) = \{1 - \mu_{\cdot}(x)\}, \qquad x \in X \qquad (8.7)$$

Equations 8.5 and 8.6 are referred to as the fuzzy minimum and maximum operators, respectively. There are many extensions to the above min–max definition to the fuzzy set operations.[5] These fuzzy set operations are often used in soil predictions.[7,8]

8.2.4 POSSIBILITY AND PROBABILITY

The fundamental difference between possibility theory and probability theory is that possibility theory is about similarity between objects or quality of a given object when compared to a prototype, while probability deals with the chance of the occurrence of an event. For example, there is a major difference between the following two statements: "It is possible that we can have another passenger on the bus" and "It is probable that we can have another passenger on the bus." The former addresses the issue of whether the bus can hold another passenger, while the latter states whether there will be another passenger coming to take this bus. The following two statements will make this distinction even more clear: "There is little chance that it will rain" and "It will rain a little." The former is a probability statement stating whether it will rain or not, while the latter is a possibility statement describing the intensity of rain, not about whether it will rain.

Fuzzy logic is about possibility, and crisp logic (Boolean logic) is about probability. Possibility is more appropriate for classifying soils. For example, the statement "The soil at this site is about 70% *similar to the prototype* of Miami silt loam" is more appropriate than "There is a 70% chance the soil at this site *is the prototype* of Miami silt loam." The reason is that the soil already exists at that site. In classification, we want to know which class the local soil belongs to by comparing the properties of the local soil with those of the prototype of the candidate class. It is not the issue of which soil class occurs at the given site. Soil classification is based on possibility, not probability. It is inappropriate to use probability to measure possibility, even though we have been doing it for a long time and are very much used to the idea.

8.3 SOIL AS A CONTINUUM AND PRESENTATION OF SOILS UNDER BOOLEAN LOGIC

Soil is a continuum both in spatial (geographic) domain and in attribute (property) domain. Soil is a result of the interaction of its formative environment factors, such as geology, climate, topography, and organism, over time, as outlined by Dokuchaiev[9] and Hilgard.[10,11] The continuous spatial variation of many of these factors causes soil to vary continuously over space, although abrupt changes of soil over space do exist at times. This spatial variation of soil causes the property values to be continuous in their attribute (property) domain. In other words, the values of a soil property are also continuous over the domain of that property.

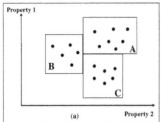 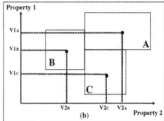

FIGURE 8.3 Discretization of soil-landscape in the attribute space. (a) Dots represent the locations of soils in the attribute space; rectangles represent the boundaries of soil classes in the attribute space. (b) Dots represent the centers of soil classes; the intervals between the projected centers on the respective axes represent the attribute resolution on these property axes.

Thus, soil classification and soil mapping must treat soil as a continuum in both spatial and attribute domains.

Traditional methods in soil classification and soil mapping take a Boolean approach and treat soil as distinct and discrete entities, rather than as a continuum. In classification, field observations on soils are grouped into types (classes) according to their diagnostic properties.[12] The outcome of this process is a list of classes (sets) with their respective boundaries defined. There are two contradicting issues associated with the classes so defined. First, each of these soil classes does not represent a pure concept with a definitive set of properties; rather, it is often a collection of soil objects with varying soil properties. As a result, the definition for each of these classes is given in terms of ranges of property value (Figure 8.3a). Thus, the concept of the given class cannot be represented by a single prototype (pedon). In fact, there can be many prototypes (pedons) for a given class. In other words, the concept of a given class is not represented as a single point in an m-dimensional property space (where m is the number of properties used to define class). Rather, it is presented by an m-dimensional cube (Figure 8.3a). The phenomenon of mixed objects in a single class is a typical drawback of classification under crisp set, as was discussed in Section 8.2.1.

The second issue is that in many applications each of these classes is actually treated as a distinct entity characterized by typical property values (the means or the modes of property values). Once a soil object is assigned to a class, it will be labeled with the typical property values of this class; thus, there is no difference in property values among the soil objects assigned to this class. Therefore, the typical soil property values are the only means of characterizing these soil classes in the soil property domain. In other words, we treat each of these classes as a pure concept or as a point in m-dimensional property space when we use the information of the class for real-world applications (such as deriving property maps for hydroecological modeling applications).[13]

These two contradicting issues result in the limited power in describing the changes of soil property values (attribute resolution) in the property domain.

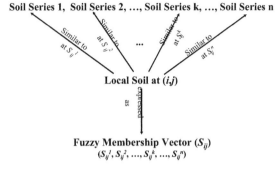

S_{ij}^k: fuzzy membership value between the soil at point (i,j) and soil series k

FIGURE 8.4 Representation of soils under fuzzy logic.

Changes in soil property are only limited to the intervals of the typical values of two adjacent soil classes (Figure 8.3b). Intermediate soil property values between two typical values of two adjacent classes often cannot be obtained. This reduction of soil attribute resolution is a drawback of soil classification based on Boolean logic and prevents the continuous variation of soils to be captured and represented (see Section 8.5 for further discussion).

8.4 REPRESENTATION OF SOILS UNDER FUZZY LOGIC

8.4.1 Fuzzy Representation of Soils

The continuous spatial variation of soil over space naturally leads itself to the employment of fuzzy logic in its representation. Under fuzzy logic a soil object at a given location (i, j) could bear partial membership (μ_{ij}^k or s_{ij}^k) in a class and bear different membership values in different classes (Figure 8.4). These different membership values in different classes can be captured and represented as an n-dimensional vector, S_{ij} ($S_{ij}^1, S_{ij}^2, \ldots, S_{ij}^k, \ldots, S_{ij}^n$), where n is the number of classes considered. In this way, subtle differences between soil objects can be reflected by difference in membership values in different soil classes. Table 8.2 shows the change of membership values along a transect in the Lubrecht study area of western Montana.[26] Two observations can be made about what is presented in this table. The first is that the change of soil from one class to another is not sudden, but rather is very mild. For example, soils at points 13 and 14 have highest membership in different classes, the soil at point 13 in Ovando and at point 14 in Elkner, but the difference in membership between the two sites is relatively small. This reflects the mild but important gradation of soil from one class to another over space. If we were to use Boolean logic to classify the soils, we would say Ovando for point 13 and Elkner for 14 and oversimplify the spatial

TABLE 8.2
Fuzzy Membership Values along a Transect in Lubrecht, Montana

Field Point	Soil Series[a]			
	Ambrant	Rochester	Elkner	Ovando
1	0.2644	0.0210	0.3541	0.3605
2	0.4595	0.0000	0.1746	0.3659
3	0.3824	0.0669	0.2467	0.3040
4	0.1617	0.0281	0.3630	0.4472
5	0.1250	0.0000	0.4928	0.3822
6	0.5239	0.3947	0.0310	0.0504
7	0.0296	0.0000	0.5389	0.4314
8	0.7156	0.0000	0.2844	0.0000
9	0.3953	0.0363	0.4628	0.1054
10	0.1898	0.1017	0.4647	0.2437
11	0.4958	0.5025	0.0011	0.0006
12	0.4901	0.4257	0.0402	0.0439
13	0.0403	0.0000	0.4406	0.5192
14	0.0708	0.0000	0.6273	0.3020

Note: Points are about 180 m apart. Obtained via a knowledge-based inference approach.[7]

[a] Ambrant, coarse-loamy, mixed, superactive, frigid Lamellic Haplustepts; Rochester, sandy-skeletal, mixed, frigid Typic Ustorthents; Elkner, coarse-loamy, mixed, superactive Lamellic Eutrocryepts; Ovando, sandy-skeletal, mixed Lamellic Cryorthents.

Source: Soil Survey Staff, Natural Resources Conservation Service, United States Department of Agriculture, Soil Series Classification Database, available at http://soils.usda.gov/technical/classification/osd/index.html.

gradation of soils. With the membership vector, we are able to capture and represent the mild but important gradation.

The second observation is that soils belonging to a same class vary subtly over space. For example, soils at points 2 and 3 have the highest membership values in Ambrant soil series, but the membership distribution in the vector is different, which can be used to reflect the subtle and gradual change of soil within a class over space. For another example, soils at points 9 and 10 have the highest membership values in Elkner, but soil at point 9 is more of an intergrade between Ambrant and Elkner, while soil at point 10 is more of an intergrade between Elkner and Ovando. This nature of in-between types (known in *Soil Taxonomy* as intergrades) cannot be captured without the use of a fuzzy membership vector.

8.4.2 MEMBERSHIP DETERMINATION AND MEMBERSHIP MAPS OF SOIL CLASSES

The grade of membership in a soil class depends on two things: the definition of the given class and the membership function that measures the similarity between the local soil object and the class definition. A collection of membership values in a soil class across a given area forms a membership map for that soil class. This section discusses the basic ideas and approaches in defining soil classes, membership computation, and derivation of membership maps.

8.4.2.1 Definition of Class Centroid

Under the notation of fuzzy logic, each class has a central concept or centroid.[14] The central concept of a class may not be a real individual but an imaginary model entity. There might be an individual that most resembles the central concept, but may not be the central concept itself. This individual is referred to as the exemplar (also known as the prototype, typical or representative soil profile, or pedon). This approach is commonly used in soil surveys. The definition of class central concept has been a challenge to practitioners in soil science due to the gradual yet complex variation of soils over space. One common approach to determination of the central concept is to use the means or modal or typical values of attributes representing the class. This approach is suited for classes with well-defined properties.

The other approach in determining central concepts is based on some version of fuzzy c-means (FCM) classification.[15,16] This approach is suited for situations when predefined soil classes are not applicable. FCM is a classifier that first optimally partitions a collection of individuals (such as a set of soil profiles) into a given set of classes.[16] It identifies the centroids of classes by minimizing the fuzzy partition error as given in Equation 8.8:[16]

$$J_m(U, v) = \sum_{i=1}^{n} \sum_{k=1}^{c} (\mu_i^{\,k})^m \left\| y_i - v_k \right\|^2 A \qquad (8.8)$$

where J_m is the fuzzy partition error and can be described as a weighted measure of the squared distance between individuals and class centroids, so it is a measure of the total squared errors as minimized with respect to each cluster[17,18]; J_m decreases as the clustering improves (meaning that pixels tend to be overall closer to their respective centroids); Y is the data (such as a set of soil profiles), and y_i is the ith feature vector; n is the number of objects (profiles) in Y; c is the number of clusters (classes) in Y; v is a vector of cluster centers; v_k is the center of cluster k; U is a fuzzy c-partition of Y; m is a weighting exponent that determines the degree of fuzziness (with $m = 1$ being nonfuzzy); A is a weighting matrix that is based on the norm under which the distance is measured; and $\mu_i^{\,k}$ is the membership of the ith object (y_i) belonging to the kth cluster and is computed as follows:

$$\mu_i^l = \frac{\|y_i - v_l\|^2 \, A}{\sum_k^c \|y_i - v_k\|^2 \, A} \tag{8.9}$$

Once the number of classes is determined, the centroid of each so-derived class is treated as the central concept of the derived class. In most cases, one does not know the number of classes that best describe the structure in the dataset. To judge the effectiveness of the clustering results generated using the above fuzzy c-means algorithm, two cluster validity measures (partition coefficient (F) and entropy (H)) are defined as[16]

$$F_c(\hat{\mu}) = \sum_{i=1}^n \sum_{k=1}^c (\hat{\mu}_i^k)^2 / n \tag{8.10}$$

$$H_c(\hat{\mu}) = -\sum_{i=1}^n \sum_{k=1}^c (\hat{\mu}_i^k \log_a(\hat{\mu}_i^k))/n \tag{8.11}$$

The partition coefficient F will take the values of $1/c$ to 1, while entropy H ranges from zero to $\log_a(c)$.[17] F measures the amount of overlap between clusters and is inversely proportional to the overall average overlap between pairs of fuzzy sets.[17] H, conversely, is a scalar measure of the amount of fuzziness in a given fuzzy partition U.[15] The best fuzzy c-partition, e.g., the number of classes that best describe the structure in the dataset, is thus the c-partition that realizes the highest $F_c(\hat{\mu})$ and the lowest $H_c(\hat{\mu})$. Note that both H and F will reach maxima and minima, respectively, at the same points, and in this sense they are essentially equivalent.[15] It is often the case that F increases and H decreases as the number of classes decreases. To determine if a fuzzy clustering can be considered optimal, the improvement in entropy or partition coefficient over adjacent partitioning is examined. If there is a significant improvement, the current partition of Y is considered to be the optimal partition of the dataset.

de Gruijter and McBratney[19] and McBratney and de Gruijter[20] noted that there is an insufficiency in identifying class centroids using the regular FCM.[15,16] The insufficiency is related to the existence of extragrades. Extragrades are those individuals falling far outside of the convex hull of class centroids[20] (Figure 8.5). These extragrades should be distinguished from the intragrades (those falling inside or close to the convex hull), but are not separated from intragrades under the regular FCM.[20,21] In order to improve the separation of extragrades from intragrades, de Gruijter and McBratney[19] and McBratney and de Gruijter[20] defined an extragrade class by modifying the FCM objective function:

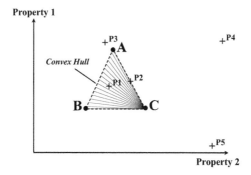

FIGURE 8.5 Extragrades and intragrades in relation to class centroids (A, B, C): points 1, 2, and 3 are intragrades, while points 4 and 5, which are way outside of the convex hull, are extragrades.

$$J_m(U, v) = \alpha \left(\sum_{i=1}^{n} \sum_{k=1}^{c} (\hat{\mu}_i^{\ k})^m \left\| y_i - v_k \right\|^2 A \right) +$$

$$(1 - \alpha) \left(\sum_{k=1}^{c} (\hat{\mu}_i^{\ *})^m \sum_{k=1}^{c} \left\| y_i - v_k \right\|^{-2} A \right) \qquad (8.12)$$

where $\mu_i^{\ *}$ is the membership value of an individual i in the extragrade class and α is the parameter determining the mean value of $\mu_i^{\ *}$.

8.4.2.2 Membership in Classes: Membership Functions

Once class centroids are defined, membership in each class can be determined in two basic ways: one based on natural grouping and the other based on knowledge of predefined classes.[3] The former again uses the results from optimizing the FCM objective function. The results consist of three outcomes: the number of naturally occurring groups (the number of classes), the set of centroids for the set of classes, and the membership values of individuals in the set of the classes. Membership values of individuals in each of the final classes are computed using Equation 8.9.

Membership computation based on expert knowledge of predefined classes is more and more widely used in application of fuzzy set theory in soil science.[3,22–25] This group of approaches can be referred to as the semantic import model (SI).[23] The basic idea is that a membership function is defined with expert knowledge of class specification and class behavior in the attribute space. Class behaviors describe how membership value changes with respect to changes of soil attributes within the attribute space. These knowledge-based membership functions are of the three basic forms shown in Figure 8.1, but the exact shape is determined by the knowledge in the application domain. Burrough[2] and Bur-

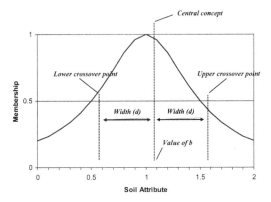

FIGURE 8.6 Metrics of a membership function.

rough et al.[23] use the following function to depict a symmetrical bell-shaped membership function:

$$\mu_{fuzzy}(x) = \frac{1}{1 + \left(\dfrac{x-b}{d}\right)^2} \tag{8.13}$$

where x is the attribute value of a soil entity, b is the attribute value representing the central concept, and d is one half of the width of the bell-shaped curve between the two crossover points (the upper and lower values where the membership values are at 0.5) (Figure 8.6). This function can be modified to represent the other two membership curves (Z-shaped and S-shaped) outlined in Figure 8.1. Z-shaped curves can be captured using Equations 8.14 and 8.15, and S-shaped with Equations 8.16 and 8.17:

$$\mu_{fuzzy}(x) = \frac{1}{1 + \left(\dfrac{x-b}{d}\right)^2} \quad \text{for } x > b \tag{8.14}$$

$$\mu_{fuzzy}(x) = 1 \quad \text{for } x \le b \tag{8.15}$$

$$\mu_{fuzzy}(x) = \frac{1}{1 + \left(\dfrac{x-b}{d}\right)^2} \quad \text{for } x < b \tag{8.16}$$

$$\mu_{fuzzy}(x) = 1 \text{ for } x \geq b \qquad (8.17)$$

Zhu[25] developed an approach using domain expert knowledge to define membership functions for fuzzy concepts. The approach is based on personal construct theory and allows the domain experts to focus on the definition of one concept at a time. The membership function for a given concept is defined by focusing on the determination of critical attribute points. For a given attribute, the expert first determines the curve type (bell-shaped or Z-shaped or S-shaped) that best describes the concept and then determines the critical attribute values for this type of curve. For example, if a bell-shaped curve is used to capture the concept, the expert will need to provide the attribute value or range of values over which the membership is at unity (1.0), the attribute values at the upper and lower crossover points, and the values (upper and lower values) at which the membership reaches zero. These critical points are then linked via a spline function to form a continuous curve. This approach was successfully applied in eliciting the knowledge of soil experts in defining the relationships between soil classes (series) and key environmental factors (soil-forming factors), which were then used to map the spatial distribution of soils under fuzzy logic.[4,25–27]

8.4.2.3 Fuzzy Membership Maps

Membership maps of soil classes can be produced using either a combination of fuzzy classification and spatial interpolation techniques[28] or a knowledge-based approach.[4,7,26] As an illustration of the former approach, Odeh et al.[28] derived fuzzy membership maps by combining FCM with kriging. They first use the modified fuzzy c-means (FCM) algorithm[20] to partition the pedons sampled from the field into intragrade and extragrade classes. Membership values of these pedons in each of the derived classes computed from the FCM classifier were then used to construct the semivariogram for each class. These semivariograms were then used in kriging to interpolate the membership distribution of different classes over the study area. The approach does not require a set of predefined soil classes, but does require the availability of a large collection of well-distributed field observations. Thus, it is more suited for soil mapping in small areas with intensive sampling.

Zhu and his colleagues developed a knowledge-based approach, Soil Land Inference Model (SoLIM), for deriving membership maps of soil series across landscape (see detailed discussion in Section 8.5). The SoLIM approach uses geographical information processing techniques[26,27] to characterize the soil formative environment and employs artificial intelligence and machine learning techniques[25,30–32] to capture the relationships between soil and its formative environment. A set of inference techniques constructed under fuzzy logic combines the environmental data with the extracted knowledge to produce membership maps of soil classes across an area.[7] Figure 8.7 shows fuzzy membership maps for two soil classes in Wisconsin: Dorerton (loamy-skeletal, mixed, active, mesic

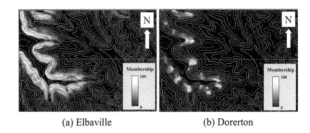

(a) Elbaville   (b) Dorerton

FIGURE 8.7 Fuzzy membership maps for soil series Elbaville (a) and Dorerton (b) over a study area in Wisconsin with a size of 1.6×2.2 km. Light tone indicates high membership values, while dark tone indicates low membership values. Elbaville occurs on steep back slope, while Dorerton occurs on noses. The spatial gradation of these two soils is clearly captured and depicted with the gradation of membership values over space.

Typic Hapludalfs) and Elbaville (fine-loamy, mixed, superactive Glossic Hapludalfs). Elbaville occurs on steep back slopes, while Dorerton occurs on nose slope positions. The transition from Elbaville on steep back slopes to Dorerton on nose positions is clearly captured and depicted with the gradation of derived fuzzy membership values. This knowledge-based approach for fuzzy membership derivation has proven to be successful and practical in soil survey over large areas, where expert knowledge of soil-environment relationships is available for all mapped soil series (or classes) (see Zhu and Band[7] and Zhu et al.[33] for discussion on evaluation).

8.4.3 USE OF FUZZY MEMBERSHIP

The application of fuzzy sets in soil science has been rather recent, and researchers are still studying the use and interpretation of fuzzy membership of soils, particularly the use and interpretation of fuzzy membership maps of soil classes.[3] Current use of fuzzy membership maps has been mainly in the following three areas: land suitability analysis, derivation of continuous maps of soil attributes, and characterization of uncertainty in soil mapping. Land suitability analysis based on fuzzy membership maps allows practitioners to make finer distinctions about different levels of suitability, rather than just the suitable vs. nonsuitable division. These finer distinctions provide users flexibility to control the quality of the final selection using an α-cut strategy. Chang and Burrough[22] applied fuzzy logic to compute suitability for apple production over an area in northeast China. Burrough[2] and Burrough et al.[23] interpret fuzzy membership as land suitability for various purposes. Dobermann and Oberthur[24] used a combination of fuzzy membership functions with Monte-Carlo simulation to produce maps of membership values for three soil fertility classes and two multivariate soil fertility qualities. These pieces of information were then assessed for suitability for intensive rice production.

Zhu et al.[33] proposed using fuzzy membership values as weights in computing soil property values intermediate to the typical or modal values prescribed to soil

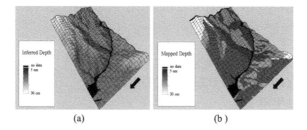

(a)   (b)

FIGURE 8.8 Soil A-horizon depth maps in Lubrecht of western Montana with a size of 4.8 × 7.5 km. (a) Derived from fuzzy membership maps using the SoLIM approach. (b) Derived from the conventional soil map.

classes. They assume that the soil property value at a given point (i, j) can be estimated from a membership-weighted combination of typical values for the prescribed soil classes. As a first approximation, they used a linear and additive function to estimate soil property values at a given point:

$$V_{ij} = \frac{\displaystyle\sum_{k=1}^{n} s_{ij}^{k} \bullet v^{k}}{\displaystyle\sum_{k=1}^{n} s_{ij}^{k}} \tag{8.18}$$

where V_{ij} is the specific soil property value at site (i, j), S_{ij}^{k} (used interchangeably with μ_{ij}^{k} in this chapter) is the membership of soil at location (i, j) in soil class k, v^{k} is the prescribed soil property value of soil series (soil type) k, and n is the total number of prescribed soil series (soil types) in the area. Using thickness of soil A-horizon in ponderosa pine/Douglas fir stands of the Lubrecht Experimental Forest of western Montana as an example, Zhu and his colleagues[33] found that the thickness map derived from fuzzy membership maps using the above equation is better in capturing the spatial variation of A-horizon depth than that derived from the conventional soil map of the area (Figure 8.8).

Zhu[34] illustrated the use of the fuzzy membership maps for portraying the uncertainty associated with assigning local soils to individual soil classes. He conceived two different types of uncertainty associated with the assignment. The first is the exaggeration uncertainty, which occurs when a local soil is assigned to a class to which the local soil does not fully belong. In other words, the membership of the local soil in that class is being exaggerated. The second type of uncertainty is the ignorance uncertainty, which occurs when a local soil is assigned to a class while potential membership in other classes is ignored. Both types of uncertainty can be estimated from the membership vector, S_{ij} (S_{ij}^{1}, S_{ij}^{2}, ..., S_{ij}^{k}, ..., S_{ij}^{n}), at the point. Using Lubrecht as an example, he showed that the uncertainty measures are useful in portraying uncertainty related to the class assignment, and spatial variation of classification uncertainty can be effectively depicted using this approach.

8.5 RECENT APPLICATIONS OF FUZZY SET CONCEPTS (OR FUZZY MODELS) IN SOIL SURVEY

Application of fuzzy set theory and fuzzy logic in soil science has experienced tremendous growth in recent years. Examples of applications range from fuzzy classification of soil pedons, to fuzzy set-derived soil interpretations, to land suitability evaluations to information retrieval to fuzzy geostatistics, to the production mode of soil survey. McBratney and Odeh[3] have provided an excellent survey of applications of fuzzy logic and fuzzy set theory in soil science. This section focuses on one of the recent applications of fuzzy set in soil survey. The discussion is focused around the SoLIM approach recently developed by Zhu and his colleagues.[4,7,25–27,31,32] We will first examine the issues facing the traditional approach to soil survey and then examine how fuzzy logic helps to address these issues.

8.5.1 Limitations of the Conventional Approach to Soil Survey

Soil survey can be considered a process of realizing a soil classification in geographic space (geographic domain). The conventional (traditional) approach to soil survey employs the area-class model[35] for capturing and representing soil spatial distribution. There are two limitations associated with soil mapping using the area-class model: the class assignment generalization and spatial generalization. Class assignment generalization, also referred to as generalization of soils in attribute space, is related to the assignment of local soils to prescribed soil classes under Boolean logic. Under Boolean logic, soil at a given point can belong to one and only one soil class, and that soil is assumed to have the soil properties of the soil class to which the soil is assigned. Under this notion, the difference in soil properties between two neighboring soil objects can either be exaggerated to be the difference of the two soil classes (when these two soil objects are assigned to two different classes) or be completely ignored (when the two soil objects are assigned into a single class) (Figure 8.9).

Spatial generalization, also referred to as generalization of soils in geographic space, is related to the map scale and the cartographic techniques employed for producing soil maps. At a certain scale, only soil objects larger than a certain size (scale dependent, called minimum map unit size) can be delineated on soil maps. Soil objects smaller than the minimum map unit size are either omitted completely or merged into the surrounding soil objects (Figure 8.10). Soil units B, C, and D in Figure 8.10a are too small to be represented in Figure 8.10b. Soil unit A in Figure 8.10b is a mixed unit consisting of soil units A, B, C, and D.

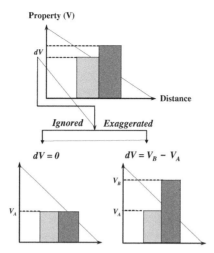

FIGURE 8.9 Generalization of soils in attribute space.

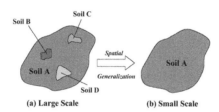

FIGURE 8.10 Generalization of soils in geographic space.

This inclusion may be noted in the map unit legend as a percentage of an inclusion, but the locations of these included units are completely lost on small-scale maps. Therefore, the spatial resolution of a soil map is the minimum mapping size, which can be a few hectares on large-scale maps to hundreds of hectares or more on small-scale maps.

Due to these two generalizations of soils, as results of the crisp logic employed and the limitation of map scales, soil spatial information produced from conventional soil survey is often incompatible with other environmental data derived from digital terrain analysis and remote sensing techniques. This incompatibility could have important implications to the interpretation of results from hydroecological models and for resource management decision making. It may be true that soil experts know the existence of the gradual gradation of soil properties over space and the inclusion of different soil objects in soil mapping units, but these cannot be mapped on soil maps due to the crisp logic employed and limitations of map scale and cartographic techniques used. Therefore, soil scientists' knowledge of soil variation cannot be fully expressed by soil maps constructed under crisp logic with the conventional cartographic techniques.

8.5.2 OVERCOMING THE LIMITATIONS: COUPLING FUZZY LOGIC WITH RASTER GIS MODEL

Zhu[27] developed a soil similarity model to overcome the two generalizations faced with the area-class model. This similarity model consists of two components, with each designed to overcome one of the two generalizations: a raster data model to minimize spatial generalization and a soil similarity vector to mitigate attribute generalization. The raster data model is better suited to representing smooth, continuous geographic features and phenomena than a vector data model. The level of gradation captured by the raster data model depends on the spatial resolution of the raster data model and is in turn limited only by the spatial resolution of the input data, rather than by arbitrary standards imposed by carto-graphic or mapping techniques. Local soil condition can be captured at pixel level, and information about small pockets of unique soil types is not eliminated. This minimizes the discrepancies between the spatial resolution of soil spatial information and other environmental data layers.[13] However, one important assumption for using the raster data model for representing soil spatial variation is that soil within a given pixel is perceived to be homogeneous; that is, the variation of soil properties within a pixel is so small that this variation can be ignored. This assumption holds if the pixel is small enough. Otherwise, the assumption is violated and the spatial generalization problem associated with the polygon model will also occur with the raster data model.

Soil similarity vector (the similarity representation of soils in the parameter domain) is based on fuzzy logic.[27] Under this fuzzy representation, the soil at a given pixel is represented by an n-element vector (*soil similarity vector* or *fuzzy membership vector*), S_{ij} (S_{ij}^1, S_{ij}^2, ..., S_{ij}^k, ..., S_{ij}^n), as described in Section 8.4.1 and shown in Figure 8.4. With this similarity representation, the local soil at a given pixel is no longer necessarily assigned to a particular class, but can be represented as an intergrade to the set of prescribed classes. This method of representation will allow the local soil to take property values intermediate to the modal (typical) values of the prescribed classes, and thus will largely cir-cumvent the problem of generalization in the parameter domain. By coupling the similarity representation with a raster geographic information systems (GIS) data model, soils in an area are represented as an array of pixels, with soil at each pixel being represented as a soil similarity vector (Figure 8.11). In this way, soil spatial variation can be represented as a continuum in both the spatial and parameter domains.

8.5.3 POPULATING THE SIMILARITY MODEL

The similarity model only provides added flexibility for representing soil spatial variation. The degree of success in using this model depends on how the model is populated or, equivalently, how the soil similarity values in the vector are determined at each pixel. The SoLIM approach takes advantage of recent devel-opments in geographic information science, artificial intelligence techniques, and

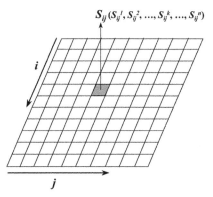

$$S_{ij}(S_{ij}^{1}, S_{ij}^{2}, ..., S_{ij}^{k}, ..., S_{ij}^{n})$$

FIGURE 8.11 The similarity model for representing detailed soil spatial information.

the classic concept of soil-landscape relationships to compute the soil similarity values at each pixel. The approach is based on the classic soil factor equation of Dokuchaiev[9] and Hilgard,[10,11] which contends that there is a relationship between soil and its formative environment, as shown in Equation 8.19:

$$S = \int f_1(E)dt \qquad (8.19)$$

In Equation 8.19, t is time and f_1 is the relationship of soil development to the formative environment, E, which generally includes variables describing climate, topography, parent material, and organisms. S is meant to be a soil class that can be expressed as a fuzzy membership value (soil similarity value). Due to the difficulty of explicitly describing integration of soil formative environmental factors over time during the course of soil formation across landscape, t is considered as part of E; thus, Equation 8.19 is simplified to

$$S = f(E) \qquad (8.20)$$

The implementation of Equation 8.20 is shown in Figure 8.12. Data on soil formative environmental conditions (E) can be derived using GIS and remote sensing techniques.[26,29,36] The soil-environmental relationships (f) can be approximated by knowledge extracted from human or nonhuman sources through a set of artificial intelligence techniques. Interactive knowledge acquisition techniques based on Kelly's person-constructed theory[37] can be used to acquire knowledge from local field soil experts.[25] For areas with abundant field observations but no experienced field soil scientists, artificial neural network (ANN) techniques can be used to extract the relationships existing in these field observations.[30] There might be situations under which there are no experienced field soil scientists, nor does there exist a large set of field observations, but descriptions of pedons with

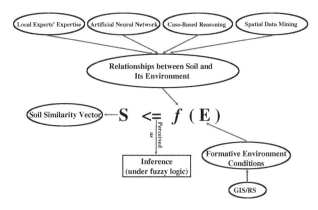

FIGURE 8.12 Implementation of SoLIM.

explicit spatial location information exist. Under these situations, a case-based approach can be applied to utilize the knowledge existing in these pedon descriptions.[32] For areas with old soil maps that need to be updated, a spatial data-mining approach can be applied.[31] Once the knowledge is obtained through the set of knowledge acquisition techniques, the acquired soil-environmental relationships can then be combined with data characterizing the soil formative environment conditions to populate the similarity model.[7,26]

Recent case studies in Wisconsin have shown that the SoLIM approach is capable of capturing and representing detailed spatial variation of soil characteristics. Not only does the approach provide more accurate and detailed soil spatial information products, but it also produces products that are not available with the conventional approach. It has been considered to be a major improvement over the traditional approach to soil survey (about 10 times faster and two thirds less of the cost).[4]

8.6 CHALLENGES AND FUTURE DIRECTIONS

Much has been accomplished in applying fuzzy logic and fuzzy set theory in soil science, but like any new field, there are many challenges facing researchers. These challenges, as alluded to in the above discussions, can be grouped into three major areas: definition of class centroids, membership definition (membership function determination), and interpretation of fuzzy membership values.

Under the notation of fuzzy logic, a class has only one central concept. It is possible that we can use FCM or local knowledge to define a set of centroids for a small set of data or over a relatively small geographic area, but these approaches are either data sensitive and *ad hoc* (such as FCM) or rather subjective (such as the local knowledge-based approach). They may not be adequate for objectively defining a set of central concepts of soil classes for an entire nation, a continent, or for the entire world. The question of how objective our current definitions of soil classes are needs to be examined. Refinements to these definitions and the creation of objective soil classes over much larger spatial scale can reduce the subjectivity and *ad hoc* nature in applying fuzzy logic in soil science. Fuzzy set theory itself might be a tool for addressing these issues. Developments in category theory, particularly prototype category theory,[38,39] might be of some value in addressing these issues.

Membership functions in soil science are still very data sensitive or *ad hoc*. Membership computed from FCM will likely change as one adds more observations to the pool of samples, since the algorithm looks for naturally occurring groups in the data. In other words, the membership computed from FCM is data dependent, and thus geographic area sensitive. Most membership functions based on expert knowledge are rather *ad hoc* and subjective, lacking any formal procedures for defining them, although efforts are under way to address these issues.[25] The issue of integrating knowledge from multiple experts during membership function definition is a challenge for researchers not only in soil science, but also in the artificial intelligence community as a whole. Using knowledge from non-human sources for membership function definition is still at its early stage, although progress has been made.[30,31] Much research is needed in this area.

Well-accepted use and interpretation of fuzzy membership maps is a major driver for a potential widespread use of fuzzy set theory in soil science. However, many researchers are too focused on the mathematics of fuzzy set theory, so use and interpretation of the products from these mathematical studies have received far less attention in the past. In particular, the use of fuzzy membership maps in assessing risk in decision making deserves much more attention.

The other research area related to the use and interpretation of fuzzy membership values is the interpretation of fuzzy membership values in the soil similarity vector. We really do not know what it means for a pixel to receive a similarity value of 0.3 for soil class A and 0.6 for soil class B. What can we say about the soil at that pixel compared to one with a similarity value of 0.2 for A and 0.8 for B? Is this difference (between 0.3 and 0.2 for soil class A) meaningful? If one wants to build a house on a site with a similarity value of 0.6 for Dorerton (loamy-skeletal, mixed, active, mesic Typic Hapludalfs), what does this mean in a practical sense? We need to identify pixels with an array of similarity values for a particular soil series, visit them on the ground, and describe and sample the soils to see what we can learn about these values and their relation to the morphology of the soil.

The work described in this chapter on using fuzzy membership values to predict topsoil thickness[33] is only a trial attempting to predict a soil property at an unsampled location. This is potentially a competitor to kriging, but substitutes soil survey landscape models for the (prohibitively) high sampling intensity and cost of model-based geostatistical strategies. If tools like SoLIM could not only assign a pixel to a soil class, but also predict some important soil properties at that location, this would be a very big development in soil survey. However, given that soils have many properties and they vary spatially in three dimensions (N-S-E-W and with depth), the relationships between membership values and different properties may be different, and the relationships may also vary along different dimensions. Research in this area could potentially be an enormous undertaking but rewarding.

ACKNOWLEDGMENTS

The support provided to the author through the One Hundred Talent Program of the Chinese Academy of Sciences is greatly appreciated. This chapter has benefited greatly from the three anonymous reviewers, and their comments and edits are appreciated.

REFERENCES

1. L Zadeh. Fuzzy sets. *Inf Control* 8: 338–353, 1965.
2. PA Burrough. Fuzzy mathematic methods for soil survey and land evaluation. *J Soil Sci* 40: 447–492, 1989.
3. AB McBratney, IOA Odeh. Application of fuzzy sets in soil science: fuzzy logic, fuzzy measurements and fuzzy decisions. *Geoderma* 77: 85–113, 1997.
4. AX Zhu, B Hudson, J Burt, K Lubich, D Simonson. Soil mapping using GIS, expert knowledge, and fuzzy logic. *Soil Sci Soc Am J* 65: 1463–1472, 2001.
5. HJ Zimmermann. *Fuzzy Set Theory and Its Applications*. Kluwer-Nijhoff Publishing, Boston, 1985, pp. 65–150.
6. GJ Klir, TA Folger. *Fuzzy Sets, Uncertainty, and Information*. Prentice Hall, Englewood Cliffs, NJ, 1988, pp. 213–240.
7. AX Zhu, LE Band. A knowledge-based approach to data integration for soil mapping. *Can J Remote Sensing* 20: 408–418, 1994.
8. X Shi, AX Zhu, JE Burt, F Qi, D Simonson. A case-based reasoning approach to fuzzy soil mapping. *Soil Sci Soc Am J* 68: 885–894, 2003.
9. KD Glinka. *The Great Soil Groups of the World and Their Development*. Edwards Bros., Ann Arbor, MI, 1927, pp. 45–57. (Translated from German by CF Marbut.)
10. H Jenny. *Factors of Soil Formation*. McGraw-Hill, New York, 1941, pp. 35–50.
11. H Jenny. *E.W. Hilgard and the Birth of Modern Soil Science*. Farallo Publication, Berkeley, CA, 1961, pp. 45–65.
12. Soil Survey Staff. *Soil Taxonomy: A Basic System of Soil Classification for Making and Interpreting Soil Surveys*, Agriculture Handbook 436. Natural Resources Conservation Service, U.S. Department of Agriculture, U.S. Government Printing Office, Washington DC, 1999, pp. 10–69.
13. AX Zhu, DS Mackay. Effects of spatial detail of soil information on watershed modeling. *J Hydrol* 248: 54–77, 2001.
14. PHA Sneath, RR Sokal. *Numerical Taxonomy*. W.H. Freeman and Company, San Francisco, 1973, pp. 110–130.
15. JC Bezdek. *Pattern Recognition with Fuzzy Objective Function Algorithms*. Plenum Press, New York, 1981, pp. 89–110.
16. JC Bezdek, R Ehrlich, W Full. FCM: the fuzzy c-means clustering algorithm. *Comput Geosci* 10: 191–203, 1984.
17. CW Ahn, MF Baumgardner, LL Biehl. Delineation of soil variability using geostatistics and fuzzy clustering analyses of hyperspectral data. *Soil Sci Soc Am J* 63: 142–150, 1999.

18. TJ Ross. *Fuzzy Logic with Engineering Applications*. McGraw-Hill, New York, 1995, pp. 67–121.
19. JJ de Gruijter, AB McBratney. A modified fuzzy k-means method for predictive classification. In *Classification and Related Methods of Data Analysis*, HH Bock, Ed. Elsevier, Amsterdam, 1988, pp. 97–104.
20. AB McBratney, JJ de Gruijter. A continuum approach to soil classification by modified fuzzy k-means with extragrades. *J Soil Sci* 43: 159–175, 1992.
21. AB McBratney, JJ de Gruijter, DJ Brus. Spacial prediction and mapping of continuous soil classes. *Geoderma* 54: 39–64, 1992.
22. L Chang, PA Burrough, Fuzzy reasoning: a new quantitative aid for land evaluation. *Soil Surv Land Eval* 7: 69–80, 1987.
23. PA Burrough, RA MacMillian, W van Deusen. Fuzzy classification methods for determining land suitability from soil profile observations and topography. *J Soil Sci* 43: 193–210, 1992.
24. A Dobermann, T Oberthur. Fuzzy mapping of soil fertility: a case study on irrigated Riceland in the Philippines. *Geoderma* 77: 317–339, 1997.
25. AX Zhu. A personal construct-based knowledge acquisition process for natural resource mapping using GIS. *Int J Geogr Inf Sci* 13: 119–141, 1999.
26. AX Zhu, LE Band, B Dutton, TJ Nimlos. Automated soil inference under fuzzy logic. *Ecol Modeling* 90: 123–145, 1996.
27. AX Zhu. A similarity model for representing soil spatial information. *Geoderma* 77: 217–242, 1997.
28. IOA Odeh, AB McBratney, DJ Chittleborough. Fuzzy c-means and kriging for mapping soil as a continuous system. *Soil Sci Soc Am J* 56: 1848–1854, 1992.
29. JP Wilson, JC Gallant, Eds. *Terrain Analysis: Principles and Applications*. John Wiley & Sons, New York, 2000, 527 pp.
30. AX Zhu. Mapping soil landscape as spatial continua: the neural network approach. *Water Resour Res* 36: 663–677, 2000.
31. F Qi, AX Zhu. Knowledge discovery from soil maps using inductive learning. *Int J Geogr Inf Sci* 17: 771–795, 2003.
32. X Shi, AX Zhu, JE Burt, F Qi, D Simonson. A cased-based reasoning approach to fuzzy soil mapping. *Soil Sci Soc Am J* 68: 885–894, 2004.
33. AX Zhu, LE Band, R Vertessy, B Dutton. Deriving soil property using a soil land inference model (SoLIM). *Soil Sci Soc Am J* 61: 523–533, 1997.
34. AX Zhu. Measuring uncertainty in class assignment for natural resource maps using a similarity model. *Photogrammetric Eng Remote Sensing* 63: 1195–1202, 1997.
35. DM Mark, F Csillag. The nature of boundaries on 'area-class' maps. *Cartographica* 27: 65–78, 1990.
36. X Shi, AX Zhu. A similarity-based method for deriving fuzzy representation of terrain positions. *Photogrammetric Eng. Sensing.* Submitted.
37. GA Kelly. A brief introduction to personal construct theory. In *Perspectives in Personal Construct Theory*, D Bannister, Ed. Academic Press, London, 1970, pp. 1–29.
38. EH Rosch. Natural categories. *Cognit Psychol* 4: 328–350, 1973.
39. EH Rosch. Principles of categorization. In *Cognition and Categorization*, E Rosch, BB Lloyd, Eds. Lawrence Erlbaum Associates, Hillsdale, NJ, 1978, pp. 145–170.

9 Modeling Spatial Variation of Soil as Random Functions

Richard Webster and Margaret A. Oliver

CONTENTS

ABSTRACT

For many years pedologists classified soil so as to predict its properties at unvisited sites. To do so efficiently, they constructed mental models of the way soil was distributed in the landscape. These classes later became strata within which to sample and predict quantitatively within the framework of classical statistics. In the last 20 years, pedologists have replaced such models by ones in which variation is regarded as continuous, stochastic, and spatially correlated; classical statistics has given way to geostatistics. In the simplest cases, an actual soil variable is regarded as the outcome of a stationary stochastic process, which is a model completely characterized by its variogram. The variogram is central to practical geostatistics. Its values must be estimated from data to give an ordered set, the sample or experimental variogram. This in turn is modeled with one or more mathematical functions, which must be such as to guarantee nonnegative variances when values are combined. There are a few families of simple functions; they include the familiar spherical and exponential functions and the unbounded power function. Unknown values are predicted by kriging and then mapped. The techniques are illustrated with data from a recent case study in which a 23-ha field in southern England was surveyed by sampling. The data are summarized and explored, after which variograms of several variables are computed and modeled. The soil's magnesium content and pH behaved as realizations of stationary processes; the electrical conductivity contained a long-range trend. The variogram of yield could be separated into short- and long-range components, and the components are mapped separately to reveal a pattern of cultivation distinct from the regional pattern in the field. Electrical conductivity and sand content are shown to be coregionalized.

9.1   INTRODUCTION

The soil mantles the land more or less continuously, except where there is bare rock and ice, and in a way so complex that no description of it can follow every detail. Further, our knowledge of the soil's properties beneath the surface is fragmentary because it derives from samples. Any representation of the mantle involves simplification and inference or prediction as to what the soil is like between sampling points with the uncertainty that they entail.

Research in the last 40 years has provided us with quantitative descriptions based on samples. There have been two main approaches. In the first the soil is divided into discrete classes that are sampled to give estimates of means and variances of individual properties of interest. This approach may be regarded as classical in both a pedological and a statistical sense. The other views soil as a suite of continuous variables and seeks to describe the continuity in terms of spatial dependence. Specifically, it treats properties as though they were the outcomes of random processes and uses geostatistics to estimate both plausible generating functions of the processes and values of the realizations at unsampled places. The two approaches are not mutually exclusive, and they can be combined.

In what follows, we deal briefly with soil classification as a model of soil variation, partly for historical reasons and partly because of its intrinsic merit. We then devote the major part to geostatistics, which we illustrate with results from a recent case study of precision farming in the south of England. No account of the study has been published previously, and so we describe the study and the results from it in a single section after presenting the theory. We link them to the theory with cross-references.

9.2 SOIL CLASSIFICATION

Formal classification of soil has its roots in 19th-century biological taxonomy and practice in geological survey. Finite circumscribed regions are divided into parcels by boundaries, which are sharp lines across which the soil changes in some sense. For any one region the outcome is a map, technically a choropleth map, showing the region tessellated into spatial classes. These constitute a general-purpose classification. The map may purport to show the classes of some predefined scheme of classification; alternatively, the boundaries on it may be drawn where the soil changes more than elsewhere, and between which the soil is relatively homogeneous. There is no underlying mathematical model, but the pedologist who makes the map often has a mental model of the way soil varies in the landscape. Perhaps the best-known and most generally useful model is the catena.

The soil map appears mathematically as a stepped function for any one soil property, as in Figure 9.1a of a transect across a region. Variation within the classes may be acknowledged, but it is not evident. The reality is more like Figure 9.1b, which is the same transect, but now with all the data on pH in the subsoil measured at 10-m intervals.[1] Some of the boundaries can still be recognized where there are large jumps in the data, but others are not obvious.

By the 1960s, taxonomists were putting numerical limits on the discriminating criteria for consistency. This helped to codify description. It did nothing, however, to quantify the variation in properties that could not be assessed readily in the field. And it was unhelpful to the mapmaker who wished to place boundaries where there were maxima in the rate of change in the landscape. Description needed a formal statistical basis, a need first recognized by civil engineers in the 1960s (see, for example, Morse and Thornburn[2] and Kantey and Williams[3]).

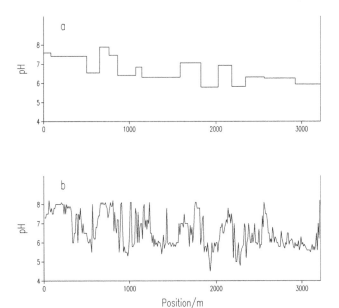

FIGURE 9.1 Subsoil pH along a transect. (a) A stepped function of mean values representing a soil classification with horizontal lines for the means within classes; the vertical lines are the class boundaries. (b) The measured values.

9.2.1 SAMPLING AND ESTIMATION

In this approach a soil property, Z, takes values at an infinity of points, $\mathbf{x}_i = \{x_{i1}, x_{i2}\}, i = 1, 2, \ldots, \infty$, in a region $\mathfrak{R}$. In principle, the points could occupy discrete positions in a third dimension (depth), but in regional survey, the depth is so small in relation to the lateral distances that it is best treated separately. The values, $z(\mathbf{x}_i)$, comprise the population, which has a mean, μ, and variance, here denoted as σ_T^2, signifying the total variance in $\mathfrak{R}$. The region is divided into K spatial strata or classes, R_k, $k = 1, 2, \ldots, K$, that are mutually exclusive and exhaustive and that are what the map displays. Each stratum has its own mean and variance, which we denote μ_k and σ_k^2, respectively. The region is then sampled, and the property at the N sampling points is measured to give data, $z(\mathbf{x}_1), z(\mathbf{x}_2), \ldots, z(\mathbf{x}_N)$, of which n_k belong in class R_k.

If sampling is unbiased, then the mean for the kth class is estimated simply by

$$\hat{\mu}_k = \bar{z}_k = \frac{1}{n_k} \sum_{i=1}^{n_k} z(\mathbf{x}_i) \quad \text{for} \quad \mathbf{x}_i \in R_k \tag{9.1}$$

The variance within R_k is estimated from the same sample by

$$\hat{\sigma}_k^2 = s_k^2 = \frac{1}{n_k - 1} \sum_{i=1}^{n_k} \left\{ z(\mathbf{x}_i) - \bar{z}_k \right\}^2 \quad \text{for} \quad \mathbf{x}_i \in R_k \tag{9.2}$$

This in turn provides the variance of the mean, the estimation variance, as

$$s_k^2(\mu) = s_k^2 / n_k \tag{9.3}$$

Further, one can use the sample mean for a class as predictor for any unsampled point, $\mathbf{X}_0$, in the same class. If $\mathbf{X}_0$ is chosen at random, then its prediction variance is

$$s_k^2(\mathbf{X}_0) = s_k^2 + s_k^2 / n_k \tag{9.4}$$

Confidence limits on variances from small samples are wide, but one can narrow them by making use of the pedologist's mental model of variation. When a soil surveyor subdivides a particular region to display the spatial distribution of the soil, he or she usually tries to create classes of the same categorical level, for example, all soil series or all soil families. Ideally, the variances within these are equal; i.e., there is a common within-class variance:

$$\sigma_W^2 = \sigma_k^2 \quad \text{for all } k \tag{9.5}$$

Then, although each class might be represented by few sampling points and the individual σ_k^2 are not well estimated, σ_W^2 is well estimated by the pooled variance s_W^2, which can replace s_k^2 in Equations 9.3 and 9.4.

9.3 THE GEOSTATISTICAL APPROACH

Imposing boundaries between classes to describe variation that is patently continuous is obviously artificial. The stepped function it creates puts into mathematical form the pedologist's mental model. Pedologists lived uncomfortably with it for many years, partly because there seemed no practicable alternative. They considered polynomials, but such functions would have to be of a very high order and could have no generality; the variation is too complex, perhaps even chaotic, as Figure 9.1b shows. Such variation looks as though it might be random, and it is this idea that provides the alternative model: if the variation appears random, then let us treat it as if it were so.[4] This is the basis of modern geostatistics and its approach to describing soil variation.

9.3.1 RANDOM VARIABLES AND RANDOM FUNCTIONS

Again, a region $\mathfrak{R}$ is regarded as comprising an infinite number of points $\mathbf{x}_i, i = 1, 2, \ldots, \infty$. Whereas in the classical approach the values of z at these points constitute the population, in the geostatistical approach this population is assumed to be just one realization of a random process or random function that could generate any number of such populations. Then at each place $\mathbf{x}$ the soil property is a random variable, $Z(\mathbf{x})$ (notice the capital Z), of many values. For a continuous variable such as the soil's strength or pH, this number is infinite, and the whole process may be regarded as a doubly infinite superpopulation. The random variable at $\mathbf{x}$ has a distribution with a mean and variance and higher-order moments, and the actual value there, $z(\mathbf{x})$, is just one drawn at random from that distribution.

In these circumstances, the quantitative description of the variation involves estimating the characteristics of what are assumed to be the underlying random processes. The characteristics include the means and variances, and perhaps higher-order moments, and, most important, the spatial covariances.

The spatial covariance between the variables at any two places $\mathbf{x}_1$ and $\mathbf{x}_2$ is given by

$$C(\mathbf{x}_1, \mathbf{x}_2) = E\left[\left\{Z(\mathbf{x}_1) - \mu(\mathbf{x}_1)\right\}\left\{Z(\mathbf{x}_2) - \mu(\mathbf{x}_2)\right\}\right] \tag{9.6}$$

where $\mu(\mathbf{x}_1)$ and $\mu(\mathbf{x}_2)$ are the means at $\mathbf{x}_1$ and $\mathbf{x}_2$, and E denotes the expected value. In practice, $C(\mathbf{x}_1, \mathbf{x}_2)$ cannot be estimated because we only ever have the one realization, and to overcome this apparent impasse we invoke assumptions of stationarity.

9.3.2 STATIONARITY

Starting with the first moment, we assume that the mean, $\mu = E[Z(\mathbf{x})]$, is constant for all $\mathbf{x}$, and so we can replace $\mu(\mathbf{x}_1)$ and $\mu(\mathbf{x}_2)$ by the single value μ, which we can estimate by repetitive sampling.

Next, when $\mathbf{x}_1$ and $\mathbf{x}_2$ coincide, Equation 9.6 defines the variance, $\sigma^2 = E[\{Z(\mathbf{x}) - \mu\}^2]$. We assume this to be finite and, like the mean, to be the same everywhere. We then generalize Equation 9.6 so that it applies to any pair of points $\mathbf{x}_i$ and $\mathbf{x}_j$ separated by a vector, or *lag*, $\mathbf{h} = \mathbf{x}_i - \mathbf{x}_j$, so that

$$\begin{aligned}
C(\mathbf{x}_i, \mathbf{x}_j) &= E\left[\left\{Z(\mathbf{x}_i) - \mu\right\}\left\{Z(\mathbf{x}_j) - \mu\right\}\right] \\
&= E\left[\left\{Z(\mathbf{x})\right\}\left\{Z(\mathbf{x}+\mathbf{h})\right\} - \mu^2\right] \\
&= C(\mathbf{h})
\end{aligned} \tag{9.7}$$

and this is also constant for any given $\mathbf{h}$. This constancy of the mean and variance and of a covariance that depends only on separation and not on absolute position constitutes *second-order stationarity*.

Equation 9.7 shows that the covariance is a function of the lag and only of the lag; it describes quantitatively the dependence between values of Z with changing lag. It is readily converted to the dimensionless *autocorrelation* by

$$\rho(\mathbf{h}) = C(\mathbf{h}) / C(\mathbf{0}) \tag{9.8}$$

where $C(0) = \sigma^2$ is the covariance at lag $\mathbf{0}$.

9.3.3 INTRINSIC VARIATION AND THE VARIOGRAM

In many instances the assumption of constant mean throughout a region is untenable, and if the mean changes, then the variance will appear to increase indefinitely with increasing area. The covariance cannot be defined then because there is no value for μ to insert in Equation 9.7. Faced with this situation, geostatisticians consider the differences from place to place and their squares, as follows. For small lag distances, the expected differences are zero,

$$E[Z(\mathbf{x}) - Z(\mathbf{x} + \mathbf{h})] = 0 \tag{9.9}$$

and the expected squared differences define the variances for those lags,

$$E\left[\left\{Z(\mathbf{x}) - Z(\mathbf{x} + \mathbf{h})\right\}^2\right] = \mathrm{var}\left[Z(\mathbf{x}) - Z(\mathbf{x} + \mathbf{h})\right] = 2\gamma(\mathbf{h}) \tag{9.10}$$

Equation (9.10) gives the variance of the difference at lag $\mathbf{h}$. The variance per point, $\gamma(\mathbf{h})$, is half of this value and is known as the semivariance. The above two equations constitute the *intrinsic hypothesis* of geostatistics.[5] Like the covariance, the semivariance depends only on the lag and not on the absolute positions $\mathbf{x}$ and $\mathbf{x} + \mathbf{h}$. As a function, $\gamma(\mathbf{h})$ is the variogram, often still called the semivariogram.

If the process $Z(\mathbf{x})$ is second-order stationary, then the semivariance and the covariance are equivalent:

$$\gamma(\mathbf{h}) = C(\mathbf{0}) - C(\mathbf{h})$$

$$= \sigma^2\{1 - \rho(\mathbf{h})\} \tag{9.11}$$

If it is intrinsic only, then the covariance does not exist, but the semivariance remains valid, and it is this validity in a wide range of circumstances that makes the variogram so useful in summarizing spatial variation.

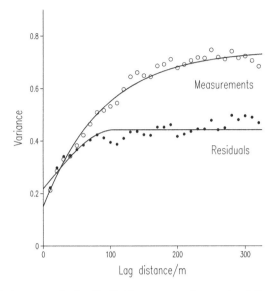

FIGURE 9.2 Variograms of subsoil pH. The upper variogram is of the raw data shown in Figure 9.1b; the lower curve is for the differences between the raw data and the means in Figure 9.1a, i.e., for the residuals.

9.3.4 ESTIMATING THE VARIOGRAM

Semivariances are readily estimated from data, $z(\mathbf{x}_1), z(\mathbf{x}_2), \ldots$, by the method of moments:

$$\hat{\gamma}(\mathbf{h}) = \frac{1}{2m(\mathbf{h})} \sum_{i=1}^{m(\mathbf{h})} \left\{ z(\mathbf{x}_i) - z(\mathbf{x}_i + \mathbf{h}) \right\}^2 \tag{9.12}$$

in which $m(\mathbf{h})$ is the number of paired comparisons at lag $\mathbf{h}$. By changing $\mathbf{h}$, we obtain a sample or experimental variogram, which can be displayed as a graph of $\hat{\gamma}$ against $\mathbf{h}$. The upper set of points in Figure 9.2 is an example in which the experimental semivariances for the data in Figure 9.1b are plotted against the lag distance, $h = |\mathbf{h}|$, for the one-dimensional transect.

The values of $\mathbf{h}$ define discrete points on the variogram, and so sampling is best planned with regular intervals along a line in one dimension or on a grid in two or three. Figure 9.2 is an example deriving from data recorded at regular intervals (10 m) on a line. Otherwise, where data are irregularly scattered, the actual separations have to be placed into bins, with limits in separating distance and also in direction if there is more than one dimension (Figure 9.3). This introduces an arbitrariness that is absent with data recorded at regular intervals. The practitioner must decide how wide to make the bins, and that decision will affect the result to some degree — see the variograms of subsoil pH in Figure 9.11. Narrow bins tend to give rise to erratic variograms (Figure 9.11a); wide

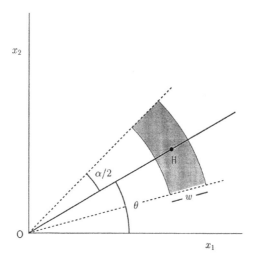

FIGURE 9.3 Discretization of the lag into bins for irregularly scattered data. The quantity w is the width of the bin and $\alpha/2$ is the angular tolerance.

bins tend to smooth and lose detail (Figure 9.11b). Inevitably, the choice is a compromise. In particular, widening the angle of the discretization loses information on any fluctuation in variance with changing direction (anisotropy) until at 180° all information is lost.

9.3.5 Models for Variograms

The underlying variogram, Equation 9.10, is a continuous function in as many dimensions as the variable $Z(\mathbf{x})$. The experimental variogram estimates it at a set of points, with more or less error and point-to-point fluctuation arising from the sampling. To obtain a variogram to describe the spatial variation in R, we fit a plausible function to the experimental values.

Figure 9.2 shows the principal features of many, if not most, experimental variograms. They are as follows.

1. The variance increases from near the ordinate with increasing lag distance.
2. It reaches a maximum at which it remains thereafter.
3. Any simple smooth line or surface placed through the points and projected to the ordinate cuts it at some value greater than 0.

The model must also be mathematically acceptable in that it cannot give rise to negative variances when random variables are combined. Technically, the covariance function, if it exists, must be positive semidefinite, and the variogram must be conditional negative semidefinite (CNSD) — see Webster and Oliver[6] for an explanation.

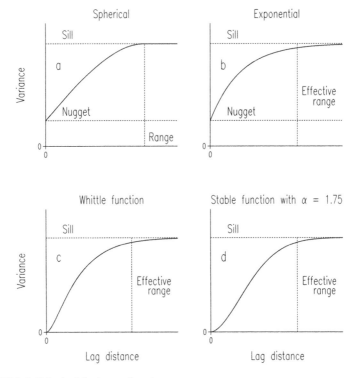

FIGURE 9.4 Principal features of variograms.

There are only a few families of simple functions that satisfy the above criteria. They can be divided into those that are bounded and those that are not. In the first group are the popular spherical and exponential models plus several other members of this family, including the circular and pentaspherical models from Matérn's[7] set, Gaussian and stable models, and Whittle's[8] function. We list them and their formulae in their isotropic forms, i.e., for $h = |\mathbf{h}|$, and illustrate them fitted to experimental values in the case study (Figure 9.10, Figure 9.17, and Figure 9.20). They are as follows.

9.3.5.1 Circular

The circular function has the following equation:

$$\gamma(h) = c_0 + c\left\{1 - \frac{2}{\pi}\cos^{-1}\left(\frac{h}{a}\right) + \frac{2h}{\pi a}\sqrt{1 - \left(\frac{h}{a}\right)^2}\right\} \quad \text{for} \quad 0 < h \leq a$$

$$= c \quad \text{for} \quad h > a \tag{9.13}$$

$$= 0 \quad \text{for} \quad h = 0$$

Here $\gamma(h)$ is the semivariance at lag h, and c is the *a priori* variance of the autocorrelated process. The quantity c_0 is the intercept on the ordinate and is known as the *nugget variance*, a term deriving from gold mining. The combined $c_0 + c$ is known as the *sill* of the model. These quantities are illustrated generically in Figure 9.4, and in Figure 9.10a we show the function fitted to the experimental variogram of topsoil magnesium, Mg_0.

The function has a distance parameter, a; this is its *range*, also known as its correlation range. It marks the limit of spatial dependence; values at places closer to one another than the distance parameter are more or less correlated, whereas those farther apart are not. It implies that all the variance in R is encountered within that distance. The function gets its name from the formula for the area of two intersecting circles, which are of diameter a.

The semivariance at lag 0 is itself zero, and for continuous processes such as most physical properties of the soil, $\gamma(h)$ should increase gradually as h increases from 0. In practice, we usually lack sufficient estimates of $\gamma(h)$ near the ordinate to fit a model through the origin (see below), and we take the conservative approach above. The nugget variance is therefore best regarded as embodying variation within the shortest sampling interval plus any measurement error.

9.3.5.2 Spherical

The equation for the spherical function is

$$\gamma(h) = c_0 + c\left\{\frac{3}{2}\left(\frac{h}{a}\right) - \frac{1}{2}\left(\frac{h}{a}\right)^3\right\} \quad \text{for} \quad h \le a$$

$$= c \quad \text{for} \quad h > a \qquad (9.14)$$

$$= 0 \quad \text{for} \quad h = 0$$

in which h, c, c_0, and a have the same meanings as before. This function gets its name from the formula for the volume of two intersecting spheres, each of diameter a. The function is illustrated in Figure 9.4a and is fitted to experimental values in Figure 9.2 and Figure 9.10b.

9.3.5.3 Pentaspherical

This function has the equation

$$\gamma(h) = c_0 + c\left\{\frac{15}{8}\left(\frac{h}{a}\right) - \frac{5}{4}\left(\frac{h}{a}\right)^3 + \frac{3}{8}\left(\frac{h}{a}\right)^5\right\} \quad \text{for} \quad h \le a$$

$$= c \quad \text{for} \quad h > a \qquad (9.15)$$

$$= 0 \quad \text{for} \quad h = 0$$

It can be thought of as the five-dimensional analog of the circular and spherical models. Figure 9.10c shows it fitted to the same experimental variogram as the circular and spherical models.

There is also a one-dimensional analog of these models; it is the bounded linear model, which increases linearly from its intersect on the ordinate to its sill, at which point it becomes constant. It is little used partly because it is valid (CNSD) in only one dimension. The circular model is valid in one and two dimensions, but not in three. The spherical and pentaspherical models are valid in one, two, and three dimensions.

9.3.5.4 Exponential

The models above, starting with the bounded linear model, show increasingly gradual curvature. A model that curves even more gradually is the exponential function. Its equation is

$$\gamma(h) = c_0 + c\left\{1 - \exp\left(-\frac{h}{r}\right)\right\} \tag{9.16}$$

in which c_0 and c have the same meanings as before, but now with a distance parameter, r. The exponential model approaches its sill asymptotically and therefore has no definite range. A working range is often taken as $a' = 3r$, at which point the function has reached 95% of c. This model is shown in Figure 9.4b and Figure 9.10d.

9.3.5.5 Models with Reverse Curvature at the Origin

Some variograms appear to approach the origin with decreasing gradients. These may be represented by the general equation

$$\gamma(h) = c\left\{1 - \exp\left(-\frac{h^\alpha}{r^\alpha}\right)\right\} \tag{9.17}$$

in which $1 < \alpha \leq 2$. If $\alpha = 2$, we have the Gaussian function. This is at the limit of acceptability and gives rise to unstable prediction. It is best replaced by stable models with $\alpha < 2$. Figure 9.4d shows the function with $\alpha = 1.75$, and an example with $\alpha = 1.95$ is shown for elevation in the case study below (Figure 9.20a). Another recommended function to describe such variation is Whittle's *elementary correlation*,[8] which embodies a Bessel function and derives from diffusion in two dimensions. It is illustrated in Figure 9.4c.

9.3.5.6 Unbounded Models

Variograms of processes that are intrinsic but not second-order stationary increase without bound as the lag distance increases. These can usually be fitted by power functions, for which the general equation including a nugget is

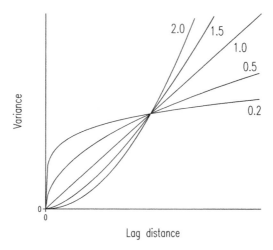

FIGURE 9.5 Curves of unbounded power functions with several values of exponents between 0 and 2.

$$\gamma(h) = c_0 + wh^{\alpha} \tag{9.18}$$

The parameter w describes the intensity of the process, and the exponent, which must lie strictly between 0 and 2 (these limits are excluded), describes the curvature. If $\alpha < 1$, then the curve is convex upwards; if it is 1, then we have a straight line; and if $\alpha > 1$, then the curve is concave upwards. The curve with $\alpha = 2$ is a parabola and describes a smoothly continuous process that is not random. Figure 9.5 illustrates the curves for several values of α.

9.3.5.7 Anisotropy

The variogram of a two-dimensional process is itself two-dimensional, and if the process is anisotropic, then so is its variogram, which is then a function of both distance, h, and direction, θ. In the simplest cases, the anisotropy is geometric, meaning that it can be made isotropic by a linear transformation of the coordinates. The transformation is defined by reference to an ellipse:

$$\Omega(\theta) = \sqrt{A^2 \cos^2(\theta - \phi) + B^2 \sin^2(\theta - \phi)} \tag{9.19}$$

where A and B are the long and short diameters, respectively, of the ellipse and ϕ is its orientation, i.e., the direction of the long axis (Figure 9.6). Equation 9.19 is embodied into the models as follows. For the bounded models, Ω replaces the distance parameter of the isotropic variogram. So, for example, in the exponential we have

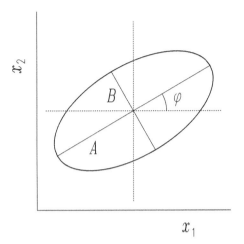

FIGURE 9.6 Ellipse showing the parameters to describe geometric anisotropy.

$$\gamma(h,\theta) = c_0 + c \left\{ 1 - \exp\left(-\frac{h}{\Omega(\theta)} \right) \right\} \qquad (9.20)$$

and in the power function,

$$\gamma(h,\theta) = c_0 + \left\{ \Omega(\theta)h \right\}^{\alpha} \qquad (9.21)$$

Figure 9.7 shows an anisotropic spherical function. Notice how the range of the model changes with changing direction.

9.3.6 COMBINING TREND AND RANDOM FLUCTUATION

The above functions describe processes that are entirely random though correlated. We can represent the processes by the general model

$$Z(\mathbf{x}) = \mu_V + \varepsilon(\mathbf{x}) \qquad (9.22)$$

in which μ_V is the mean, i.e., constant in some neighborhood V, and $\varepsilon(\mathbf{x})$ is the autocorrelated variance as defined in Equation 9.10. It often happens that such models are unacceptable, either because there is an evident long-range trend across a region or because over short distances the variation appears smooth. In these circumstances μ_V cannot be treated as constant, but must be replaced by a deterministic term, say $u(\mathbf{x})$, that depends on the position $\mathbf{x}$. The model then becomes

$$Z(\mathbf{x}) = u(\mathbf{x}) + \varepsilon(\mathbf{x}) \qquad (9.23)$$

FIGURE 9.7 Perspective diagram of a geometrically anisotropic spherical variogram.

If $u(\mathbf{x})$ can describe the variation over the whole of R simply, then it is called a trend. If it is local only, then it is known as a drift. In either event, it is usually represented by a low-order polynomial, so that Equation 9.23 becomes

$$Z(\mathbf{x}) = \sum_{j=0}^{J} a_j f_j(\mathbf{x}) + \varepsilon(\mathbf{x}) \tag{9.24}$$

in which the a_j are unknown coefficients and the $f_j(\mathbf{x})$ are known functions of our choosing.

Trend can be gradual, such as the smooth predictable change associated with the water table or an inclined or undulating land surface, or it can occur as abrupt transitions from one soil class to another. Gradual trend may be evident in a display of the data on a map. It may also be revealed in the experimental variogram, in which the points appear to lie on a line of ever-increasing gradient, especially if the curve fitted is of a power function with exponent exceeding 2. Figure 9.20a in the case study shows an experimental variogram that signifies the presence of trend.

It is fairly easy, even if somewhat arbitrary, to separate any long-range trend from the short-range apparently random fluctuation, and to estimate the parameters of the two components separately. We illustrate the procedure below with an example from the case study. It is not as easy to separate short-range drift; these circumstances involve a full structural analysis, effectively a process of trial and error.[9]

9.3.7 FITTING MODELS

Having described some of the mathematical functions for variograms, we now turn to fitting them to the experimental or sampling estimates. The matter is controversial, but we cannot devote space to the controversy here. The usual approach is to fit the simplest model that looks reasonable and makes sense. Nevertheless, there are difficulties arising largely from the following attributes:

1. The accuracy of the observed semivariances is not constant.
2. The variation may be anisotropic.
3. The experimental variogram can contain a lot of point-to-point fluctuation.
4. Most of the models are nonlinear in one or more parameters.

The first three items make fitting by eye unreliable; items 1 and 2 impair one's intuition, and item 3 means that several plausible models might be drawn through the estimates. Large deviations about a model can lead to unstable mathematical solutions where nonlinear parameters must be found by iteration (item 4).

These difficulties are perhaps the most serious facing the practitioner, yet they must be overcome because all subsequent tasks depend on the model finally chosen. We recommend a procedure that embodies both visual inspection and statistical fitting, the latter having been tried, tested, and refined over the years as statistical computation has advanced. So, first inspect a graph of the experimental variogram. Then choose one or more CNSD models that have shapes that match the principal features of the graph. Then fit each model in turn by weighted least squares, i.e., by minimizing the sum of the squares of the deviations, suitably weighted, between the experimental and fitted values. Then draw each fitted model to check that the fit does indeed seem reasonable. If all the chosen models seem to fit well, then you might finally choose the one for which the residual sum of squares is least.

9.3.7.1 Complexity

The models we have mentioned are all simple in that they have no more than three parameters. If none of them fits well, we might try fitting more complex models, i.e., ones with more parameters. It is almost always possible to improve the fit of a model by adding parameters. One might be tempted to continue adding parameters until the fit is perfect, but that is not sensible; we need to compromise between simplicity, or parsimony, with few parameters and close fit with more parameters. Akaike's information criterion[10] enables us to do this, and we describe and illustrate its use in Webster and Oliver.[6]

9.3.7.2 Weights

Item 1 above drew attention to the fact that the $\hat{\gamma}(\mathbf{h})$ are not equally reliable. In particular, they are based on different numbers of paired comparisons, $m(\mathbf{h})$ in Equation 9.12. So as a first step, one should weight the experimental values in proportion to m.

We also know that reliability of an estimate is in general inversely related to the variance. This led Cressie[11] to include the fitted semivariances, $\gamma^*(\mathbf{h}_j), j = 1,2,\ldots$, in the weights. Cressie's proposed weights are

$$w_{\text{Cressie}}(\mathbf{h}_j) = m(\mathbf{h}_j) / \gamma^{*2}(\mathbf{h}_j) \qquad (9.25)$$

This scheme tends to give more weight to estimates at the shorter lags than weighting on pairs alone, because in most instances the variogram is an increasing function, and so the fit is better there. This is usually desirable for kriging (see below); it might be less desirable, however, if you want a more nearly equal weighting to estimate the distance parameters of the variograms.

A complication in Cressie's scheme is that you cannot know the $\gamma^{*2}(\mathbf{h}_j)$ until you have fitted the model. Therefore, you have to iterate. We have found, however, that the fitting converges rapidly and there is little change after the first iteration.

9.4 COMBINING CLASSIFICATION WITH GEOSTATISTICS

In some instances neither a classification nor a variogram alone can represent spatial variation in soil properties. The choropleth map implies abrupt changes, whereas the variogram is based on a model of random but continuous fluctuation. If there appear to be both kinds of variation, then the two approaches can be combined. Figure 9.2 shows such a situation. By recognizing the class boundaries, as in Figure 9.1a, and analyzing the variance (Table 9.1), we can obtain residuals from the class means. Their variance is the residual mean square, and a portion of this is likely to be autocorrelated and have its own variogram. The lower set of points in Figure 9.2 is an example of a within-class variogram obtained by superimposing the classification of Figure 9.1a on the data of Figure 9.1b.

The variogram of the residuals differs from the variogram of the original data in two important respects:

TABLE 9.1
Analysis of Variance of pH of Subsoil

Source	Degrees of Freedom	Mean Square	F Ratio
Between classes	14	8.1456	19.27
Within classes	306	0.4228	
Total	320	0.7607	

1. The sill of the fitted model is less by an amount approximately equal to the between-class variance, as expected.
2. The range of the model is much less. This is because the class-to-class variation, which evidently dominated the variation over the whole transect, has been removed to leave only the short-range correlation.

9.5 SIMULTANEOUS VARIATION IN TWO OR MORE VARIABLES

Any two variables, say z_u and z_v, may be correlated, and in particular linearly correlated. That relation is conventionally expressed by the product–moment correlation coefficient:

$$\rho = \frac{\text{cov}[uv]}{\sqrt{\text{var}[u] \times \text{var}[v]}} \quad (9.26)$$

i.e., the covariance of z_u and z_v divided by the product of their standard deviations. It is also known as the Pearson correlation coefficient.

Two spatial random variables, $Z_u(\mathbf{x})$ and $Z_v(\mathbf{x})$, may also be *spatially* intercorrelated in that each is spatially correlated both with itself, i.e., autocorrelated, and with the other. The two are then said to be cross-correlated. In these circumstances, the two variables have in addition to their autovariograms, as defined by Equation 9.10 and for present purposes denoted $\gamma_{uu}(\mathbf{h})$ and $\gamma_{vv}(\mathbf{h})$, a cross-variogram, $\gamma_{uv}(\mathbf{h})$, defined by

$$\gamma_{uv}(\mathbf{h}) = \frac{1}{2}\mathbb{E}\left[\{Z_u(\mathbf{x}) - Z_u(\mathbf{x} + \mathbf{h})\}\{Z_v(\mathbf{x}) - Z_v(\mathbf{x} + \mathbf{h})\}\right] \quad (9.27)$$

If both variables are second-order stationary with means μ_u and μ_v, then both will have covariance functions, $C_{uu}(\mathbf{h})$ and $C_{vv}(\mathbf{h})$, as defined in Equation 9.7, and a cross-covariance:

$$C_{uv}(\mathbf{h}) = \mathbb{E}\left[\{Z_u(\mathbf{x}) - \mu_u\}\{Z_v(\mathbf{x} + \mathbf{h}) - \mu_v\}\right] \quad (9.28)$$

There is also a cross-correlation coefficient, ρ_{uv}, given by

$$\rho_{uv}(\mathbf{h}) = \frac{C_{uv}(\mathbf{h})}{\sqrt{C_{uu}(\mathbf{0})C_{vv}(\mathbf{0})}} \qquad (9.29)$$

This is effectively the extension of the Pearson product–moment correlation coefficient of Equation 9.26 into the spatial domain, and when $\mathbf{h} = 0$, it is the Pearson coefficient.

The cross-covariance is in general not symmetric, i.e.,

$$\mathrm{E}\Big[\{Z_u(\mathbf{x}) - \mu_u\}\{Z_v(\mathbf{x}+\mathbf{h}) - \mu_v\}\Big] \neq \mathrm{E}\Big[\{Z_v(\mathbf{x}) - \mu_v\}\{Z_u(\mathbf{x}+\mathbf{h}) - \mu_u\}\Big] \quad (9.30)$$

In words, the cross-covariance between $Z_u(\mathbf{x})$ and $Z_v(\mathbf{x})$ in one direction is different from that in the other, or expressed another way,

$$C_{uv}(\mathbf{h}) \neq C_{uv}(-\mathbf{h}) \text{ or equivalently } C_{uv}(\mathbf{h}) \neq C_{vu}(\mathbf{h}) \qquad (9.31)$$

since

$$C_{uv}(\mathbf{h}) = C_{vu}(-\mathbf{h})$$

One can envisage asymmetry between two soil properties at different depths on a slope as result of creep or solifluction. The subsoil will tend to lag behind the topsoil. Similarly, irrigation by flooding always from the same end of a field might distribute salts differentially in the direction of flow. Asymmetric covariances have not been reported in the soil literature as far as we know, however.

The cross-variogram and the cross-covariance function (if it exists) are related by

$$\gamma_{uv}(\mathbf{h}) = C_{uv}(\mathbf{0}) - \frac{1}{2}\{C_{uv}(\mathbf{h}) + C_{uv}(-\mathbf{h})\} \qquad (9.32)$$

This quantity contains both $C_{uv}(\mathbf{h})$ and $C_{uv}(-\mathbf{h})$ and, in consequence, loses any information on asymmetry; it is an even function, i.e., it is symmetric:

$$\gamma_{uv}(\mathbf{h}) = \gamma_{vu}(\mathbf{h}) \text{ for all } \mathbf{h}$$

Cross-semivariances can be estimated in a way similar to that of the autosemivariances,

$$\hat{\gamma}_{uv}(\mathbf{h}) = \frac{1}{2m(\mathbf{h})} \sum_{i=1}^{m(\mathbf{h})} \left\{ z_u(\mathbf{x}_i) - z_u(\mathbf{x} + \mathbf{h}) \right\} \left\{ z_v(\mathbf{x}_i) - z_v(\mathbf{x} + \mathbf{h}) \right\} \qquad (9.33)$$

and the sample cross-variogram is formed simply by incrementing $\mathbf{h}$. There is an equivalent formula for computing the cross-covariances. Notice that there must be numerous places where both z_u and z_v have been measured.

9.5.1 Modeling the Cross-Variogram

The cross-variogram can be modeled in the same way as the autovariogram, and the same restricted set of functions is available. There is one additional constraint. Any linear combination of the variables is itself a regionalized variable, and its variance cannot be negative. This is assured by adopting the *linear model of coregionalization*. In it the variable $Z_u(\mathbf{x})$ is assumed to be the sum of independent (orthogonal) random variables $Y_j^k(\mathbf{x})$,

$$Z_u(\mathbf{x}) = \sum_{k=1}^{K} \sum_{j=1}^{k} a_{uj}^k Y_j^k(\mathbf{x}) + \mu_u \qquad (9.34)$$

in which the superscript k is an index, not a power. There is a similar assumption for $Z_v(\mathbf{x})$. If the assumptions hold, then the pair of variables has a cross-variogram:

$$\gamma_{uv}(\mathbf{h}) = \sum_{k=1}^{K} \sum_{j=1}^{k} a_{uj}^k a_{vj}^k g^k(\mathbf{h}) \qquad (9.35)$$

The products in the second summation can be replaced by b_{uv}^k to give

$$\gamma_{uv}(\mathbf{h}) = \sum_{k=1}^{K} b_{uv}^k g^k(\mathbf{h}) \qquad (9.36)$$

The quantities are the variances and covariances, e.g., nugget and sill variances, for the independent components of a spherical model. For two variables there are the three nugget variances b_{uu}^1, b_{vv}^1, and b_{uv}^1, and similarly, three sills of the correlated variances. The coefficients $b_{uv}^k = b_{vu}^k$ for all k, and for each matrix of coefficients,

$$\begin{bmatrix} b_{uu}^k & b_{uv}^k \\ b_{vu}^k & b_{vv}^k \end{bmatrix}$$

must be positive definite. Since the matrix is symmetric, it is sufficient that $b_{uu}^k \geq 0$ and $b_{vv}^k \geq 0$, and that its determinant is positive or zero; this leads to

$$\left| b_{uv}^k \right| = \left| b_{vu}^k \right| \le \sqrt{b_{uu}^k b_{vv}^k} \ \ \forall k$$

This is Schwarz's inequality.

Any number of regionalized variables may be embodied in the linear model of coregionalization. If there are V of them, then the full matrix of coefficients, $[b_{ij}^k]$, will be of order V, and its determinant and all its principal minors must be positive or zero.

Schwarz's inequality has the following consequences:

1. Every basic structure, $g^k(\mathbf{h})$, present in the cross-variogram must also appear in the two autovariograms; i.e., $b_{uu}^k \ne 0$ and $b_{vv}^k \ne 0$ if $b_{uv}^k \ne 0$. As a corollary, if a basic structure $g^k(\mathbf{h})$ is absent from either autovariogram, then it may not be included in the cross-variogram.
2. Structures may be present in the autovariograms without their appearing in the cross-variogram; i.e., b_{uv}^k may be zero when $b_{uu}^k > 0$ or $b_{vv}^k > 0$.

Parameters of the linear model of coregionalization with the above constraints can be fitted by iteration. One usually obtains the distance parameters first by fitting models independently to the experimental variograms and choosing good compromise values from these. Then with the distance parameter fixed, the values of the b_{uv}^k are found to minimize the sums of the squares of the residuals, subject to the condition that the solution guarantees nonnegative variances, i.e., is CNSD. One can check the validity of the resulting model by plotting it on a graph of the experimental cross-semivariances plus the limiting values that would hold if the correlation between the variables were perfect. These limits constitute the hull of perfect correlation, which is obtained from the coefficients b_{uu}^k and b_{vv}^k by

$$\mathrm{hull}\left[\gamma(\mathbf{h}) \right] = \pm \sum_{k=1}^{K} \sqrt{b_{uu}^k b_{vv}^k} \, g^k(\mathbf{h})$$

The line should fit close to the experimental values for the model. It must also fall within the hull to be acceptable. If it lies close to the hull, then the cross-correlation is strong; if in contrast it is far from the bounds, then the cross-correlation is weak. This is illustrated below in the case study by the coregionalization between the soil's apparent electrical conductivity, EC_a, and topsoil sand, $sand_0$ (Figure 9.24).

9.6 KRIGING: IMPLEMENTING THE GEOSTATISTICAL MODEL

Although the emphasis in this contribution is on modeling the spatial variation in soil, we should remember that the force driving the development of geostatistics was economic. In Russia, meteorologists wanted to interpolate atmospheric variables from sparse recording stations; in South Africa, miners wanted to estimate the gold contents of ores locally from drill cores; elsewhere, petroleum engineers wished to estimate oil reserves from logged boreholes; and all wanted their estimates to be unbiased with minimum variance. Kolmogorov[12] worked out the theory for doing these in the 1930s, but without a computer, no one could implement it. The advent of computers gave mining and petroleum engineers the opportunity. This combination of theory and technology was also the breakthrough that soil scientists had sought; it enabled them to predict spatially without having to classify the soil first, with all the controversy and dissatisfaction that incurred.

Spatial prediction with the above properties of unbiasedness and minimum variance is termed *kriging* by geostatisticians in recognition of the pioneering work by D.G. Krige in the South African goldfields.[13] It can take several forms. In the simplest cases, the estimator is a linear sum of data, a weighted average:

$$\hat{Z}(\mathbf{x}_0) = \sum_{i=1}^{N} \lambda_i z(\mathbf{x}_i) \tag{9.37}$$

In this sum, $z(\mathbf{x}_1), z(\mathbf{x}_2), ..., z(\mathbf{x}_N)$ are the measured values of z at places $\mathbf{x}_1, \mathbf{x}_2, ..., \mathbf{x}_N$, and the λ_i are the weights. The weights sum to 1 to ensure unbiasedness,

$$\sum_{i=1}^{N} \lambda_i = 1 \tag{9.38}$$

and the expected error is $E\left[\hat{Z}(\mathbf{x}_0) - Z(\mathbf{x}_0)\right] = 0$. The prediction variance is

$$\begin{aligned}
\mathrm{var}\left[\hat{Z}(\mathbf{x}_0)\right] &= E\left[\left\{\hat{Z}(\mathbf{x}_0) - Z(\mathbf{x}_0)\right\}^2\right] \\
&= 2\sum_{i=1}^{N} \lambda_i \gamma(\mathbf{x}_i, \mathbf{x}_0) - \sum_{i=1}^{N}\sum_{j=1}^{N} \lambda_i \lambda_j \gamma(\mathbf{x}_i, \mathbf{x}_j)
\end{aligned} \tag{9.39}$$

where $\gamma(\mathbf{x}_i, \mathbf{x}_j)$ is the semivariance of Z between sampling points $\mathbf{x}_i$ and $\mathbf{x}_j$, and $\gamma(\mathbf{x}_i, \mathbf{x}_0)$ is the semivariance between the ith sampling point and the target point $\mathbf{x}_0$.

More generally, we may want to predict the average value of Z in a block B that is larger than the support of the data. The ordinary kriged estimate is still a weighted average of the data,

$$\hat{Z}(B) = \sum_{i=1}^{N} \lambda_i z(\mathbf{x}_i) \tag{9.40}$$

but now with variance

$$\mathrm{var}\left[\hat{Z}(B)\right] = \mathrm{E}\left[\left\{\hat{Z}(B) - Z(B)\right\}^2\right]$$

$$= 2\sum_{i=1}^{N} \lambda_i \overline{\gamma}(\mathbf{x}_i, B) - \sum_{i=1}^{N}\sum_{j=1}^{N} \lambda_i \lambda_j \gamma(\mathbf{x}_i, \mathbf{x}_j) - \overline{\gamma}(B, B) \tag{9.41}$$

The quantity $\overline{\gamma}(\mathbf{x}_i, B)$ is the average semivariance between the ith sampling point and the target block B, and $\overline{\gamma}(B, B)$ is the average semivariance within B, the within-block variance.

Equation 9.39 for a target point leads to a set of $N+1$ equations in $N+1$ unknowns:

$$\sum_{i=1}^{N} \lambda_i \gamma(\mathbf{x}_i, \mathbf{x}_j) + \psi(\mathbf{x}_0) = \gamma(\mathbf{x}_j, \mathbf{x}_0) \quad \text{for all } j$$

$$\sum_{i=1}^{N} \lambda_i = 1 \tag{9.42}$$

This is the ordinary punctual kriging system, in which the quantity $\psi(\mathbf{x}_0)$ is a Lagrange multiplier. The system can be represented in matrix form as follows:

$$\begin{bmatrix} \gamma(\mathbf{x}_1,\mathbf{x}_1) & \gamma(\mathbf{x}_1,\mathbf{x}_2) & \cdots & \gamma(\mathbf{x}_1,\mathbf{x}_N) & 1 \\ \gamma(\mathbf{x}_2,\mathbf{x}_1) & \gamma(\mathbf{x}_2,\mathbf{x}_2) & \cdots & \gamma(\mathbf{x}_2,\mathbf{x}_N) & 1 \\ \cdot & \cdot & \cdots & \cdot & \cdot \\ \cdot & \cdot & \cdots & \cdot & \cdot \\ \cdot & \cdot & \cdots & \cdot & \cdot \\ \gamma(\mathbf{x}_N,\mathbf{x}_1) & \gamma(\mathbf{x}_N,\mathbf{x}_2) & \cdots & \gamma(\mathbf{x}_N,\mathbf{x}_N) & 1 \\ 1 & 1 & \cdots & 1 & 0 \end{bmatrix} \begin{bmatrix} \lambda_1 \\ \lambda_2 \\ \cdot \\ \cdot \\ \cdot \\ \lambda_N \\ \psi(\mathbf{x}_0) \end{bmatrix} = \begin{bmatrix} \gamma(\mathbf{x}_1,\mathbf{x}_0) \\ \gamma(\mathbf{x}_2,\mathbf{x}_0) \\ \cdot \\ \cdot \\ \cdot \\ \gamma(\mathbf{x}_N,\mathbf{x}_0) \\ 1 \end{bmatrix} \tag{9.43}$$

In matrix notation it is simply

$$\mathbf{A}\boldsymbol{\lambda} = \mathbf{b} \tag{9.44}$$

where $\mathbf{A}$ is the augmented matrix of semivariances among the data points on the left-hand side of Equation 9.43, $\mathbf{b}$ is the augmented vector of semivariances between the data points and the target, and $\boldsymbol{\lambda}$ is the vector of weights and the Lagrange multiplier. One solves the equation by inverting matrix $\mathbf{A}$ and postmultiplying the inverse by $\mathbf{b}$ to obtain the kriging weights:

$$\boldsymbol{\lambda} = \mathbf{A}^{-1}\mathbf{b} \qquad (9.45)$$

The weights, λ_i, are then inserted into Equation 9.37 to give the estimate of Z at $\mathbf{x}_0$. The estimation variance (prediction variance or specifically kriging variance) is

$$\sigma^2(\mathbf{x}_0) = \sum_{i=1}^{N} \lambda_i \gamma(\mathbf{x}_i, \mathbf{x}_0) + \psi(\mathbf{x}_0) \qquad (9.46)$$

and it is obtained from the matrix equation as

$$\hat{\sigma}^2(\mathbf{x}_0) = \mathbf{b}^{\mathrm{T}}\boldsymbol{\lambda} \qquad (9.47)$$

If a target point, $\mathbf{x}_0$, happens to be one of the data points, say $\mathbf{x}_j$, then $\sigma^2(\mathbf{x}_0)$ is minimized when $\lambda(\mathbf{x}_j) = 1$ and all of the other weights are 0. In fact, $\sigma^2(\mathbf{x}_0) = 0$, and by inserting the weights into Equation 9.37, we obtain the recorded value, $z(\mathbf{x}_j)$, as our estimate of $z(\mathbf{x}_0)$. Punctual kriging is thus an exact interpolator.

The equivalent kriging system for blocks is

$$\sum_{i=1}^{N} \lambda_i \gamma(\mathbf{x}_i, \mathbf{x}_j) + \psi(B) = \bar{\gamma}(\mathbf{x}_j, B) \quad \text{for all } j$$

$$\sum_{i=1}^{N} \lambda_i = 1 \qquad (9.48)$$

with the associated variance obtained as

$$\sigma^2(B) = \sum_{i=1}^{N} \lambda_i \bar{\gamma}(\mathbf{x}_i, B) + \psi(B) - \bar{\gamma}(B, B) \qquad (9.49)$$

The only differences between it and the punctual kriging system is in the right-hand side vector, which is now

$$
\mathbf{b} =
\begin{bmatrix}
\overline{\gamma}(\mathbf{x}_1, B) \\
\overline{\gamma}(\mathbf{x}_2, B) \\
\cdot \\
\cdot \\
\cdot \\
\overline{\gamma}(\mathbf{x}_N, B) \\
1
\end{bmatrix}
$$

and the additional within-block variance appears in the expression for the kriging variance:

$$
\hat{\sigma}^2(B) = \mathbf{b}^T \lambda - \overline{\gamma}(B, B) \tag{9.50}
$$

Long-range trend need not complicate the analysis because the kriging is done in fairly small neighborhoods centered on each $\mathbf{x}_0$ or B in turn. However, if one wants to adhere to the random model, then one can first remove the trend, compute and model the variogram of the residuals from the trend, krige with that model and the residuals, and finally add the trend back to the kriged estimates. This combination is illustrated in the case study for EC_a, with results in Figure 9.19 and Figure 9.20. The estimates will be unbiased, but the kriging variances will be for the residuals and not the estimation variances. More advanced techniques are needed to obtain them and are beyond the scope of this chapter.

9.6.1 WEIGHTS

When the kriging equations are solved to obtain the weights, λ_i, in general, the only large weights are those of the points near to the point or block to be kriged. The nearest 4 or 5 might contribute 80% of the total weight, and the next nearest 10 almost all of the remainder. The weights also depend on the configuration of the sampling. We can summarize the factors affecting the weights as follows:

1. Near points carry more weight than more distant ones. Their relative proportions depend on the positions of the sampling points and on the variogram: the larger is the proportion of the nugget variance to the total variance, the smaller are the weights of the points nearest to target point or block.
2. The relative weights of points also depend on the block size: as the block size increases, the weights of the nearest points decrease and

those of the more distant points increase, until the weights become nearly equal.

3. Clustered points carry less weight individually than isolated ones at the same distance.
4. Data points can be screened by ones lying between them and the target.

These effects are all intuitively desirable, and the first shows that kriging is local. They also have practical implications, perhaps the most important of which is that because only the nearest few data points to the target carry significant weight, matrix $\mathbf{A}$ in the kriging system need never be large and its inversion will be swift. We can replace N in Equations 9.42 and 9.48 by some much smaller number, say $n \ll N$. Typically $n = 20$ is enough.

9.7 FACTORIAL KRIGING ANALYSIS

We mentioned above that some experimental variograms are best fitted by a combination of two or more simple models. Figure 9.17a, for example, of yield, shows a variogram fitted by two isotropic spherical functions plus a nugget. The full model is

$$\gamma(h) = c_0 + c_1 \left\{ \frac{3h}{2a_1} - \frac{1}{2}\left(\frac{h}{a_1}\right) \right\} + c_2 \left\{ \frac{3h}{2a_2} - \frac{1}{2}\left(\frac{h}{a_2}\right) \right\} \quad \text{for} \quad 0 < h \le a_1$$

$$= c_0 + c_1 + c_2 \left\{ \frac{3h}{2a_2} - \frac{1}{2}\left(\frac{h}{a_2}\right) \right\} \quad \text{for} \quad a_1 < h \le a_2 \qquad (9.51)$$

$$= c_0 + c_1 + c_2 \quad \text{for} \quad h > a_2$$

$$= 0 \quad \text{for} \quad h = 0$$

where c_1 and c_2 are the *a priori* variances of the two spatially dependent components of the models, a_1 and a_2 are the corresponding ranges, and c_0 is the nugget variance.

The above equation effectively describes the variance of a random process, of range a_1, nested within another of longer range, a_2, plus the variance of a spatially uncorrelated process, the nugget variance. It is an example of a general situation in which the variogram of $Z(\mathbf{x})$ is a combination of, say, S individual variograms; thus,

$$\gamma(\mathbf{h}) = \gamma^1(\mathbf{h}) + \gamma^2(\mathbf{h}) + \cdots + \gamma^S(\mathbf{h}) \qquad (9.52)$$

where the superscripts refer to the component variograms. If we assume that the processes represented by these components are uncorrelated, then we can write Equation 9.52 as

$$\gamma(\mathbf{h}) = \sum_{k=1}^{S} b^k g^k(\mathbf{h}) \tag{9.53}$$

where $g^k(\mathbf{h})$ is the kth basic variogram function and b^k is a coefficient that measures the relative contribution of the variance of $g^k(\mathbf{h})$ to the sum.

The components on the right-hand side of Equation 9.53 correspond to S random functions that in sum form $Z(\mathbf{x})$; thus,

$$Z(\mathbf{x}) = \sum_{k=1}^{S} Z^k(\mathbf{x}) + \mu \tag{9.54}$$

in which μ is the mean of $Z(\mathbf{x})$. Each $Z^k(\mathbf{x})$ has expectation zero, and the squared differences are

$$\frac{1}{2}\mathrm{E}\left[\left\{Z^k(\mathbf{x}) - Z^k(\mathbf{x}+\mathbf{h})\right\}\left\{Z^{k'}(\mathbf{x}) - Z^{k'}(\mathbf{x}+\mathbf{h})\right\}\right] = b^k g^k(\mathbf{h}) \quad \text{if } k = k'$$

$$= 0 \text{ otherwise} \tag{9.55}$$

The last component, $Z^S(\mathbf{x})$, may be intrinsic only, so that $g^S(\mathbf{h})$ in Equation 9.53 is unbounded with a gradient b^S.

Equation 9.55 expresses the mutual independence of the S random functions in Equation 9.54. Together they define the linear model of regionalization, which we may see as a model of the real world of the soil, in which factors such as rock type, tree-throw, burrowing animals, and farmers' divisions of land into fields each operates on its own characteristic scale and with its particular form, independent of the others.

The model also enables the values of the contributing processes to be estimated separately by factorial kriging analysis, a technique devised by Matheron.[14] Each spatial component $Z^k(\mathbf{x})$ is estimated as a linear combination of the observations, $z(\mathbf{x}_i), i = 1, 2, ..., N$:

$$\hat{Z}^k(\mathbf{x}_0) = \sum_{i=1}^{N} \lambda_i^k z(\mathbf{x}_i) \tag{9.56}$$

As in ordinary kriging, N is usually replaced by much smaller n points near to $\mathbf{x}_0$.

The λ_i^k are weights assigned to the observations, but now they sum to 0, not to 1, to accord with Equation 9.54. Subject to this condition, they are chosen to minimize the kriging variance, and this leads to the kriging system

$$\sum_{j=1}^{n} \lambda_j^k \gamma(\mathbf{x}_i, \mathbf{x}_j) - \psi^k(\mathbf{x}_0) = b^k g^k(\mathbf{x}_i, \mathbf{x}_0) \text{ for all } i = 1, 2, \dots, n$$

(9.57)

$$\sum_{j=1}^{n} \lambda_j^k = 0$$

The quantity $\psi^k(\mathbf{x}_0)$ is the Lagrange multiplier.

The above system of equations is solved for each spatial component, k, to find the weights λ_j^k, which are then inserted into Equation 9.56 to estimate that component. Figure 9.18a and b shows the long- and short-range components of yield mapped separately, based on the respective distance parameters, a_2 and a_1, of the fitted variogram function.

We need to krige the local mean at $\mathbf{x}_0$, which is again a linear combination of the data:

$$\hat{\mu}(\mathbf{x}_0) = \sum_{j=1}^{n} \lambda_j^{\text{mean}} z(\mathbf{x}_j)$$

(9.58)

The weights for this are obtained from the following system:

$$\sum_{j=1}^{n} \lambda_j^{\text{mean}} \gamma(\mathbf{x}_i, \mathbf{x}_j) - \psi^{\text{mean}}(\mathbf{x}_0) = b^k g^k(\mathbf{x}_i, \mathbf{x}_0) \text{ for all } i = 1, 2, \dots, n$$

(9.59)

$$\sum_{j=1}^{n} \lambda_j^{\text{mean}} = 1$$

We again have a Lagrange multipler, $\psi^{\text{mean}}(\mathbf{x}_0)$.

Estimating the long-range component can be affected by the size of the moving neighborhood — see Galli et al.[15] To estimate a spatial component with a given range, the distance across the neighborhood should be at least equal to that range. However, it usually happens that sampling must be dense for a short-range component to be distinguished in the variogram. When the data are subsequently used for kriging, only a small proportion of them are retained for a kriging system, and those are all near to the target. Although we could make n large, and even include all N data in modern computers, the inversion of such large matrices can be unstable and is not recommended. Further, only the nearest few data to the target would contribute to the estimate because they would screen the more distant data; it would mean kriging in an effective neighborhood smaller than the neighborhood specified, so that the range of the component estimated would be smaller than the range determined from the structural analysis. Galli et al.[15] suggested a way of overcoming the shortcoming by selecting only a proportion of the data within the specified neighborhoods. Such a selection is arbitrary, and Jaquet[16] proposed what seems a better alternative. His proposal involves adding to the estimated long-range component the estimate of the local mean; this is the solution that we have adopted — see Oliver et al.[17] and below.

9.8 CASE STUDY: SOIL VARIATION AT YATTENDON

We illustrate the most important aspects of the theory with results from a recent case study. No description of the study has been published previously, and the results are new. Therefore, we devote a section to the background, the sample surveys planned to obtain data at a suitable intensity, and the exploration of those data before analyzing them formally.

The study was part of a project on precision farming for the British Home Grown Cereals Authority.[18] Its aim was to explore factors in the soil and in the environment more generally that might cause variation in cereal yield within individual fields. The particular field, National Grid Reference SU 458174, covers 23 ha on the Yattendon Estate. It is part of the Chalk hill country of southern England and has the typical undulating topography of this region. The soil is a Luvisol, its texture varies from sandy loam to clay loam, and it is moderately to well drained. The parent material is the Reading Beds, comprising sediments of varied particle size of Tertiary age, which rest on the Upper Chalk (soft limestone of Cretaceous age).

Several surveys were done during 2002; they included observations of the following:

1. Attributes of the topsoil (0 to 15 cm) and subsoil (30 to 60 cm); Table 9.2 lists the ones analyzed here.
2. Electromagnetic induction (EMI) in which the apparent electrical conductivity (EC_a) of the soil was measured at field capacity.
3. Attributes of the growing crop (winter wheat).

Yield data from previous crops of winter wheat (1995, 1997, 1999, and 2001) were used to guide the initial soil sampling. The variograms of yield were all bounded and could be modeled as combinations of two spatial components, one with an average range of about 40 m and the other of about 100 m. Based on this information, we sampled as follows: at the nodes of a 30 m × 30 m grid, with additional observations at 15-m intervals along short transects from randomly selected grid nodes. In this way, we ensured that the variation apparent in the yield data would be represented adequately and efficiently. In addition, we aimed to have at least 100 sampling sites to ensure that our variograms would be reliable.[19]

From the full set of data we have selected the variables listed in Table 9.2 for illustration.

TABLE 9.2
Summary Statistics of Properties Recorded at Yattendon

Variable	Number of Sites	Mean	Minimum	Maximum	Variance	Standard Deviation	Skewness
EC_a/mS m^{-1}	3275	21.1	6.5	82.5	9.354	87.5	2.25
$Log_{10}EC_a$	3275	1.290	0.813	1.916	0.166	0.028	0.43
Elevation/m	3555	92.2	84.0	113.0	35.05	5.92	0.03
Leaf area index	195	1.79	0.35	3.60	0.371	0.609	0.40
Mg_0/mg kg^{-1}	230	67.3	12.8	136.4	524.63	22.9	0.29
Mg_{30}/mg kg^{-1}	122	53.18	14.73	146.3	455.32	21.3	1.47
$Log_{10}Mg_{30}$	122	1.693	1.168	2.165	0.029	0.171	−0.32
pH_{30}	122	7.176	6.140	8.220	0.241	0.491	0.35
$Sand_0$/%	230	50.84	14.00	83.00	207.41	14.4	0.02
Yield/t ha^{-1}	5896	7.08	1.50	11.8	4.273	2.07	−0.52

9.9 EXPLORING THE DATA

Table 9.2 summarizes the data. The distributions of all but two of the variables are fairly symmetric; the two exceptions are the soil's apparent electrical conductivity, EC_a, and subsoil magnesium, Mg_{30}, for which the skewness coefficients exceed 1. These large coefficients were not caused by outliers, and we might have transformed the data to their common logarithms to make the distributions more nearly normal and stabilize the variances. This transformation would have diminished the skewness to 0.43 for EC_a and to −0.32 for Mg_{30}. In the event, we have done the analyses on the original data for ease of interpretation.

Experimental variograms were computed by Equation 9.12 in four directions to reveal any anisotropy in the variation. The results for topsoil magnesium, Mg_0, are shown in Figure 9.8 as an example for the directions 0°, 45°, 90°, and 135°. There is little divergence among the different directions, and we can treat the variation as isotropic. Likewise, there was no evidence of anisotropic variation in the other properties.

The shapes of the variograms of several properties suggest the presence of regional trend in the variation, for example, that of EC_a (Figure 9.20a). In these circumstances $\hat{\gamma}(\mathbf{h})$ computed from the data by Equation 9.12 does not estimate $\gamma(\mathbf{h})$ of the random residual. The experimental values follow a curve that is concave upwards, and this is the form that several computed variograms had. We return to the example below.

9.10 MODELING THE VARIOGRAM AND CROSS-VALIDATION

The experimental variogram may be computed to any desired lag, provided there are sufficient comparisons, and one must choose a suitable maximum lag to

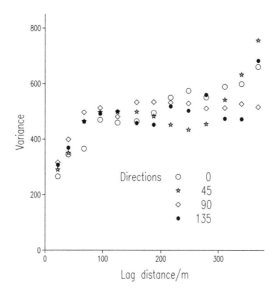

FIGURE 9.8 Experimental variogram of topsoil magnesium, Mg_0, computed in four dirctions, 0°, 45°, 90°, and 135°.

which to fit the model. We illustrate this with the variogram of Mg_0 computed to 400 m (Figure 9.9). The sample values lie close to a monotonic nondecreasing curve, compatible with second-order stationarity to about 300 m. Thereafter, the sample values increase markedly. A simple bounded model fitted to the whole set of values clearly cannot represent the upturn beyond 300 m. We could force a fit with a circular model (Figure 9.9), but it is a poor representation of the whole sequence.

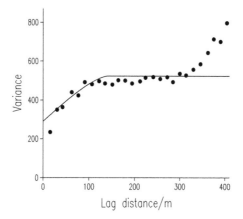

FIGURE 9.9 Experimental variogram of topsoil magnesium, Mg_0, computed to lag 400 m and the best-fitting circular function.

TABLE 9.3
Number of Pairs of Comparisons for Each Lag
Interval on the Experimental Variogram for
Topsoil Mg

Lag	Counts	Lag	Counts	Lag	Counts
15.7	88	164.4	1014	313.8	838
31.0	522	179.6	1354	330.1	1028
43.0	410	194.5	1412	345.6	749
62.9	1192	212.0	1403	362.1	727
77.2	345	224.0	987	375.3	493
91.5	1300	241.0	1673	390.3	582
107.0	780	256.3	979	404.4	388
122.5	1256	272.2	1343		
135.2	910	285.0	767		
151.3	1343	300.1	1112		

We know from experience that the semivariances at the longer lags become increasingly unreliable as the number of comparisons from which they are computed decreases. Table 9.3 shows that the number of paired comparisons starts to decrease at the distance at which the semivariances start to increase. If we want the fitted model for kriging, then usually the form of the variogram at the long lag distances is of no consequence.

By limiting the experimental variogram to a maximum lag distance of 300 m, we could fit several of the simple bounded models, as Figure 9.10 shows; Table 9.4 lists the parameters of four plausible models and the residual mean squares (RMSs). The RMS is least for the exponential function (Figure 9.10d), which provides the best fit in this sense. In addition, the table gives the percentage of the variance accounted for by each model; it is 93.7% for the exponential function. The function has a smaller nugget variance than the other models (Table 9.4 and

TABLE 9.4
Parameters of Models Fitted to the Experimental Variogram of
Topsoil Mg at Yattendon

Model Type	Variances		Distance Parameters			Percentage of
	c_0	c	a/m	r/m	RMS[a]	Variance
Circular	226.7	271.5	99.5		1.178	91.3
Spherical	215.6	282.8	110.0		1.145	91.6
Pentaspherical	205.3	293.5	130.0		1.107	91.9
Exponential	103.2	400.3		35.7	0.911	93.7

[a] RMS is the residual mean square.

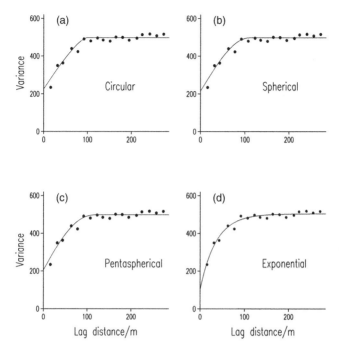

FIGURE 9.10 Experimental variogram of topsoil magnesium, Mg_0 and fitted models: (a) circular, (b) spherical, (c) pentaspherical, and (d) exponential, which fits best.

Figure 9.10d); it is approximately half. This has implications for kriging because the larger the nugget variance is, the greater is the smoothing of the predicted values. The distance parameter of the function is 35.7 m, giving an effective range of spatial dependence of 107 m, which is similar to the ranges of the three other models.

We mentioned above that the size of bins and the lag interval affect the form of the experimental variogram computed from irregularly scattered data. We show here how that carries through to the modeling. Figure 9.11a shows the experimental variogram of subsoil pH, pH_{30}, with a lag interval and bin width of 15 m and the best-fitting pentaspherical function, which accounted for 95.5% of the variance. Figure 9.11b shows the experimental variogram computed afresh with a lag interval and bin width of 30 m. In this instance, the exponential function fitted best, accounting for 94.3% of the variance. Notice that the experimental semivariances increase in a smoother progression with wider bins, but that the sill variance and effective range are much the same. The pentaspherical model has a substantially larger nugget variance, and one might wonder whether this would have an important effect on subsequent kriging. A comparison by cross-validation shows there to be little difference in the accuracy of the predictions made with them (Table 9.5). The mean error, mean squared error, and MSDR are very similar for the two models. The MSDR, mean-squared deviation ratio, is the mean of the squared errors divided by the kriging variances,[6] and a value of 1 indicates a well-chosen model for kriging. Provided the experimental variogram

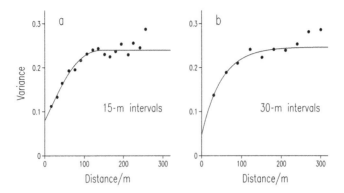

FIGURE 9.11 Experimental variogram of subsoil pH, pH_{30}. (a) Computed with bins 15 m wide and fitted with a pentaspherical function. (b) Computed with a bin width of 30 m and fitted with an exponential model.

TABLE 9.5
Results of Cross-Validation for Subsoil pH

Model Type	Mean Error	Mean Squared Error	Mean Squared Deviation Ratio
Exponential	–0.01449	0.1495	1.045
Pentaspherical	–0.01374	0.1492	1.006

is reliable (this one was computed from 122 values) and the model fits well, the initial lag interval selected does not appear to have much effect even though the narrower bins produced a more irregular variogram. Further, the similarity of the predictions is evident in the maps (Figure 9.15a and b).

We fitted the following functions to the variograms of other variables: circular to Mg_{30} (Figure 9.12a) and spherical to topsoil sand (Figure 9.12b); Table 9.6 lists their model parameters. The stable exponential function was fitted to the variogram of elevation computed on the residuals from a linear trend, which was fitted by standard least squares regression. Figure 9.22a shows it. The model approaches the origin in an upwardly concave way, but the exponent must be less than 2. It expresses smooth local variation that is not predictable, as in the case of local trend or drift. A power function was fitted to the residuals from a linear trend for leaf area index (LAI) (Figure 9.23a). This function is intrinsic only and expresses the increase in variance as the area studied increases.

The experimental variogram for yield (1995) was fitted by a double spherical function (Figure 9.17a). The nugget and short- and long-range components of the model are also shown separately. If the components are assumed to be independent, so that we have a linear model of regionalization, then their contributions can be estimated separately by factorial kriging analysis, as described

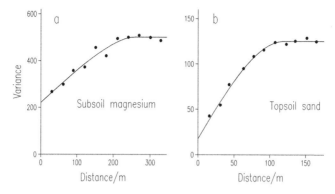

FIGURE 9.12 Experimental variograms and fitted models of (a) subsoil magnesium, Mg_{30} (circular), and (b) topsoil sand, $sand_0$ (spherical).

TABLE 9.6
Parameters of Models Fitted to Experimental Variograms at Yattendon

Variable	Model Type	Sills			Distance Parameters		
		c_0	c_1	c_2	a_1/m	a_2/m	r/m
		Bounded Variances					
EC_a^{res}	Exponential	9.15	41.1		26.9		
$Elevation^{res}$	Stable*	0	9.54		95.9		
Mg_0	Exponential	103.2	400.3				35.7
Mg_{30}	Circular	221.9	278.4		250.3		
pH_{30}	Pentaspherical	0.0796	0.160		154.4		
pH_{30}	Exponential	0.0466	0.200				49.70
$Sand_0$	Spherical	17.68	107.5		120.1		
Yield	Double spherical	0.6176	1.022	1.003	31.13	108.8	

		Unbounded Variances		
		c_0	Intensity (w)	Exponent (α)
Leaf area index	Power function	0	0.0410	0.389

Note: The superscript res means residuals from a fitted trend function.

* The exponent of the stable model is 1.965.

above. We show the results in Figure 9.18, after illustrating ordinary kriging at the end of the following section.

9.11   KRIGING

We have used the models fitted to the experimental variograms (Table 9.6) with the data to krige at unsampled places and have mapped the results. The examples illustrate the commonly used models, and the maps of the kriged predictions show the kinds of patterns that these models describe. Predictions were made by ordinary kriging over blocks of 10 m × 10 m at the nodes of a 5 m × 5 m grid.

Figure 9.10d shows the experimental variogram and the exponential model fitted to Mg_0. The parameters of the model were used to krige the values, which were then mapped (Figure 9.13a). The irregular shape and size of the patches of large and small values express the randomness of the exponential process in space; the structures have what appear to be random extents. The maximum distance across the field is about 600 m. The average extent of the patches is a measure of the approximate distance across the patches and of the extent of spatial dependence ($a' = 3r$) of about 100 m (Table 9.6).

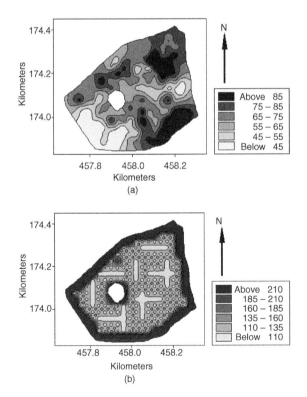

FIGURE 9.13 Block kriged map of (a) topsoil Mg and (b) its associated map of the kriging variances. (See color version on accompanying CD.)

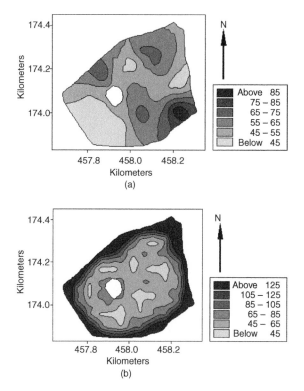

FIGURE 9.14 Maps of (a) kriged estimates of subsoil magnesium, Mg_{30}, and (b) the associated kriging variances. (See color version on accompanying CD.)

Figure 9.13b is the map of the kriging variances. The large variances around the field margins and the copse in the central part of the field show the edge effects, where there were fewer data from which to predict. The dark linear patches in the center are of the small kriging variances where the sampling was more intensive on the short transects.

Figure 9.12a shows the experimental variogram and the fitted circular model of Mg_{30}. The range of spatial variation for subsoil Mg is more than twice as long as that for the topsoil. This is expressed clearly in the kriged map (Figure 9.14a), where the spatial structures have a much larger extent than those for topsoil Mg. They are also more regular, though the two variables have similar variances. Figure 9.14b is the map of kriging variances — there were fewer parts of the field with more intensive sampling in the subsoil.

Figure 9.12b shows the experimental variogram and the fitted spherical model to topsoil sand content, $sand_0$. This has a range of spatial dependence similar to that of topsoil Mg. The kriged map of $sand_0$ (Figure 9.16) shows a similar pattern of variation to that of Mg_0, but with large values where the sand content is small. The spatial structures in the map of $sand_0$ are more regular than those for Mg_0; the patchiness in the variation is distinct.

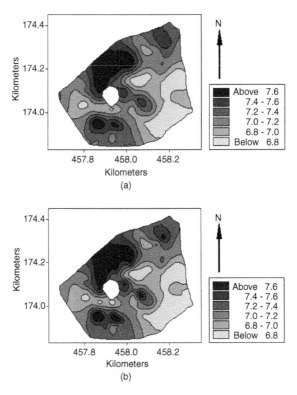

FIGURE 9.15 Subsoil pH, pH_{30}, map of block kriged predictions using (a) the pentaspherical model and (b) the exponential function. (See color version on accompanying CD.)

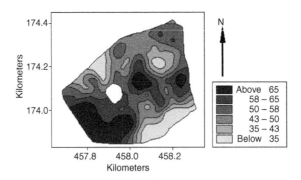

FIGURE 9.16 Block kriged map of topsoil sand. (See color version on accompanying CD.)

For properties that have two spatial scales of variation, factorial kriging can be used to explore them separately. Figure 9.17a shows the double spherical function fitted to the variogram of the yield for 1995. The function was used for ordinary kriging first; Figure 9.17b is the map of predictions. The pattern of variation appears complex because of the long- and short-range components of

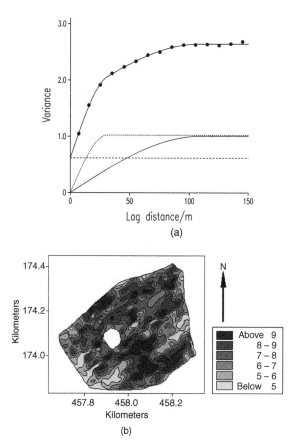

(a)

(b)

FIGURE 9.17 (a) Variogram of yield in 1995 fitted, with nugget and short- and long-range components shown separately. (b) Map of ordinary kriged predictions made with this model. (See color version on accompanying CD.)

variation present. These were then extracted separately and predicted by factorial kriging. The map of the long-range predictions (Figure 9.18a) is similar to that from ordinary kriging, except that it is less noisy. The regions with large and small yields are clear in both maps. In the southwest, north, and southeast of the field, there is an inverse relation between yield and topsoil sand content (Figure 9.16), but the relation changes in the northeast and central parts of the field. The map of the short-range prediction (Figure 9.18b) shows a much smaller scale of variation with a strong regular pattern. This component of the variation appears to relate to the lines of management in the field. These were in a northeast-to-southwest direction, and the larger values are probably in the zones between the tramlines, where the soil has suffered less compaction from machinery. There is also some evidence of variation perpendicular to these lines, perhaps the effects of operations in a different direction. These management effects were not evident in the map of ordinary kriged predictions.

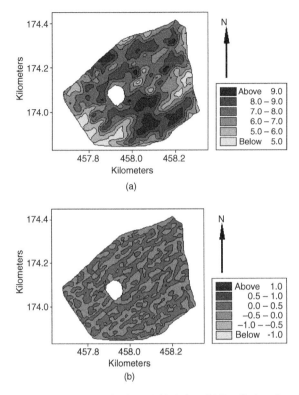

FIGURE 9.18 Maps of yield made by factorial kriging. (a) Predictions based on the long-range component of the variogram, a_2. (b) Predictions based on the short-range component of the variogram, a_1. (See color version on accompanying CD.)

9.11.1 EXAMPLES OF TREND

Figure 9.19a is a map of the raw values of EC_a. The linear feature with large EC_a crossing the field diagonally is where a pipeline was placed some 6 years previously. Otherwise, the values have a patchy distribution, and there is no clear indication of the presence of trend. The semivariances computed from these data increase monotonically following an approximately exponential form to a lag of about 125 m (Figure 9.20a), and thereafter increase more markedly. This behavior is symptomatic of trend, though not conclusive evidence of it. Practitioners should not assume that trend can be detected from the appearance of the variogram alone. They should realize that the estimated semivariances at long lags can be unreliable, as in Figure 9.9, which behaves similarly.

Linear and quadratic trend functions were fitted on the coordinates of EC_a; they accounted for 18 and 42%, respectively, of the variance. The latter percentage is strong evidence for a quadratic trend. The variogram of the residuals from the trend (Figure 9.20b) differs substantially from that of the raw data; it is now

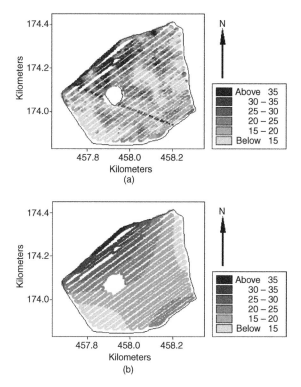

FIGURE 9.19 Pixel maps of apparent electrical conductivity of the soil, EC_a, of (a) raw values and (b) the quadratic trend. (See color version on accompanying CD.)

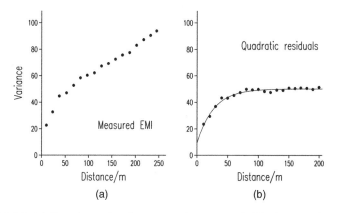

FIGURE 9.20 (a) Experimental variogram of the measured EC_a. (b) Experimental variogram of the residuals from the quadratic trend and the fitted model.

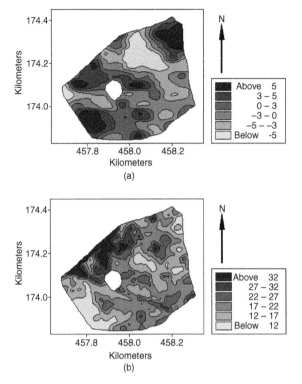

FIGURE 9.21 (a) Map of block kriged residuals of EC_a from the quadratic trend surface. (b) Map of block kriged predictions with the trend added back. (See color version on accompanying CD.)

bounded, and it can be fitted by an exponential function. Table 9.6 lists the parameters of the model.

The quadratic trend in EC_a is evident in Figure 9.19b, in which the residuals have been removed. It shows an upwardly convex surface with small values on either side of the central zone of large values. This surface relates to both the physiography of this field (Figure 9.22b) and the soil texture (Figure 9.16, the map of $sand_0$).

Kriging was done on the residuals from the quadratic trend with the parameters of the fitted exponential function (Table 9.6). Figure 9.21a shows the contour map of the kriged residuals. Unlike the map of the raw data, this map shows the patchy deviations from the trend, which is as we should expect from the bounded variogram.

To map EC_a on its original scale of measurement, we added the quadratic trend surface to the predicted residuals. Figure 9.21b shows the interpolated map of EC_a. The large values are on the plateau, where the soil contains more clay. The large values associated with the pipeline are still evident. The pattern of variation has remained patchy, but it no longer resembles that of the map of the residuals.

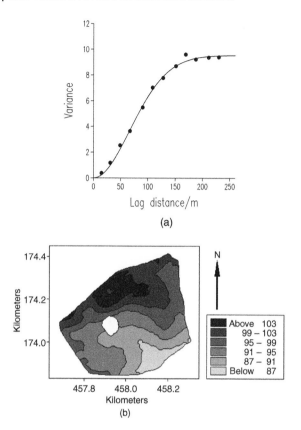

FIGURE 9.22 Elevation. (a) Experimental variogram and fitted stable model with exponent 1.965 for the residuals from a linear trend. (b) Map of kriged predictions made with this model with the trend added back. (See color version on accompanying CD.)

The same approach was applied to predict elevation. More than 68% of its variation was accounted for by a linear trend surface. Figure 9.22b shows the kriged map of elevation; the residuals were kriged and the linear trend was added back to the predictions. The map shows the smoothly undulating topography of this field. It also shows some similarity with the quadratic trend surface fitted to the EC_a data, which suggests that the trend in those data might be linked to the field's physiography.

For leaf area index (LAI) more than 36% of its variation was accounted for by a linear trend surface. The map of kriged predictions (with the trend added back) shows that the variation is not patchy — there is continuous change from large to small values across the field (Figure 9.23b). This is what we should expect from an unbounded model; it shows that the full extent of the variation has not been encompassed and that the variance would continue to increase if we were to increase the extent of our study.

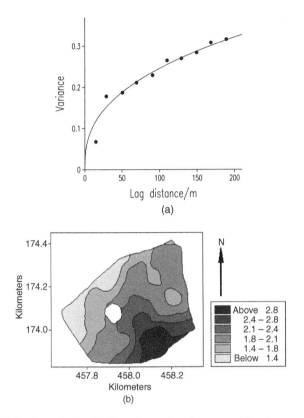

FIGURE 9.23 Leaf area index. (a) Experimental variogram and fitted power function for the residuals from a linear trend. (b) Map of kriged predictions made with this model with the trend added back. (See color version on accompanying CD.)

9.12 COREGIONALIZATION

Finally, we use the data from the case study to illustrate coregionalization. The electrical conductivity, EC_a, and the sand content of the topsoil, $sand_0$, are inversely correlated; the Pearson correlation coefficient is -0.67. We analyzed the data to see whether this correlation extends spatially in that the geographic distributions are similar. We computed the experimental cross-variogram of the two properties with Equation 9.33, in addition to the two autovariograms. We then fitted the linear model of coregionalization (Equation 9.34) to them. Figure 9.24 shows the three variograms with the fitted spherical model with the common range 96.6 m. The sill values are listed in Table 9.7. In the graph of the cross-variogram we have added the hull of perfect correlation. The cross-variogram lies about halfway between the zero of the variance scale and the hull, showing that the moderate correlation represented in the Pearson coefficient extends into the spatial domain.

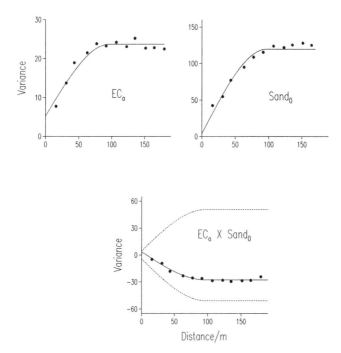

FIGURE 9.24 Model of coregionalization with autovariograms of the quadratic residuals of EC_a and topsoil sand, and the cross-variogram and fitted model; the dashed lines in the bottom graph form the hull of perfect correlation.

TABLE 9.7
Model Parameters of the Auto- and Cross-Variograms for the Coregionalization of EC_a^{res} and Topsoil Sand

Variable	c_0	c
EC_a^{res}	5.253	18.46
$EC_a^{res} \times Sand_0$	3.550	−31.21
$Sand_0$	3.801	115.8

Note: The basic variogram is spherical with a range of 96.6 m.

9.13 CONCLUSION

Geostatistics now has a huge repertoire, and a single chapter cannot cover everything. Here we concentrate on the basic linear techniques and their underlying theory to enable a practitioner to get started. Our recent investigation at Yattendon illustrates the techniques in action. It also shows that those techniques cannot be applied automatically, but rather, a situation needs a lot of thought, an exploration of the data, and understanding. Only then should the data be analyzed formally for local estimation and mapping.

ACKNOWLEDGMENTS

The results described in the case study were from a project funded by the U.K.'s Home Grown Cereal Authority (HGCA Project 2298), and we thank it for its support. We also thank Dr. Z.L. Carroll and Dr. S.J. Baxter for their work on this project and the resulting data analysis.

REFERENCES

1. R Webster, HE de la Cuanalo. Soil transect correlograms of north Oxfordshire and their interpretation. *J Soil Sci* 26: 176–194, 1975.
2. RK Morse, TH Thornburn. Reliability of soil units. In *Proceedings of the 5th International Conference on Soil Mechanics and Foundation Engineering*, Vol. 1. Dunod, Paris, 1961, pp. 259–262.
3. BA Kantey, AAB Williams. The use of soil engineering maps for road projects. *Trans South Afr Inst Civ Eng* 4: 149–159, 1962.
4. R Webster. Is soil variation random? *Geoderma* 97: 149–163, 2000.
5. G Matheron. *Les variables régionalisées et leur estimation*. Masson, Paris, 1965.
6. R Webster, MA Oliver. *Geostatistics for Environmental Scientists*. John Wiley & Sons, Chichester, U.K., 2001.
7. B Matérn. Spatial variation: stochastic models and their applications to problems in forest surveys and other sampling investigations. *Meddelanden från Statens Skogforskningsinstitut* 49: 1–144, 1960.
8. P Whittle. On stationary processes in the plane. *Biometrika* 41: 434–449, 1954.
9. RA Olea. *Optimum Mapping Techniques Using Regionalized Variable Theory*, Series in Spatial Analysis 2. Kansas Geological Survey, Lawrence, 1975.
10. H Akaike. Information theory and an extension of maximum likelihood principle. In *Second International Symposium on Information Theory*, BN Petrov, F Csáki, Eds. Akadémia Kiadó, Budapest, 1973, pp. 267–281.
11. N Cressie. Fitting variogram models by weighted least squares. *Math Geol* 17: 563–586, 1985.
12. AN Kolmogorov. Sur l'interpolation et l'extrapolation des suites stationaires. *C R Acad Sci Paris* 208: 2043–2045, 1939.
13. DG Krige. Two-dimensional weighted moving average trend surfaces for ore evaluation. *J South Afr Inst Mining Metall* 66: 13–38, 1966.
14. G Matheron. *Pour une analyse krigeante de données régionalisées*, Note N-732 du Centre de Géostatistique. Ecole des Mines de Paris, Fontainebleau, 1982.
15. A Galli, F Gerdil-Neuillet, C Dadou. Factorial kriging analysis: a substitute to spectral analysis of magnetic data. In *Geostatistics for Natural Resources Characterization*, G Verly, M David, A Journel, A Marechal, Eds. D. Reidel, Dordrecht, Netherlands, 1984, pp. 543–557.

16. O Jaquet. Factorial kriging analysis applied to geological data from petroleum exploration. *Math Geol* 21: 683–691, 1989.
17. MA Oliver, R Webster, K Slocum. Filtering SPOT imagery by kriging analysis. *Int J Remote Sensing* 21: 735–752, 2000.
18. MA Oliver, ZL Carroll. *Description of Spatial Variation in Soil to Optimize Cereal Management*, project report 330. London, HGCA, 2004.
19. R Webster, MA Oliver. Sample adequately to estimate variograms of soil properties. *J Soil Sci* 43: 177–192, 1992.

10 Stochastic Simulation of Soil Variations

Jean-Paul Chilès and Denis Allard

CONTENTS

ABSTRACT

Conditional simulations are a means for representing the vertical and lateral variations of soil properties in a stochastic framework. Used as input to a Monte-Carlo approach, they also allow the prediction of nonlinear functionals such as the exceedance of a concentration threshold or a breakthrough time. Several methods can be used to simulate variables with a Gaussian distribution. They can be generalized to simulate other continuous variables that can be considered transforms of Gaussian variables. Varied models and methods are used to simulate indicator or categorical variables such as soil type: sequential indicator simulation, truncated plurigaussian simulation, Boolean model, etc. The basic methods can be extended to more general situations: simulation of block values, multivariate case, and nonstationarity. The general presentation of the methods is followed by an illustration of conditional simulations to model the variability of the water content and mineral nitrogen of a soil, as part of a precision agriculture project.

10.1 INTRODUCTION AND DEFINITIONS

Soil properties vary vertically and laterally in a complex fashion, so that a statistical approach is often used to describe them.[1] Geostatistics proposes tools for characterizing that spatial variability and deriving optimal linear interpolators and predictors from scattered data, as shown in Chapter 9. Here we present methods for visualizing that spatial variability and predicting nonlinear functionals of the variables under study by Monte-Carlo techniques.

In this chapter we consider a variable z depending on a location with coordinate vector $\mathbf{x}$ in the two- or three-dimensional space or, in practice, in a bounded domain of that space. Typical variables are variables with a continuous distribution, such as soil thickness, yield, nutrient content, metal concentration, or salinity, and categorical variables such as soil type. To make its dependence on location explicit, we call it a regionalized variable and denote it as $z(\mathbf{x})$. This function is known at a finite number of data points, denoted as $\mathbf{x}_{\alpha}$, $\alpha = 1, \ldots, N$, and we are interested in z at other locations or in quantities that are functions of z. We consider situations where the spatial variability of the variable is too complex to be described analytically, so that probabilistic models are more appropriate. The regionalized variable $z(\cdot)$ is thus considered a realization of some random function

$Z(\cdot)$, and predictors developed for Z are applied to z, as shown in Chapter 9. Kriging is an easy answer to the interpolation or prediction of a series of quantities: value at an unsampled location, average value in a given block or domain, gradient at a given point, and, more generally, any linear functional of $Z(\cdot)$. An easy answer because kriging considers linear predictors and thus only requires knowledge of second-order moments of $Z(\cdot)$ — in practice, in the framework of a global or local stationary model, the covariance function, or the variogram — which is not very demanding. Moreover, the associated kriging variance provides a measure of the magnitude of the prediction error.

But problems do exist that cannot be expressed by linear functionals of Z. A simple one is the determination of the probability that $Z(\mathbf{x}_0)$ exceeds a given threshold s. The answer requires more information about the random function Z than its sole variogram. A solution is to notice that the problem amounts to the prediction of the indicator function $I(\mathbf{x}) = 1_{Z(\mathbf{x})>s}$, and to work with this indicator function rather than with $Z(\mathbf{x})$. But $I(\mathbf{x})$ carries much less information than $Z(\mathbf{x})$. For example, if we have a sample point $\mathbf{x}_1$ close to $\mathbf{x}_0$, we expect a large probability for $Z(\mathbf{x}_0)$ to exceed s if $Z(\mathbf{x}_1)$ is much larger than s, and a medium probability if $Z(\mathbf{x}_1)$ is only slightly larger than s. When working with the indicator $I(\mathbf{x})$, we will make no distinction between these two situations, and hence obtain a less accurate answer.

We can have suitable answers to the question of exceedance of a threshold in specific situations, such as for Gaussian random functions with known mean. In such a case, simple kriging (see Section 10.2.1) coincides with the conditional expectation and the kriging error is a Gaussian random variable, so that we have

$$\Pr\left\{Z_0 > s \mid Z(\mathbf{x}_1) = z_1, \ldots, Z(\mathbf{x}_N) = z_N\right\} = \Pr\left\{Z_0 - z^* > s - z^* \mid z^*\right\}$$

$$= 1 - G\left(\frac{s - z^*}{\sigma_K}\right)$$

where Z_0 stands for either an unknown point value $Z(\mathbf{x}_0)$ or a weighted average of $Z(\cdot)$, Z^* is its simple kriging estimator, z^* its value conditional on the data values, σ_K^2 the corresponding kriging variance, and G the standard normal cumulative distribution function. A more general answer, at least in the stationary case, is given by disjunctive kriging, which only requires knowledge of the bivariate distributions of pairs of values.[2]

A more complex problem is to know whether or not there is some connectivity between zones with high values, which will, for example, define flow paths if Z represents hydraulic conductivity — or on the contrary, if there are low conductivity flow barriers. Such problems are highly nonlinear and cannot be solved by disjunctive kriging. In such situations, conditional simulations come into play. They are the tool of choice to evaluate the impact of spatial uncertainty on the results of complex procedures, such as numerical modeling of a dynamic system or economic optimization of the development of a natural resource. The first application domain was the design of mining exploitation scenarios.[3] They are now increasingly used in soil science applications, for example, to analyze the land use suitability for pasture,[4] to rationalize the classification of panels of an industrial site as polluted or not,[5] to evaluate the soil remediation cost,[6,7] or to estimate the spatial uncertainty of topsoil texture.[8]

Another use of conditional simulations is to obtain realistic pictures of spatial variability. In fact, the structure of a kriged map reflects both the structure of the phenomenon and the density of the data: in areas with numerous data, the kriged map gives a quite good representation of reality (the kriging variance is low); in the presence of scarce data only, the kriged map interpolates smoothly between the few data points and does not reflect the small-scale variations of reality (the kriging variance is large).

10.1.1 Definition of Conditional Simulations

What exactly is a (conditional) simulation? To explain it, we have to go back to the above sentence: $z(\cdot)$ is considered a realization of some random function $Z(\cdot)$. Indeed, a random function can be considered a collection of possible outcomes, named realizations. To make the dependency on the realization apparent, we can represent the random function as a function $Z(\mathbf{x}, \omega)$, where ω represents the outcome. Reality is considered a particular outcome for some value ω_0 of ω: $z(\mathbf{x}) = Z(\mathbf{x}, \omega_0)$. A (nonconditional) simulation of the random function $Z(\cdot)$ is simply a realization $Z(\mathbf{x}, \omega)$ for a value ω randomly selected in the set Ω of all possible outcomes.

Usually the random function Z has an infinite number of realizations. Among them, some assume at the sample points the same values as those observed, and thus can be considered to better represent the regionalized variable $z(\mathbf{x})$. They will be called conditional simulations. A conditional simulation is therefore a realization randomly selected from the subset of realizations that match the sample points. Equivalently, it is a realization of a random function with a conditional spatial distribution.

10.1.2 Importance of the Spatial Distribution

The counterpart of this very powerful technique is that it requires full knowledge of the spatial distribution of the random function, i.e., of all finite-dimensional distributions, information that we cannot fully characterize from a limited dataset. Indeed, we usually have a single realization and measurements at scattered sample points $\{\mathbf{x}_\alpha: \alpha = 1, ..., N\}$. We assume stationarity so that statistical characteristics can be obtained by considering similar point configurations (e.g., all pairs of points with a given separation vector to get a sample variogram value). If bivariate distributions can be obtained in that way, multidimensional distributions attached to n-points configurations, $n > 2$, can hardly be inferred from the data. An exception is the situation of image analysis, where the complete image is available and allows, to some degree, an inference of the multivariate distributions. We will thus have to choose a random function model from limited statistics. Fortunately, we usually have some guidance from the physics of the phenomenon under study, and geostatistical data analysis tools facilitate this choice and enable its validation.

To stress the importance of a correct choice of the random function model, Figure 10.1 shows nonconditional simulations of two very different stationary random functions, which share the same histogram and the same covariance function. There is no need to be an expert to understand that the conclusions drawn with regard to connectivity, for example, will not be the same for these two simulations.

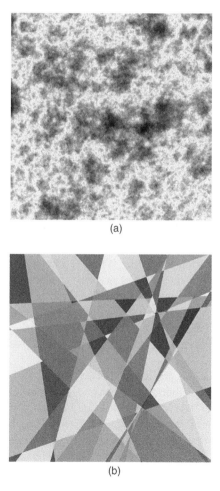

(a)

(b)

FIGURE 10.1 Simulations showing the same histogram and the same exponential covariance: (a) Gaussian RF, (b) mosaic RF with Gaussian marginal (tessellation by Poisson polygons). (See color version on accompanying CD.)

10.1.3 USE OF CONDITIONAL SIMULATIONS

If we consider a stationary field in a domain that is much larger than the correlation length, the local variations in two distant subdomains evolve independently. A single simulation over the whole domain can thus give a view of a variety of possible local situations. This is often sufficient, for example, to assess an agricultural practice scenario, depending on the local variability of soil properties. Conversely, when studying a nonstationary field such as a local contamination, or a global problem such as a breakthrough time, a single simulation provides a single answer in terms of volume of plume or flow and transport. It is then necessary to build several simulations if we want to assess the range of the possible results (typically 100 simulations may be needed).

These simulations are often ranked on the basis of a regional quantity computed from the realization (e.g., the volume of contaminated soil or the breakthrough time) to obtain the quantiles of the distribution of that regional quantity. Since the simulations represent independent drawings from the multivariate conditional distribution, the regional quantities obtained from the simulations are a representative sample of the conditional distribution of the regional quantity. And since the distribution of the ranks is by definition uniform, conditional simulations are often said to be equally probable or equally likely when one looks for quantiles. But of course, realizations corresponding to percentiles P5 or P95 usually look extreme by comparison with an average realization like P50.

10.1.4 OUTLINE OF THE CHAPTER

Generating a simulation means choosing a random function model, at least implicitly, and a simulation algorithm. Whereas a huge variety of random function models can be defined, a limited number of models are used in practical applications, and we will focus on them. There are often several algorithms to generate a nonconditional simulation of a given model. Again, we will not try to be exhaustive and simply present those that are most used or most efficient. The interested reader is referred to Chilès and Delfiner[2] and Lantuéjoul[9] for a thorough presentation.

Conditional simulations are applied to two main kinds of variables: (1) those that have a continuous distribution, such as grade, moisture content, and hydraulic conductivity, which can take on any real value between a minimum and a maximum, and (2) categorical variables, namely, discrete variables representing elements of a classification, such as soil types. Simulation algorithms are rather different in the two cases, even if there are some bridges. The next two sections concern these two types of variables. They are followed by a section describing an application to the modeling of soil properties.

For the sake of simplicity, we will focus on stationary random functions and simply mention extensions to intrinsic random functions. In what follows we will abbreviate random function by RF and stationary random function by SRF.

10.2 SIMULATION OF A CONTINUOUS VARIABLE

Among stationary random functions, Gaussian SRFs constitute the prototype of RFs with a continuous marginal distribution. The finite-dimensional distributions of a Gaussian SRF are all normal and completely defined by the mean and the covariance function — or the variogram — of the SRF, which makes the statistical inference easy. There are a series of algorithms to generate a simulation of a Gaussian SRF. Moreover, phenomena resulting from the addition of a large number of independent and identically distributed perturbations tend to be Gaussian by virtue of the central limit theorem. Last, transforming a Gaussian SRF $Y(\mathbf{x})$ into $Z(\mathbf{x}) = \varphi(Y(\mathbf{x}))$ yields a non-Gaussian SRF, thus extending the scope of Gaussian SRFs. We will therefore focus our presentation on the simulation of (possibly transformed) Gaussian SRFs and simply mention extensions and other models at the end of the present section.

10.2.1 GAUSSIAN TRANSFORMATION

Few variables conform to a Gaussian spatial distribution, and even simply to a Gaussian marginal. But if the SRF of interest $Z(\cdot)$ has a continuous marginal cumulative distribution function (cdf) $F(z)$, it can be transformed into an SRF $Y(\cdot)$, with a Gaussian marginal cdf. Indeed, denoting the standard normal cdf as $G(y)$, the RF defined by $Y(\mathbf{x}) = G^{-1}(F(Z(\mathbf{x})))$ has a Gaussian marginal distribution. Hence, $G^{-1}(F(\cdot))$ is called the *normal-score transform*. Its inverse, namely, the back-transform $\varphi(\cdot) = F^{-1}(G(\cdot))$, transforms a Gaussian variable into a variable with the marginal cdf of interest. Notice that these transformations simply amount to relating y and z values corresponding to the same quantile: $F(z) = G(y)$.

So the methods presented hereafter can be applied to variables that are not Gaussian, but can be considered transforms of a Gaussian SRF. To this end, the data are transformed into Gaussian data at the beginning of the study, and the back-transformation φ is applied at the end of the simulation process.

The transform φ is determined once the distribution F is known. In practice, it can be defined by a graphical fitting of F on the empirical distribution (the cumulative histogram of the data). It can also be expressed by an expansion with Hermite polynomials whose coefficients derive from the empirical distribution.

The Gaussian transformation only guarantees that the marginal distribution of Y is Gaussian. To be a Gaussian SRF, Y must also have normal finite-dimensional distributions. It is thus advisable to at least check that the bivariate distributions of pairs $(Y(\mathbf{x}), Y(\mathbf{x}'))$ are normal and, in the negative case, consider other simulation algorithms.[10,11] A simple and global control consists in checking that the sample variogram, which is an average of squared increments, is proportional to the square of the order one variogram, defined as an average of absolute increments.

Note that the specification of the transform φ implies knowledge of the marginal distribution of Z, and consequently of its mean. Similarly, Y is an SRF with known mean, more specifically a zero mean. As will be seen hereafter, conditional simulation algorithms for a Gaussian SRF Y include a kriging step.

This kriging is slightly different from the ordinary kriging, or kriging with unknown mean, presented in Chapter 9. Since Y has zero mean, there is no need to introduce the constraint that the weights sum up to 1 to ensure unbiasedness. Conversely, the covariance $C(\mathbf{h})$, and not solely the variogram $\gamma(\mathbf{h})$, shall exist. Simple kriging, or kriging with a known mean, in the case of a zero mean is the estimator

$$Y^* = \sum_{\alpha=1}^{N} \lambda_\alpha Y(\mathbf{x}_\alpha)$$

where the weights λ_α are solutions of the simple kriging system:

$$\sum_{\beta=1}^{N} \lambda_\beta \, C(\mathbf{x}_\beta - \mathbf{x}_\alpha) = C(\mathbf{x}_0 - \mathbf{x}_\alpha) \qquad \alpha = 1, ..., N$$

The simple kriging variance is then

$$\sigma_K^2 = C(\mathbf{0}) - \sum_{\alpha=1}^{N} \lambda_\alpha \, C(\mathbf{x}_0 - \mathbf{x}_\alpha)$$

Below in this section, the preliminary Gaussian transformation, if necessary, is assumed to have been done, and Z denotes a Gaussian SRF.

10.2.2 SEQUENTIAL GAUSSIAN SIMULATION

For a Gaussian RF *with known mean*, kriging coincides with conditional expectation. The distribution of $Z(\mathbf{x}_{N+1})$ conditional on $Z(\mathbf{x}_\alpha) = z_\alpha$, $\alpha = 1, ..., N$ is thus Gaussian, with mean z^* and variance σ_K^2, where z^* is the simple kriging estimate associated with the kriging estimator Z^* of $Z(\mathbf{x}_{N+1})$ from the $Z(\mathbf{x}_\alpha)$ values, and σ_K^2 the associated kriging variance. Simulating $Z(\mathbf{x}_{N+1})$ conditionally on the data thus amounts to selecting a random value in the normal distribution with mean z^* and variance σ_K^2.

By repeating this algorithm, it is possible to sequentially simulate Z at M new locations. The algorithm is the following:

1. Compute the simple kriging estimate z^* of $Z(\mathbf{x}_{N+1})$ from $Z(\mathbf{x}_1) = z_1, ...,$ $Z(\mathbf{x}_N) = z_N$, and the corresponding kriging variance σ_K^2.
2. Select a random value Y with standard normal distribution.
3. Assign $z^* + \sigma_K Y$ to $Z(\mathbf{x}_{N+1})$.

4. Include this outcome in the conditioning dataset by increasing N by 1 and decreasing M by 1.
5. If $M > 0$, go to step 1.

The statistical properties of the simulation are independent of the order in which the M new points are scanned. This method is perfect for small grids, but equivalent algorithms need less computing time because they are based on a single decomposition (e.g., a Cholesky decomposition) of the global covariance matrix of the $N + M$ variables $Z(x_\alpha)$.

The sequential Gaussian simulation method requires approximations for medium and large grids, for it is not possible to solve the kriging system in a satisfactory manner when $N + M$ is large (the upper limit is about 5000, but in practice, due to computing time, it is much lower). In this situation, each kriging is performed from a subset of the data (true data plus already simulated points). The choice of the scanning sequence of the simulated points then becomes critical, as well as the selection of the subsets.[12]

10.2.3 CONDITIONING BY KRIGING

In the rest of this section several simulation algorithms are presented. Most of them do not provide conditional simulations, but only nonconditional simulations, that is, realizations that reproduce spatial variability as modeled by the covariance, but otherwise do not honor the data. Fortunately, a procedure based on kriging allows us to pass from a nonconditional simulation $S(\cdot)$ to a conditional simulation $T(\cdot)$ that, while retaining the structural features of the former, is calibrated on the sample data.

Its principle, due to G. Matheron, is quite simple. Consider an RF $Z(\cdot)$ known at N sample points x_α, $\alpha = 1, 2, \ldots, N$. Let $Z^*(x)$ denote the kriging estimator of $Z(x)$ at the point x based on the data $Z(x_\alpha)$, and let us start from the trivial decomposition

$$Z(x) = Z^*(x) + [Z(x) - Z^*(x)]$$

The kriging error $Z(x) - Z^*(x)$ is of course unknown since $Z(x)$ is not known. Let us now assume that we have a nonconditional simulation $S(\cdot)$ independent of $Z(\cdot)$, *with the same covariance as* $Z(\cdot)$. Now consider the same equality as above for $S(x)$, where $S^*(x)$ is the kriging estimator obtained as if the simulation were known only at the sample points x_α:

$$S(x) = S^*(x) + [S(x) - S^*(x)]$$

Since the nonconditional simulation can be computed everywhere, the true value $S(x)$ is known and so is the error $S(x) - S^*(x)$. Hence, the idea of substituting, in the decomposition of $Z(x)$, the unknown error by the simulation of this error; this gives $T(x)$ defined by

$$T(\mathbf{x}) = Z^*(\mathbf{x}) + [S(\mathbf{x}) - S^*(\mathbf{x})]$$

Since kriging is an exact interpolator, at a sample point we have $Z^*(\mathbf{x}_\alpha) = Z(\mathbf{x}_\alpha)$ and $S^*(\mathbf{x}_\alpha) = S(\mathbf{x}_\alpha)$, so that $T(\mathbf{x}_\alpha) = Z(\mathbf{x}_\alpha)$. And because the kriging estimator and the kriging error are not correlated, $T(\cdot)$ preserves the covariance of $Z(\cdot)$ and $S(\cdot)$.

The proof that the substitution of errors preserves the covariance is valid for kriging with a global neighborhood. If the number of data is too large, it is necessary to use local neighborhoods. A careful design of the neighborhood search algorithm is needed to avoid the introduction of spurious discontinuities due to neighborhood changes.

When considered conditional on the $Z(\mathbf{x}_\alpha)$, $T(\cdot)$ is no longer stationary. The method of construction entails that the mean of a large number of independent conditional simulations at a given point tends to the kriging estimate, and their variance tends to the kriging variance. In figurative terms, a conditional simulation "vibrates" in between the data points within an envelope defined by the kriging standard error.

A conditional simulation is meant to behave like the real field, but not to estimate it. As an estimator of $Z(\mathbf{x})$, a simulation $T(\mathbf{x})$ would perform very poorly: it is easily shown that the corresponding estimation variance, namely, the variance of $T(\mathbf{x}) - Z(\mathbf{x})$, would be twice the kriging variance.

We now have to examine how to generate nonconditional simulations. These have to be computed at the location of the points where the final simulation is requested and at the location of the data points. So, we need algorithms able to simulate at any location within a domain encompassing all these points.

10.2.4 Turning Bands

Some covariance models may be simulated directly in the plane. But it is often simpler to use the turning bands method, which enables the construction of simulations in plane from simulations on lines.[13] This method consists of adding up a large number of independent simulations defined on lines scanning the plane.

More specifically, consider a system of n_D lines emanating from the origin of space and scanning the plane regularly; the angle between two adjacent lines is π/n_D. We denote by $\theta_t \in [0, \pi]$ the angle of the line D_t with the x-axis; by $\mathbf{u}_t$ the unit vector of D_t, with components $\cos\theta_t$ and $\sin\theta_t$; and by s_t the abscissa on D_t, centered at the origin.

Independent zero-mean nonconditional simulations $S_t(s_t)$ with covariance $C_1(h)$ are generated on the lines D_t. Let us consider a point $\mathbf{x} = (x, y)$ in the plane. Its projection on D_t is a point with abscissa

$$s_t = <\mathbf{x}, \mathbf{u}_t> = x \cos\theta_t + y \sin\theta_t$$

The simulation at $\mathbf{x}$ is then defined by

$$S(\mathbf{x}) = \frac{1}{\sqrt{n_D}} \sum_{t=1}^{n_D} S_t(s_t)$$

The elementary simulations being independent and with the same covariance $C_1(h)$, the covariance of $S(\mathbf{x})$ is

$$C_2(\mathbf{h}) = \frac{1}{n_D} \sum_{t=1}^{n_D} C_1(<\mathbf{h}, \mathbf{u}_t>)$$

If the number of lines is large enough, the simulation is approximately Gaussian and the discrete sum expressing $C_2(\mathbf{h})$ is an approximation of the integral

$$C_2(r) = \frac{1}{\pi} \int_0^{\pi} C_1(r \cos \theta) \, d\theta$$

with $r = |\mathbf{h}|$. C_2 is an isotropic covariance. In practice, C_2 is given and we have to use the inverse relation that gives C_1 as a function of C_2; that is, for $h \geq 0$,

$$C_1(h) = C_2(0) + h \int_0^{\pi/2} \frac{dC_2}{dh} (h \sin \theta) \, d\theta$$

Gneiting[14] derives C_1 explicitly for the most commonly used covariances C_2. Notice that when C_2 has a finite range, this is not necessarily the case for C_1. The turning bands method can be generalized to the three-dimensional space (and even to any dimension). The covariance C_3 obtained is isotropic. The relationship between C_1 and C_3 is much simpler than between C_1 and C_2:

$$C_3(r) = \frac{1}{r} \int_0^r C_1(u) \, du \qquad C_1(r) = \frac{d}{dr} \left[r \, C_3(r) \right]$$

In particular, if C_3 has a finite range, C_1 has the same range, a property that can be useful for some simulation algorithms. So, it can be easier to consider a two-dimensional simulation as a planar section of a three-dimensional simulation. This, of course, assumes that the covariance to simulate in two-dimensional space is a valid model in three-dimensional space, which is the case in most applications.

In practice, the simulations along each line are often discretized so that the same value $S_t(s_t)$ is assigned to a whole band perpendicular to D_t and containing s_t. Hence, the name turning bands is given to the method.

The remaining problem is now to simulate SRFs on lines with the given covariance $C_1(h)$. A number of methods are available for that, but none is both fully general and always efficient. The sequential Gaussian method, for example, is general, but it requires a discretization on the lines and is not efficient if the number of discretization points becomes large. Some simulation algorithms directly generate a Gaussian SRF, but many others can be used because the addition, properly scaled, of a large number of independent simulations of a non-Gaussian SRF tends to a Gaussian SRF. Specifically, if $S_1(\cdot)$, $S_2(\cdot)$, ..., is a sequence of independent zero-mean, finite-variance simulations with the same spatial distribution, by virtue of the central limit theorem the spatial distribution of the random function $T_k(\cdot)$ defined by

$$T_k(\mathbf{x}) = \frac{S_1(\mathbf{x}) + \ldots + S_k(\mathbf{x})}{\sqrt{k}}$$

tends to a Gaussian RF with the same mean and covariance as the S_i as $k \to \infty$.

10.2.5 Generalizations

10.2.5.1 Simulation in the Presence of a Nugget Effect

The basic simulation methods do not include a possible nugget effect. This can be easily simulated separately since it is an independent and purely random component. Since the nugget effect represents measurement errors or microscale fluctuations, we are often not interested in reproducing it. If we intend to use a method with a separate conditioning, we must nevertheless simulate the nugget effect at the simulated data points so that the simulated data have the same statistical characteristics as the true data; kriging shall be replaced by a cokriging that filters the nugget effect (also known as *factorial kriging*). If a Gaussian transformation is used, remember, however, that this transformation is defined from data that include the nugget effect component.

10.2.5.2 Simulating Nested and Anisotropic Covariances

The sequential Gaussian algorithm can be used with any covariance model, but the turning bands algorithm provides simulations with isotropic covariances. The main types of anisotropy can be handled easily. For example, a geometric anisotropy can be reduced to the isotropic case by means of a linear transformation of the coordinates (see Chapter 9).

The covariance we want to simulate is often modeled as a sum of several basic models. Some methods for generating nonconditional simulations can be applied directly to the global model. Others require the independent generation of a simulation of each component and the summing of them.

10.2.5.3 Simulation with a Change of Support

We may be interested in simulating the average value of Z in a block B that is much larger than the support of the data. If Z is a Gaussian SRF, this can be easily done because the block value is also Gaussian and we remain within a multivariate Gaussian framework. It suffices, for example, to generate a nonconditional simulation of the block value and to condition it by block kriging (see Chapter 9).

This was an upscaling problem. We may also be interested, like Kumar,[15] in a downscaling problem: simulate point-support values, or average values in small blocks, from data with different supports, for example, point-support data, average values in 20×20 m units, and pixels of remote sensing observations. In the Gaussian case, the solution is similar to that of the upscaling problem.

These problems call for more complex solutions if Z is not a Gaussian SRF but the transform of a Gaussian SRF. The block values are then considered transforms of SRFs, but the transform φ_B for the block B is not the transform φ for the point-support data: a change-of-support model, for example, the discrete Gaussian change-of-support model, is used to determine φ_B and all the necessary covariances from φ and the covariance function associated with the point support.[2]

10.2.5.4 Intrinsic Random Functions with an Unbounded Variogram

As shown in Chapter 9, there are phenomena that do not have a central value like the mean and display unbounded variations, but can be represented by intrinsic random functions (IRFs). Such random functions are characterized by an unbounded variogram $\gamma(\mathbf{h})$, the prototype being the power variogram $\gamma(\mathbf{h}) = b |\mathbf{h}|^\alpha$, $0 < \alpha < 2$, $b > 0$, and among this family the linear variogram $\gamma(\mathbf{h}) = b |\mathbf{h}|$.

The various methods presented above for simulating Gaussian SRFs have extensions to Gaussian intrinsic functions:

- Conditioning by a separate kriging shall be done with ordinary kriging; it preserves the variogram as well as the joint distribution of increments.
- The turning bands formulas relating C_1 to C_2 or C_3 remain valid with variograms instead of covariances. As a consequence, the turning bands operator does not alter a power variogram, but simply affects its multiplicative constant.
- The basic nonconditional simulation methods have extensions to intrinsic random functions. In particular, a one-dimensional Gaussian IRF with a linear variogram is a Brownian motion, whose simulation is straightforward.

The main difference with SRFs is that it is not possible to define the marginal distribution (the histogram) of an IRF because it is defined by its increments. In particular, its mean is defined up to a constant (IRF with a bounded variogram), or simply does not exist (IRF with an unbounded variogram). So the Gaussian transformation presented for SRFs has no counterpart with IRFs, so that the simulation method is limited to Gaussian IRFs. An *ad hoc* solution, if a Gaussian transformation is necessary and the data are representative of the study domain, is to consider the IRF as an SRF with a very large sill and a very large range.[16]

The conditioning procedure can be extended as well to nonstationary models with a trend, for example, to the universal kriging model, where the RF is considered the sum of an IRF and a polynomial drift with unknown coefficients.

10.2.5.5 Multivariate Simulations

We often have to simulate several correlated variables. Let us denote as Z_i, $i = 1, ..., p$ the various SRFs, which we will assume Gaussian (after a possible transformation). A simple case is the intrinsic model or *proportional covariance model*, where all direct and cross-covariances are proportional:

$$C_{ij}(\mathbf{h}) = \text{Cov}(Z_i(\mathbf{x}), Z_j(\mathbf{x} + \mathbf{h})) = b_{ij}\, C(\mathbf{h})$$

This model is valid provided that the matrix $\mathbf{B} = [b_{ij}]$ is positive definite. In that case, the Gaussian SRFs Z_i can be expressed as linear combinations of at most p independent Gaussian SRFs Y_j with the same covariance $C(\mathbf{h})$:

$$Z_i(x) = \sum_{j=1}^{p} a_{ij}\, Y_j(x)$$

The coefficients a_{ij} must of course satisfy

$$\sum_{k=1}^{p} a_{ik}\, a_{jk} = b_{ij} \qquad i, j = 1, ..., p$$

or in matrix terms, $\mathbf{A}\,\mathbf{A}^\mathsf{T} = \mathbf{B}$, where $\mathbf{A}$ is the matrix of the a_{ij} and $\mathbf{A}^\mathsf{T}$ is the transpose of $\mathbf{A}$ (there are solutions because $\mathbf{B}$ is positive definite). A nonconditional simulation of Z_i is then derived from a nonconditional simulation of the factors Y_j, which is simply a set of p independent nonconditional simulations with covariance $C(\mathbf{h})$. The conditioning step shall then be done by cokriging instead of series of krigings.

The intrinsic model is somewhat restrictive. A more general model is the *linear model of coregionalization*, which is simply a sum of independent proportional covariance models (see Chapter 9, Equation 9.34 to Equation 9.36, or Chilès and Delfiner[2] and Wackernagel,[17] for a more detailed presentation). Its nonconditional simulation is obvious since it amounts to adding independent simulations of the various proportional covariance models, and again the conditioning step can be done by cokriging.

10.2.6 OTHER MODELS

Gaussian RFs have the property to maximize entropy: among RFs sharing the same covariance, their realizations are those where the zones with high values present the lowest continuity. The same observation applies to zones with low values. If the simulated variable represents permeability, this implies a poor connectivity of the low permeability zones as well as of the high permeability zones, which can be a desirable property for a porous medium, but would not adequately represent a medium with flow barriers or conduits such as faults. Since the use of a Gaussian transform would not change these observations, there is clearly a need for other random function models than Gaussian ones.

Random tessellations (e.g., Voronoi polygons or Delaunay triangles associated with a random point process, or Poisson polygons) are the basis of mosaic random functions that have a constant value in nonoverlapping cells subdividing the space, and independent values in distinct cells. Figure 10.1b is a realization of a tessellation by Poisson polygons valued according to a normal distribution (the marginal distribution of the random function is normal, but not its multivariate distributions).

Boolean random functions generalize the Boolean model, presented hereafter by assigning to random objects a value that is not necessarily 0 or 1, nor an integer value, but a real value, or even some (random) function whose support coincides with the object.

Émery[10] proposes an extension of the sequential simulation algorithm to nongaussian random functions with isofactorial bivariate distributions, i.e., models that are intermediate between Gaussian RFs and mosaic RFs based on random tessellations.

10.3 SIMULATION OF A CATEGORICAL VARIABLE

The simplest categorical variable is one that only assumes the values 0 or 1, that is, the indicator of a set, which we consider here as random. But the covariance function is an extremely poor tool for describing the geometric properties of these very special random functions. For example, the covariance does not give any information on the connectivity of the medium. In fact, the covariance is the same for the random set considered and its complement (e.g., grains and pores), while their connectivities are generally very different. Figure 10.2 shows simulations of three very different indicator RFs, which share the same mean 1/2 and the same exponential covariance. Figure 10.2a is obtained by thresholding a Gaussian random function. This is not the case for Figure 10.2b, which derives from a tessellation by Poisson polygons.

Richer tools than the covariance have been developed in mathematical morphology, but these can be determined only if we have a large continuous image of a realization of the random set. We refer the reader to the literature on mathematical morphology and stochastic geometry.[18,19] This presentation is limited to basic random-set models that can be of use for geostatistical simulations

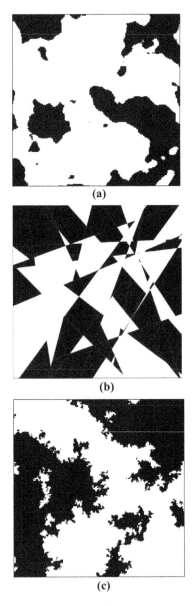

FIGURE 10.2 Simulations of indicator RFs with the same mean and the same exponential covariance: (a) truncated Gaussian RF, (b) mosaic RF with Gaussian marginal, (c) sequential indicator simulation.

and to simple generalizations allowing for the representation of m-valued indicators ($m > 2$; e.g., soil types).

Conversely, some indicator simulation methods derive from standard geostatistical methods, since they involve an underlying continuous variable. This continuous variable can have a physical meaning, for example, when we study the indicator associated with a given concentration threshold. It can also be a conventional feature of the model; for example, nested indicators (e.g., lithological facies) can be obtained by slicing a Gaussian variable at successive levels.

It is generally easy to simulate a random set once its type and parameters are selected. However, there is no general method for conditioning a simulation of a random set. The conditioning problem must be examined case by case. In this section, we first consider the sequential method, as adapted to indicators, which allows a direct construction of conditional simulations.

10.3.1 SEQUENTIAL INDICATOR SIMULATION

10.3.1.1 Simulation of a Single Indicator Variable

The sequential indicator simulation algorithm[20,21] is similar to the sequential Gaussian Simulation algorithm in Section 10.2.2. Let us consider the case of a binary variable $I(\cdot)$ with mean p, variance $\sigma^2 = p(1-p)$, and covariance $C(\mathbf{h})$, which is known at N points, $N \geq 0$, and shall be simulated at M other points. The algorithm is the following:

1. Compute the simple kriging estimate I^* of $I(\mathbf{x}_{N+1})$ from $I(\mathbf{x}_1), \ldots, I(\mathbf{x}_N)$.
2. Select a random value Y with Bernoulli distribution with mean I^* (i.e., $Y = 1$ with probability I^*, and $Y = 0$ with probability $1 - I^*$).
3. Assign Y to $I(\mathbf{x}_{N+1})$.
4. Include this outcome in the conditioning dataset by increasing N by 1 and decreasing M by 1.
5. If $M > 0$, go to step 1.

It can be shown, in the case of a known mean p, that this algorithm reproduces the mean and the indicator covariance, provided that kriging produces estimates that always lie in the interval [0, 1]. This is, however, not the case, except for some very specific situations (e.g., a pure nugget effect), so that the covariance is not reproduced exactly. This is due to the fact that, contrary to the Gaussian case, I^* does not coincide with the conditional expectation of I. It is often a very poor approximation of that conditional expectation, so that the sequential indicator algorithm performs poorly.[11] The approximation seems not so bad for a mosaic model. However, the simulation shown in Figure 10.2c, obtained with the sequential indicator algorithm, does not really look mosaic. So it is not easy to say when the sequential indicator algorithm can be recommended.

10.3.1.2 Generalization to a Categorical Variable

The method can easily be extended to a categorical variable $Z(\mathbf{x})$ that can take on the integer values 1, 2, ..., p by working with the nested indicators $I_i(\mathbf{x}) = 1_{Z(\mathbf{x}) \leq i}$. The categorical variable can correspond to slices between successive thresholds of an underlying continuous variable, or to a subdivision in facies or soil types without reference to any other variable. Nested indicators satisfy the order relation

$$0 \leq I_1(\mathbf{x}) \leq I_2(\mathbf{x}) \leq \ ... \ \leq I_p(\mathbf{x}) \leq 1$$

They can be estimated at the point $\mathbf{x}_{N+1}$ by cokriging. Their estimates define an approximation to the cumulative distribution function for $Z(\mathbf{x}_{N+1})$, from which the value of the simulation can be drawn.

The method has some advantages: the possibility to represent a more continuous spatial structure of the extreme values of the categorical variable than of the medium values, and the possibility to make use of soft information represented by indicator data between 0 and 1.[21,22] The status of the simulated random function is not clearer than that of a unique indicator function. Being nonparametric, the method is supposed to be able to reproduce the behavior of the categorical variable, and, for example, to reproduce a mosaic behavior where transitions are observed from any value to any other, as well as a diffusive behavior where transitions are possible only from one value to the preceding or the next one. But nothing in the algorithm can guarantee one of these behaviors to be honored exactly.

This method is sometimes used to simulate a continuous variable whose variation range has been divided in several classes, but that approach cannot be recommended in general — see a comparison of that approach with sequential Gaussian simulations for simulating soil texture in Druck Fuks and Voltz.[8]

10.3.2 Truncated Gaussian Simulation

Consider an indicator or a series of indicators that originate from applying one or more thresholds to a standard Gaussian SRF $Y(\mathbf{x})$,

$$I_i(\mathbf{x}) = 1_{y_{i-1} \leq Y(\mathbf{x}) < y_i} \qquad \text{with} \qquad -\infty = y_0 < y_1 < \cdots < y_{i-1} < y_i < \cdots < y_m = +\infty$$

In most applications the indicators represent geological facies or soil types, but the RF Y does not correspond to an actual regionalized variable. The thresholds y_i are thus chosen so as to match the proportions p_i of the various facies:

$$y_i = G^{-1}(\sum_{j=1}^{i} p_j) \qquad i = 1, ..., m - 1$$

Once the thresholds are fixed, the direct and cross-covariances of the various indicators derive from the correlogram $\rho(\mathbf{h})$ of the underlying Gaussian $Y(\mathbf{x})$, but the relations cannot be easily inverted. In applications, $\rho(\mathbf{h})$ is chosen such that the theoretical variograms of the indicators deduced from $\rho(\mathbf{h})$ fit the sample variograms well.

The simulation of I_i reduces to that of Y. Figure 10.2a was obtained by thresholding a Gaussian simulation at $y = 0$.

The nonconditional case is thus straightforward. In the conditional case, we do not know the exact value of Y at data point $\mathbf{x}_\alpha$, but simply to which interval $[y_{i-1}, y_i]$ belongs $Y(\mathbf{x}_\alpha)$. Let B_α denote that interval. The construction of a conditional simulation of Y is performed in two steps:

1. Simulation of the $Y(\mathbf{x}_\alpha)$ conditional on the intervals B_α.
2. Simulation of the whole grid conditional on these simulated $Y(\mathbf{x}_\alpha)$.

The second step is an ordinary conditional simulation, so only the first step needs to be addressed here. An exact iterative method has been proposed in Freulon and de Fouquet,[23] which is a direct application of the Gibbs sampler method.[24] An initial state, consistent with the data but not with the spatial structure, is obtained by assigning a value within B_α for each site $\mathbf{x}_\alpha$. The spatial structure is progressively introduced by iterating the following sequence:

1. Select a site $\mathbf{x}_\alpha$ (randomly or by scanning these sites).
2. Ignore the value at this site, and estimate it by kriging from the values at all the other sites; also compute the corresponding kriging variance.
3. Replace the value at this site by the kriged value, plus a Gaussian residual with variance equal to the kriging variance, randomly selected so that the new value belongs to the interval B_α.

This method can be generalized to nonstationary models where the proportions of the different facies, and thus the corresponding thresholds, vary in space.[25] Truncated Gaussian simulations are of diffusive type, in the sense that the facies i can be surrounded only by facies $i-1$ and $i+1$, as shown in Figure 10.3a. The method has been extended to facies that do not follow one another in a fixed order by using two Gaussian SRFs, independent or correlated, to define the facies, as shown in Figure 10.3b: truncated plurigaussian simulations.[26,27] Finally, there are adaptations for including connectivity constraints.[28]

10.3.3 OBJECT-BASED SIMULATION: BOOLEAN MODELS

We are considering here models obtained by combining *objects* placed at random points. These models can also be considered *marked point processes*, in the sense

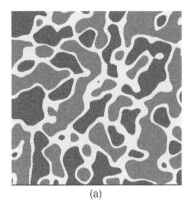 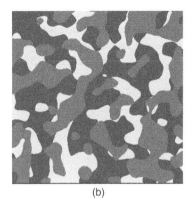

(a)   (b)

FIGURE 10.3 Examples of simulation of three soil types: (a) truncated Gaussian simulation (two thresholds applied to the same Gaussian RF — the brown and green soil types cannot be in contact); (b) truncated plurigaussian simulation (two Gaussian RFs, each one with its own threshold — each soil type can be in contact with the other two).

that they are based on a point process and marks (here the objects) attached to the points of the process.[19]

10.3.3.1 Poisson Point Process

The Poisson point process in $\mathbb{R}^n$ corresponds exactly to the intuitive idea of points distributed in space at random. The Poisson point process with *intensity* λ ($\lambda > 0$) is characterized by the following properties:

1. The number $N(V)$ of points inside a domain V is a Poisson random variable with parameter $\lambda |V|$, where $|V|$ represents the measure of V (length, surface, or volume):

$$\Pr\left\{N(V) = k\right\} = e^{-\lambda|V|} \frac{(\lambda|V|)^k}{k!}$$

$$\mathrm{E}\left[N(V)\right] = \mathrm{Var}\left[N(V)\right] = \lambda|V|$$

2. If V_i, $i = 1, 2, \ldots, p$, are disjoint domains, the random variables $N(V_i)$ are mutually independent.

The Poisson point process has an important conditional property that corresponds to the notion of random points: given that the number of points $N(V)$ inside the domain V is equal to n_0, these points are independently and uniformly distributed over V.

Thus, a Poisson point process with intensity λ can be simulated within a bounded domain V as follows:

1. Draw the number of points n_0 from a Poisson distribution with mean $\lambda |V|$.
2. Draw the n_0 points independently from a uniform distribution within V.

10.3.3.2 Boolean Random Set

A Boolean random set corresponds to the intuitive idea of the union of randomly located objects. Two independent ingredients are required for its construction:

1. A set of germs X_i in $\mathbb{R}^n$, $i = 1, 2, \ldots$, corresponding to a Poisson point process with intensity λ
2. A family of independent and identically distributed random objects A_i, $i = 1, 2, \ldots$ (e.g., discs in two dimensions, or balls in three dimensions, with random diameters)

The union of the A_i shifted to points X_i constitutes by definition a Boolean random set X:

$$X = \bigcup_i \tau_{X_i} A_i$$

where $\tau_{\mathbf{h}}$ denotes the operator of translation by a vector $\mathbf{h}$. The space $\mathbb{R}^n$ is thus divided into two phases: the union of objects X and its complement, the background. The appearance of a Boolean set depends considerably on the intensity and the shape of the objects, as shown in Figure 10.4.

Sometimes observations are done through cross-sections, whereas reality is three-dimensional, or through scan lines, whereas reality is two- or three-dimensional. A useful property of Boolean random sets is that their intersection by a line or a plane is still a Boolean random set (the reverse is not true). A number of results are available about the properties of Boolean models and the inference

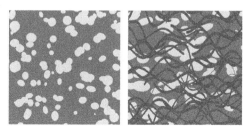

FIGURE 10.4 Realizations of Boolean random sets.

of their parameters, especially from images or scan lines.[18,19,29] In particular, the mean and covariance of the indicator of X are

$$p = 1 - e^{-\lambda K(0)} \qquad C(h) = e^{-2\lambda K(0)}\left(e^{\lambda K(h)} - 1\right)$$

where $K(\mathbf{h}) = \mathrm{E}\,|A \cap A_{-\mathbf{h}}|$ is the mean geometric covariogram of the primary objects (in particular $K(\mathbf{0})$ is their mean volume). Another interesting property is that when the objects are convex, the distribution of background intercepts is exponential.

10.3.3.3 Conditional Simulation of a Boolean Random Set

An iterative Markov procedure gives a general solution provided that the objects have a strictly positive surface (two-dimensional simulation) or volume (three-dimensional) and the number of conditioning points, known to belong to the background or the objects, is finite.[9] To simplify the presentation, we consider the generation of a simulation in a bounded domain V_S in the case where the objects, before being shifted to the X_i, are all enclosed in a sphere with radius R. Let V denote the domain, including V_S and all points at a distance of V_S shorter than R. Any point belonging to V can be the germ of an object intersecting V_S. In the generation of the simulation we will thus consider the restriction of the Poisson process to V, rather than a restriction to V_S. The expected number of germs in that domain is $\nu = \lambda\,|V|$.

Clearly a Boolean random set is the union of a Poisson number (with mean value ν) of independent objects centered in V. The principle of the algorithm is the following:

1. Start with a pattern of objects that is simply compatible with the data (but does not claim to represent a Boolean random set); denote the number of objects by N.
2. Add a random object at a random location with probability $\nu/(\nu + N)$, or remove an existing object — randomly selected — with probability $N/(\nu + N)$; keep this new pattern if it remains consistent with the data and update N; otherwise, forget it.
3. Go to step 2.

Any means can be used to achieve the first step. Lantuéjoul[9] provides an iterative algorithm, with a finite number of operations, for that. The second step of the algorithm tends to equalize N with ν: when $N < \nu$, objects are added more often than removed, and the other way around when $N > \nu$. The main algorithm never ends but, in practice, must be stopped after a finite number of iterations.

10.3.3.4 Extensions

There are a lot of extensions of this simple model. Let us mention two of them:

- Extension to other point processes: Regionalization of the Poisson intensity, cluster process, hard-core models, etc.
- Boolean random function: This makes it possible to simulate several soil types. A categorical variable Y_i that can take on the values 1, 2, ..., p is assigned to each object X_i: this variable can be a deterministic function of the object (e.g., the type of the object X_i if the objects are selected among several types), or an independent random variable, or any intermediate situation. A categorical random function is then obtained by assigning at any point $\mathbf{x}$ the value 0 if $\mathbf{x}$ belongs to the background, or the largest Y_i among those associated with objects overlapping $\mathbf{x}$.

Conditional simulation algorithms have been developed for these models too.[9,30,31]

10.3.4 SIMULATION FROM TRAINING IMAGES

When we have training images, a shortcut to the modeling of the RF is to use the sample multivariate distributions of the images. The method, very similar to sequential indicator simulation, is presented by Strebelle.[32] When simulating at $\mathbf{x}_{N+1}$ conditional on the values at $\mathbf{x}_1, ..., \mathbf{x}_N$, instead of kriging the indicator function and selecting in the corresponding Bernoulli distribution, we search in the training images for all configurations similar to the present one — i.e., geographical configuration identical to that of $\mathbf{x}_1, ..., \mathbf{x}_N, \mathbf{x}_{N+1}$, up to a translation, and same values $z_1, ..., z_N$. These configurations define an experimental conditional distribution for Z_{N+1}, which is used to simulate its value. The training images must be large enough so that these experimental conditional distributions capture the essence of the phenomenon rather than anecdotic situations. These images can also be simulations of unconditional realizations of models that go beyond a simple covariance for characterizing spatial structure, for example, stochastic process-based models — cf. Lopez[33] for the three-dimensional simulation of meandering channels.

10.4 APPLICATION TO A SOIL DATASET

10.4.1 INTRODUCTION

In this section, we illustrate the use of conditional simulations for continuous soil variables. The data considered here were collected in an agricultural field (ca. 10 ha) in Chambry, Northern France, in February 2001, as part of a precision agriculture project.[34] Precision agriculture aims at defining location-dependent management within an agricultural field, for nitrogen fertilization, for example. For assessing the soil variability, permanent variables (proportions of sand, silt and clay, Ca content) were sampled once, in the year 2000. From these samples, soil maps were hand drawn by soil scientists. Nonpermanent variables (water content and mineral nitrogen) were sampled two to three times a year, according to agricultural practice. At each node of a grid, given some economical and environmental criterion, the local optimal nitrogen fertilization amount is then computed according to the local value of these variables using a plant growth model. The mesh of the grid on which these variables must be provided depends on the mechanical properties of the devices used for the fertilization.

The aim of the geostatistical study is thus to model the spatial variability of these two variables and to provide maps of their local values as input values of the plant growth model for studying agriculture scenarios.

The dataset contains 152 sampling points, among which 84 belong to a pseudoregular grid with a 36-m mesh. The remaining 68 points belong to four sampling crosses that were located in the four main soil types, as depicted in Figure 10.5a. Soil water content (WC, measured in equivalent mm of water) and mineral nitrogen (N, in kg/ha) were measured on soil cores up to 120 cm. Each soil sample is itself a mixture of three cores taken at 50 cm distance. WC ranges from 135 to 463 mm; its average is 337 mm and its standard deviation is 60 mm. Nitrogen ranges from 15 to 68 kg/ha, with an average of 36 kg/ha and a standard deviation of 11 kg/ha. The correlation ($\rho = 0.71$) between the two variables is good, but the cross-plot (Figure 10.5b) shows that their relationship is not linear. Their histograms (Figure 10.5c and d) are not Gaussian distributed.

The experimental variograms (Figure 10.6) show a significant nugget effect for nitrogen (about 20% of the variance) and a negligible one for water content, which is usual in agricultural datasets. A linear model of coregionalization (see Chapter 9) is fitted with two structures: a nugget effect and an exponential structure with a scale parameter of 50 m (i.e., a practical range of 150 m) (cf. Equation 9.16 in Chapter 9). The sills are estimated using a weighted least squares procedure, as described in Goulard and Voltz.[35] The fitted model is

$$\gamma_{WC,WC}(\mathbf{h}) = 6 \text{ nug} + 3955 [1 - \exp(-|\mathbf{h}|/50)]$$

$$\gamma_{WC,N}(\mathbf{h}) = -11 \text{ nug} + 464 [1 - \exp(-|\mathbf{h}|/50)]$$

$$\gamma_{N,N}(\mathbf{h}) = 19 \text{ nug} + 83 [1 - \exp(-|\mathbf{h}|/50)]$$

where nug represents a unit nugget effect component, equal to 0 if $\mathbf{h} = \mathbf{0}$ and to 1 otherwise. Note that the nugget effect on the cross-variogram is negative, indicating a slightly larger correlation between WC and N at distinct locations ($\rho(\mathbf{h}) = 0.81$ when $|\mathbf{h}| \gg 50$ m) than at the same location ($\rho(\mathbf{0}) = 0.71$).

Ordinary cokriging of these two variables on a 5×5 m grid is displayed in Figure 10.7. In accordance with the model above, the same long-range structures are visible on both maps: values are generally higher in the top of the image and a zone of very low values of both WC and N is visible near the right border of the field. This zone corresponds to a very different soil type than the rest of the field. It is a summit sandy area where chalk and silt have been eroded: water and

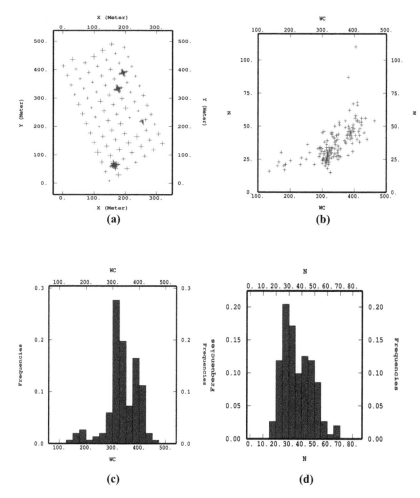

FIGURE 10.5 (a) Location of the data in the agricultural field. Four sampling crosses are located in the main soil types. (b) Cross-plot of nitrogen vs. water content. (c) Histogram of WC. (d) Histogram of N.

nitrogen are thus drained. Notice that these two maps are very smooth, a typical property of kriging. They do not reflect the true variability of the soil, and hence are not suitable for the purpose of simulating plant growth at a small scale. For reproducing small-scale variability, it is thus necessary to provide maps of simulated values instead of maps of interpolated values. Because it is important to honor the correlation between WC and N, the variables will be cosimulated. In addition to the variograms, these maps must of course honor the data at the sample points, and thus we will perform conditional simulations. Figure 10.5c and d has shown that the data are not Gaussian distributed. Since most of the available algorithms for simulating random fields end up simulating Gaussian SRFs, we

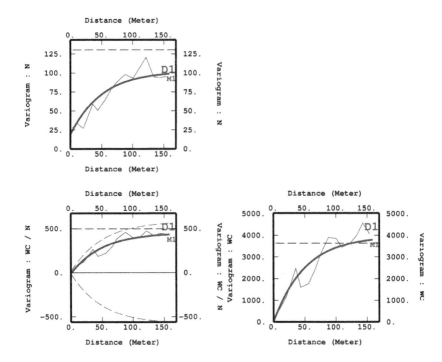

FIGURE 10.6 Variogram fitting of raw data. Thin line, experimental variogram; thick line, fitted model; dashed line, variance or covariance value; dashed curves, envelope of authorized cross-variogram models.

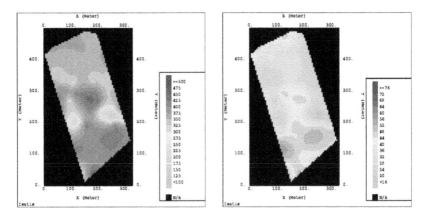

FIGURE 10.7 Kriging maps of (a) WC and (b) N. (See color version on accompanying CD.)

will first perform a Gaussian transformation of each variable, then conditionally cosimulate the transformed fields on a grid, and at the end back-transform the images for providing conditional cosimulations of the raw variables. We now detail and illustrate the different steps.

10.4.2 GAUSSIAN TRANSFORMATION

The first step is to perform a Gaussian transformation of the data and to check the bigaussian hypothesis on the transformed variables. The Gaussian transformation is performed for each variable as explained in Section 10.2.1. The transforms φ_{WC} and φ_N are modeled by expansions with Hermite polynomials.

Let Y_{WC} and Y_N denote the transformed variables corresponding to WC and N. By definition, they have a Gaussian marginal distribution. Since we intend to build simulations with algorithms designed for Gaussian random functions, it is advisable to at least check the Gaussian character of the distributions of pairs of values such as $(Y_{WC}(\mathbf{x}), Y_{WC}(\mathbf{x}'))$, $(Y_N(\mathbf{x}), Y_N(\mathbf{x}'))$, or $(Y_{WC}(\mathbf{x}), Y_N(\mathbf{x}'))$. The crossplot of $Y_{WC}(\mathbf{x}_\alpha)$ vs. $Y_N(\mathbf{x}_\alpha)$ at the same data point, depicted by Figure 10.8a, shows that the pair of variables can be marginally modeled as a bigaussian vector. Concerning the bigaussian character of pairs of values at separate locations, we use the simple and global way of checking based on the comparison of the order-one variogram with the usual variogram. Indeed, if the SRF Y is Gaussian, its order-one variogram,

$$\gamma_1(\mathbf{h}) = \tfrac{1}{2}\mathrm{E}\left[\left|Y(\mathbf{x}+\mathbf{h})-Y(\mathbf{x})\right|\right]$$

is proportional to the square root of its usual variogram (which can be considered a variogram of order 2). Figure 10.8b and c depicts the ratio $\sqrt{\gamma(\mathbf{h})}\,/\,\gamma_1(\mathbf{h})$ as a function of $|\mathbf{h}|$, for both variables. In the Gaussian case this ratio should be constant and equal to $\sqrt{\pi} = 1.77$. This condition is well verified for N, but there are some small variations for WC. The general shape of the curve, however, is not far from being a constant. The variograms of (Y_{WC}, Y_N) are modeled using a linear model of coregionalization with the same structures as for the raw data, namely, a nugget effect component and an exponential model (Figure 10.9). Its parameters are

$$\gamma_{YWC,YWC}\,(\mathbf{h}) = 0.09 \text{ nug} + 0.91\,[1 - \exp(-|\mathbf{h}|/50)]$$

$$\gamma_{YWC,YN}\,(\mathbf{h}) = 0.70\,[1 - \exp(-|\mathbf{h}|/50)]$$

$$\gamma_{YN,YN}\,(\mathbf{h}) = 0.01 \text{ nug} + 0.99\,[1 - \exp(-|\mathbf{h}|/50)]$$

In the bigaussian case, it is possible to compute the variogram of Z corresponding to a given variogram for Y using the function φ.[2] The variograms obtained for Z_{WC} and Z_N should fit the variograms computed on the raw data.

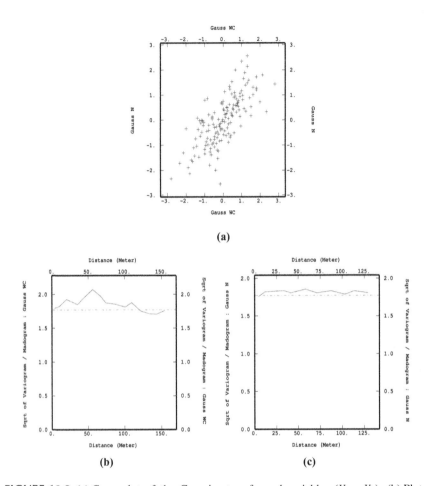

FIGURE 10.8 (a) Cross-plot of the Gaussian-transformed variables (Y_{WC}, Y_N). (b) Plot of $\sqrt{\gamma(\mathbf{h})} \, / \, \gamma_1(\mathbf{h})$ as a function of $|\mathbf{h}|$ for WC. (c) Same plot for N.

This provides an additional way for checking the validity of the bigaussian model. On the data, the agreement is good, but in order to save place, the corresponding figures have not been reported here. In conclusion, we can consider that the transformed variables conform to a bigaussian model, and we make the assumption that multipoint statistics are also Gaussian, so that (Y_{WC}, Y_N) is a Gaussian pair of stationary random functions.

10.4.3 CONDITIONAL COSIMULATION

We are now ready to perform one or several conditional cosimulations of the Gaussian pair (Y_{WC}, Y_N). Each cosimulation necessitates a nonconditional cosim-

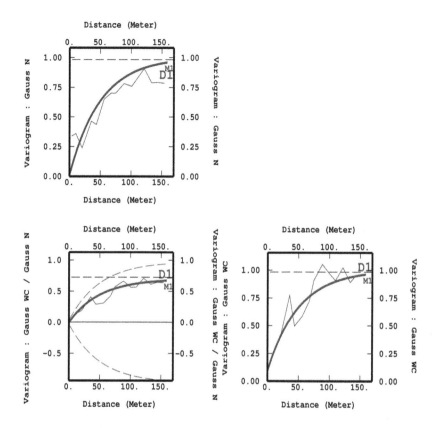

FIGURE 10.9 Variogram fitting of the transformed variables. Thin line, experimental variogram; thick line, fitted model; dashed line, variance or covariance value; dashed curves, envelope of authorized cross-variogram models.

ulation followed by conditoning by cokriging. Conditional simulations of the raw variables (WC, N) will then be obtained by back-transformation of the simulations of (Y_{WC}, Y_N).

The specificity of a cosimulation of two or more variables by comparison with separate simulations of each variable is to reproduce the cross-structures. In the present case, a nonconditional simulation of the pair of variables (Y_{WC}, Y_N) is easy. Let us begin with the continuous component (with an exponential variogram). Starting from two independent Gaussian simulations Y_1 and Y_2 with zero mean, unit variance, and an exponential variogram with a parameter scale of 50 m, we obtain correlated simulations with exponential direct and cross-variograms by linear combinations of the form

$$S_1(\mathbf{x}) = a_{11} Y_1(\mathbf{x}) + a_{12} Y_2(\mathbf{x})$$

$$S_2(\mathbf{x}) = a_{21} Y_1(\mathbf{x}) + a_{22} Y_2(\mathbf{x})$$

The sills of these variograms being $a_{11}^2 + a_{12}^2$ for $\gamma_1(\mathbf{h})$, $a_{11} a_{21} + a_{12} a_{22}$ for the cross-variogram $\gamma_{12}(\mathbf{h})$, and $a_{21}^2 + a_{22}^2$ for $\gamma_2(\mathbf{h})$, it suffices to select a_{11}, a_{12}, a_{21}, and a_{22} so that these sills equate 0.91, 0.70, and 0.99, respectively, for $S_1(\mathbf{x})$ and $S_2(\mathbf{x})$ to represent nonconditional simulations of the continuous component. Here we simulate Y_1 and Y_2 by using the turning bands algorithm (with 500 bands).

The nuggets of Y_{WC} and Y_N are independent, since the cross-variogram of Y_{WC} and Y_N has no nugget effect; so they can be simulated as two independent Gaussian white noises $\varepsilon_1(\mathbf{x})$ and $\varepsilon_2(\mathbf{x})$ with variances 0.09 and 0.01, respectively. A nonconditional simulation (S_{WC}, S_N) of (Y_{WC}, Y_N) is then obtained by adding the nugget and the exponential components:

$$S_{WC}(\mathbf{x}) = S_1(\mathbf{x}) + \varepsilon_1(\mathbf{x})$$

$$S_N(\mathbf{x}) = S_2(\mathbf{x}) + \varepsilon_2(\mathbf{x})$$

Note that the method presented here can be easily generalized to any linear model of coregionalization with several variables and several variogram components, as explained in Section 10.2.6.4.

A conditioning by cokriging followed by a back-transformation then provides a conditional cosimulation of WC and N, which are a consistent image of their possible spatial variations.

10.4.4 RESULTS

Figure 10.10a and b shows a conditional simulation of WC and N, respectively. Clearly, the conditional simulations show small-scale variability that was not in the kriging maps, but the same large-scale patterns are still. Figure 10.10c and d depicts the histogram of N and the cross-plot (WC, N). One can see that they are very similar to the same figures computed on the data.

It is of course possible to perform as many independent realizations as desired. This allows us to explore the variability of different scenarios for nitrogen fertilization and more generally for crop management.

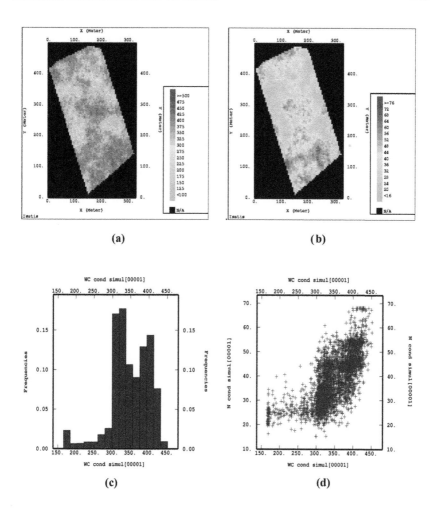

FIGURE 10.10 Conditional simulations of (a) WC and (b) N. (c) Histogram of WC. (d) Cross-plot of simulated N vs. WC. (See color version on accompanying CD.)

REFERENCES

1. GBM Heuvelink, R Webster. Modelling soil variation: past, present, and future. *Geoderma* 100: 269–301, 2001.
2. JP Chilès, P Delfiner. *Geostatistics: Modeling Spatial Uncertainty.* Wiley, New York, 1999.
3. AG Journel. Geostatistics for conditional simulation of orebodies. *Econ Geol* 69: 673–687, 1974.
4. MFP Bierkens, PA Burrough. The indicator approach to categorical soil data. II. Application to mapping and land use suitability analysis. *J Soil Sci* 44: 369–381, 1993.

5. PC Kyriakidis. Selecting panels for remediation in contaminated soils via stochastic imaging. In *Geostatistics Wollongong '96*, Vol. 2, EY Baafi, NA Schofield, Eds. Kluwer, Dordrecht, Netherlands, 1997, pp. 973–983.

6. MJ Broos, L Aarts, CF van Tooren, A Stein. Quantification of the effects of spatially varying environmental contaminants into a cost model for soil remediation. *J Environ Manage* 56: 133–145, 1999.

7. P Goovaerts. Geostatistical modelling of uncertainty in soil science. *Geoderma* 103: 3–26, 2001.

8. S Druck Fuks, M Voltz. Comparison of geostatistical simulation approaches for estimating the spatial uncertainty of soil texture. In *geoENV III: Geostatistics for Environmental Applications*, Vol. 1, P Monestiez et al., Eds. Kluwer, Dordrecht, Netherlands, 2001, pp. 463–474.

9. C Lantuéjoul. *Geostatistical Simulation: Models and Algorithms*. Springer, Berlin, 2002.

10. X Émery. Conditional simulation of non-Gaussian random functions. *Math Geol* 34: 77–100, 2002.

11. X Émery. Simulation conditionnelle de modèles isofactoriels. Ph.D. dissertation, E.N.S. des Mines de Paris, Paris, 2004.

12. JJ Gómez-Hernández, EF Cassiraga. Theory and practice of sequential simulation. In *Geostatistical Simulations*, M Armstrong, PA Dowd, Eds. Kluwer, Dordrecht, Netherlands, 1994, pp. 111–124.

13. G Matheron. The intrinsic random functions and their applications. *Adv Appl Prob* 5: 439–468, 1973.

14. T Gneiting. Closed form solutions of the two-dimensional turning band equation. *Math Geol* 30: 379–390, 1998.

15. P Kumar. Multiple scale conditional simulation. In *Scaling Methods in Soil Physics*, Y Pachepsky, DE Radcliffe, HM Selim, Eds. CRC Press, Boca Raton, FL, 2003, pp. 179–192.

16. CV Deutsch, AG Journel. *GSLIB: Geostatistical Software Library and User's Guide*, 2nd ed. Oxford University Press, New York, 1998.

17. H Wackernagel. *Multivariate Geostatistics*, 3rd ed. Springer, Berlin, 2003.

18. J Serra. *Image Analysis and Mathematical Morphology*. Academic Press, London, 1982.

19. D Stoyan, WS Kendall, J Mecke. *Stochastic Geometry and Its Applications*, 2nd ed. Wiley, Chichester, U.K., 1995.

20. F Alabert. Stochastic Imaging of Spatial Distributions Using Hard and Soft Information. M.Sc. dissertation, Stanford University, Stanford, CA, 1987.

21. AG Journel. Imaging of spatial uncertainty: a non-Gaussian approach. In *Geostatistical, Sensitivity, and Uncertainty Methods for Ground-water Flow and Radionuclide Transport Modeling*, BE Buxton, Ed. Battelle Press, Columbus, OH, 1989, pp. 585–599.

22. AG Journel, F Alabert. Non-Gaussian data expansion in the earth sciences. *Terra Nova* 1: 123–134, 1989.

23. X Freulon, C de Fouquet. Conditioning a Gaussian model with inequalities. In *Geostatistics Tróia '92*, Vol. 1, A Soares, Ed. Kluwer, Dordrecht, Netherlands, 1993, pp. 201–212.

24. G Casella, EI George. Explaining the Gibbs sampler. *Am Stat* 46: 167–174, 1992.

25. G Matheron, H Beucher, C de Fouquet, A Galli, D Guérillot, C Ravenne. Conditional Simulation of the Geometry of Fluvio-Deltaic Reservoirs, SPE paper 16753, Society of Petroleum Engineers, Richardson, TX, 1987.

26. A Galli, H Beucher, G Le Loc'h, B Doligez, Heresim Group. The pros and cons of the truncated Gaussian method. In *Geostatistical Simulations*, M Armstrong, PA Dowd, Eds. Kluwer, Dordrecht, Netherlands, 1994, pp. 217–233.

27. M Armstrong, A Galli, G Le Loc'h, F Geffroy, R Eschard. *Plurigaussian Simulations in Geosciences*. Springer, Berlin, 2003.

28. D Allard. Simulating a geological lithofacies with respect to connectivity information using the truncated Gaussian model. In *Geostatistical Simulations*, M Armstrong, PA Dowd, Eds. Kluwer, Dordrecht, Netherlands, 1994, pp. 197–211.

29. G Matheron. *Éléments pour une théorie des milieux poreux*. Masson, Paris, 1967.

30. MNM Van Lieshout. *Markov Point Processes and Their Applications*. Imperial College Press, London, 2000.

31. A Brix, WS Kendall. Simulation of cluster point processes without edge effects. *Adv Appl Prob* 34: 267–280, 2002.

32. S Strebelle. Conditional simulation of complex geological structures using multipoint statistics. *Math Geol* 34: 1–22, 2002.

33. S Lopez. Modélisation de réservoirs chenalisés méandriformes: approche génétique et stochastique. Ph.D. dissertation, E.N.S. des Mines de Paris, Paris, 2003.

34. B Mary, N Beaudoin, JM Machet, C Bruchou, F Ariès. Characterization and analysis of soil variability within two agricultural fields: the case of water and mineral N profiles. In *Proceedings of the 3rd European Conference on Precision Agriculture*, G. Grenier, S. Blackmore, Eds., Montpellier, France, 2001, pp. 431–436.

35. M Goulard, M Voltz. Linear coregionalization model: tools for estimation and choice of multivariate variograms. *Math Geol* 24: 269–286, 1992.

11 Pedometrical Techniques for Soil Texture Mapping at Different Scales

Marc van Meirvenne and Ingrid van Cleemput

CONTENTS

ABSTRACT

Soil texture is a key property influencing most soil processes. Therefore, it has been characterized and mapped intensively. Traditionally, soil texture, being a distribution of grain sizes, was classified twice: in soil textural fractions and in textural classes. The consequence was the introduction of boundaries between grain sizes and on maps. With pedometrical techniques we can avoid the last.

However, maps are scale dependent, so this chapter focuses on methods to map soil texture fractions continuously at different scales. First, we considered a nested sampling with spatial intervals between observation ranging from a few meters to several tens of kilometers. The resulting variogram displayed three levels with a very low nugget effect, illustrating the increasing variance as the spatial scale of the investigation increased. Second, soil texture was mapped on a regional scale. Therefore, an area of approximately 3000 km^2 in Belgium was used. A database of 4887 textural analyses taken inside this area was available. The area was stratified and ordinary kriging of each textural fraction independently was compared with compositional ordinary kriging of the three fractions simultaneously, guaranteeing that their sum equals 100, by taking the nature of the boundaries into account. The three resulting layers, representing the textural fractions clay, silt, and sand, form three basic geographic information systems (GIS) layers, which can be used to produce classified maps according to any texture classification. We illustrated this by applying the U.S. Department of Agriculture (USDA) textural triangle as an alternative to the Belgian triangle, which was used during the Belgian soil survey. This analysis was completed by a sensitivity analysis, resulting in an indication of the mapping quality of the resulting maps. This chapter illustrates a new use of existing data, mapping texture as texture fractions rather than classes. In this way, one of the two classifications of soil texture can be avoided.

11.1 INTRODUCTION

Soil texture is one of the most important soil properties influencing most physical, chemical, and biological soil processes. Hence, it is a key property for soil management. Ever since soil has been characterized, some indication of soil texture has been included. But since grain sizes of soil particles can vary widely at any location, the most complete way to characterize soil texture would be to use a particle-size frequency distribution.[1] However, determining the full (equivalent) soil particle diameter distribution is a cumbersome activity that was hardly possible until recently, when more sophisticated measurement methods, like laser diffractometry,[2] became available. As a result, soil scientists turned to the use of classes. Texture classes were defined on the basis of three textural fractions: clay, silt, and sand. Today, the most widespread delineations of these fractions are 0 to 2 μm for clay, 2 to 50 μm for silt, and 50 to 2000 μm for sand. Particles larger than 2 mm are not considered to belong to the fine earth, although their impact on soil management can be substantial if they are common. The three textural fractions can be presented as a three-dimensional graph, but due to the limitation that their sum must be 100%, this graph reduces to a triangle. Within this triangle texture classes are defined by covering a certain proportion of the area of the triangle. Most countries have a national soil texture triangle with national texture classes, but internationally the U.S. Department of Agriculture (USDA) soil texture triangle is most widely used, including by international institutions such as the Food and Agriculture Organization (FAO).[3]

The double classification into texture fractions and texture classes made soil survey a lot more feasible. However, it introduced another simplification of reality: boundaries and soil map polygons. This approximation was considered a price soil scientists had to pay to reduce and manage natural complexity.

With the advance of data management technology and the development of pedometrical techniques, we are now in the position to turn one step back to the natural occurrence of soil texture: mapping textural fractions numerically and avoiding the use of texture classes and their mapping boundaries. However, we are just at the beginning of exploring the possibilities of mapping the soil particle diameter distributions in a spatially continuous way. Only then will we have avoided the simplifications and boundary problems associated with classification of a continuous phenomenon, both in spatial and in attribute dimensions.[4]

This chapter focuses on some of the pedometrical techniques (like variogram analysis and compositional ordinary kriging) we currently have at our disposal to map soil textural classes in a spatially continuous way over a range of scales. We illustrate how these layers of information can be used to characterize soil spatial variability over different scales and to obtain maps of soil textural classes on a regional scale in East Flanders, Belgium. We also evaluated the sensitivity of such a classification.

11.2 CHARACTERIZING SOIL TEXTURE VARIABILITY FROM WITHIN FIELD TO A REGIONAL SCALE

Soil, as we encounter it at a given instant in time and space, is the result of several soil-forming processes, each of which acts at a specific rate and spatial scale (resolution or grain). Climatological factors can be very dynamic in time, but their average properties usually behave quite uniformly over large areas. Biological effects, on the other hand, can be extremely local (e.g., the effect of burrowing animals). The same is true for soil properties: some are very variable in time and space (e.g., nitrate–nitrogen), whereas others are quite permanent and vary in a more structured way over a landscape (e.g., thickness of a windblown sediment). As a result, the spatial structure of soil variation can be very complex, and in turn, this structure can be time-dependent.

The structure of the spatial variability is typically characterized by the variogram,[5] calculated as

$$\gamma(\mathbf{h}) = \frac{1}{2N(\mathbf{h})} \sum_{\alpha=1}^{N(\mathbf{h})} \{z(\mathbf{x}_\alpha) - z(\mathbf{x}_\alpha + \mathbf{h})\}^2$$

with $\gamma(\mathbf{h})$ being the variogram for the spatial vector $\mathbf{h}$, separating observations $z(\mathbf{x}_\alpha)$ and $z(\mathbf{x}_\alpha + \mathbf{h})$ ($\alpha = 1, \ldots, N$), N being the total number of observations of soil property Z, and $N(\mathbf{h})$ is the number of couples separated by $\mathbf{h}$.

The sill parameter of the variogram model approximately reflects the total variation encountered by the observations, whereas the nugget parameter represents the unstructured (random) part of it. The spatial scale over which the structure extends is given by the range. The latter is an important parameter to analyze the spatial dimension over which two observations are interrelated (autocorrelated). Although the variogram can be bounded at one spatial scale, it might become nested when the extent of the investigation increases, as more soil-forming factors contribute as other scales are included. Thus, variograms are scale dependent (i.e., investigation specific), despite the somewhat misleading interpretation given to the range (and sill), as if it delineated a fixed extent of influence of an observation.

Soil texture is the result of physicochemical processes acting on rocks and minerals (*in situ* or after transportation), influenced by external factors like climate, topography, and living organisms. Within a time frame relevant to soil management, texture can be considered to be time invariable, making it an interesting property to map.

To investigate the scale dependency of soil textural fractions, we combined three sampling schemes and analyzed the sand fraction. The first sampling covered an entire geomorphological unit: the polder area in the northwest of East Flanders, Belgium (Figure 11.1). It covers 6840 ha and consists of 32 polders recovered from the sea, mostly during the 16th century. The topsoil is dominantly loamy, but the texture of the area is known to display a large variation.[6] At this scale, representing a sampling interval on the order of 100 to 1000 m, 115 topsoil samples were taken.

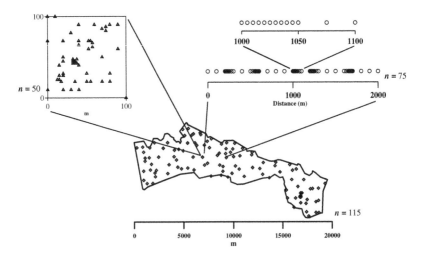

FIGURE 11.1 Polder area showing the three sampling schemes of topsoil sand: entire area (below), transect with nested sampling (top right), and 1-ha field with clustered samples (top left), indicating the number of samples taken in each scheme (*n*).

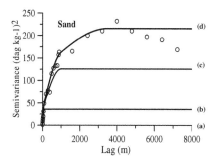

FIGURE 11.2 Nested variogram model of the three sampling schemes in Figure 11.1 combined (curve d) and the nugget effect (a) and the two models of the field and transect sampling (b and c, respectively).

Near to the center of this polder area, a 2-km-long transect was sampled according to a nested design (Figure 11.1). The basic sampling distance was 100 m, but five intervals of 50 m were sampled at 5-m intervals. Additional samples were taken at 25 m, resulting in 75 samples in total. These samples cover a spatial scale ranging from 5 to 100 m.

Within the area crossed by the transect, a 1-ha part of an agricultural field was sampled at 50 locations, with a minimum interval of 2.5 m (Figure 11.1). These locations were obtained from a random selection out of 247 grid points used to monitor nitrogen leaching.[7] This sampling covers a spatial scale between 2.5 and 10 m.

When the variogram is calculated for each of these three sampling schemes, it shows a leveling off and a bounded variogram model could be fitted (Figure 11.2). When all data points were pooled (240 data items), a nested model with a nugget effect and three simple models was used to model the increasing spatial variability with increasing spatial dimensions:

$$\gamma(h) = \gamma_0(h) + \gamma_1(h) + \gamma_2(h) + \gamma_3(h)$$

with:

$$\gamma_0(h) = \begin{cases} 0 & \text{if } h = 0 \\ C_0 & \text{if } h > 0 \end{cases}$$

$$\gamma_1(h) = C_1\left(\frac{3h}{2a_1} - \frac{1}{2}\left(\frac{h}{a_1}\right)^3 \right) \qquad 0 < h \le a_1$$

$$\gamma_1(h) = C_1 \qquad h > a_1$$

$$\gamma_2(h) = C_2\left(\frac{3h}{2a_2} - \frac{1}{2}\left(\frac{h}{a_2}\right)^3\right) \qquad h \le a_2$$

$$\gamma_2(h) = C_2 \qquad h > a_2$$

$$\gamma_3(h) = C_3\left(\frac{3h}{2a_3} - \frac{1}{2}\left(\frac{h}{a_3}\right)^3\right) \qquad h \le a_3$$

$$\gamma_3(h) = C_3 \qquad h > a_3$$

with C_0 the nugget effect (intercept on Y-axis), C_1, C_2, and C_3 the structured components of the three spherical models, $C_0 + C_1 + C_2 + C_3$ the total sill (plateau value), a_1, a_2, and a_3 the three ranges of spatial autocorrelation corresponding to the sills of every model, respectively, and h the spatial lag (separation vector) for which the variogram $\gamma(\mathbf{h})$ is calculated. For the fitted model, $C_0 = 1.5\%^2$, $C_1 = 34\%^2$, $C_2 = 90\%^2$, and $C_3 = 90\%^2$, so the sill is $215.5\%^2$ of which the random, unstructured variability represents only 0.7%. This indicates that as the separation distance reduces, the nugget effect approaches zero, its theoretical value. The fitted ranges are $a_1 = 90$ m, $a_2 = 990$ m, and $a_3 = 3500$ m, showing the increasing nature of the spatial scale as the extent of the sampling area increases, and each level contributes in a significant amount to the total variation encountered at the largest scale. It could be hypothesized that if we extend the sampling further, to cover increasingly larger areas, new levels of variation could be added, resulting in new sill values. The tendency of the variogram to increase can also be observed by taking the logarithm of both axes of Figure 11.2; the result is shown in Figure 11.3. It illustrates the almost log-linear increase of spatial variation as the spatial dimension increases. Therefore, one could wonder, as McBratney[8] did: "Does the variogram really have a sill (or a range)?"

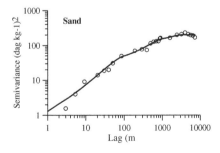

FIGURE 11.3 Nested variogram of Figure 11.2 shown on logarithmically transformed axes.

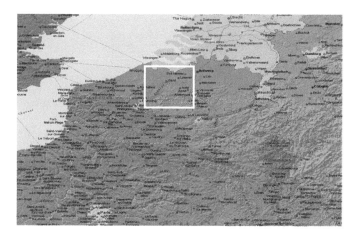

FIGURE 11.4 The study area (province of East Flanders) within Belgium.

11.3 SOIL TEXTURE MAPPING AT A REGIONAL SCALE

11.3.1 REGIONAL SCALE

We define a regional scale as a resolution that is commonly used to map at a scale of 1:50,000 to 1:100,000. This is the scale at which most national soil maps have been produced, although in some countries, like Belgium, a more detailed scale was used (mostly 1:20,000 to 1:25,000). The advantage of a regional scale is that it produces both a geological overview and agronomic detail. Regional soil maps mostly have soil texture as one of the central soil properties in their legends.

11.3.2 STUDY AREA

The application of pedometrical techniques is illustrated on an area covering 3000 km² located in Belgium: the province of East Flanders (Figure 11.4). This area was selected because a regional soil texture map was available at a scale of 1:100,000 covering variable soil parent materials (from north to south): two polder areas, a medium sand area, a fine sand area, a sandy loam area, and a silty area (Figure 11.5). Within these areas alluvial deposits associated with the major rivers are found. The sediments of the polders and river alluvia were deposited by water; all other material was wind transported (eolian sediments). Figure 11.5 shows a simplified regional soil texture map derived from a more detailed map made by Van Ruymbeke and colleagues in 1965.[9] More details can be found in Van Cleemput.[10]

   The limitations associated with a polygon soil map representing soil texture classes of soil texture fractions are as follows:

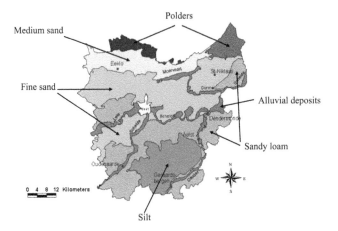

FIGURE 11.5 Generalized soil map of province of East Flanders.

1. The discrete model of spatial variation, in which soil units are represented as polygons that are internally homogeneous with abrupt boundaries, is conceptually outdated.
2. Due to the classification using the Belgian soil texture triangle, reclassification into, e.g., the internationally accepted USDA texture triangle is impossible. This creates problems of interchangeability between neighboring countries.
3. The map legend provides qualitative information, whereas increasingly quantitative information is required to be analyzed in geographic information systems (GIS)-related models and pedotransfer models.
4. No information about the accuracy of the map is provided.

For these reasons, a need existed to construct a new soil texture inventory of the area.

11.3.3 DATASET, STRATIFICATION, AND VARIOGRAMS

Within the study area we had access to 4887 topsoil (0- to 30-cm) samples, of which the three textural fractions — sand, silt, and clay (according to the definition given above) were analyzed using the conventional sieve-and-pipette method. The samples were distributed quite evenly over the province (Figure 11.6), but some areas (like the silty area in the south and the alluvial area) were less frequently sampled than the others. In some parts, detailed short-distance samples were available (like the transect sampling discussed above), allowing the characterization of the spatial structure over short intervals.

Based on the simplified soil texture map, these data were stratified according to the seven different soil texture areas presented in Figure 11.5. For each of the three stratified textural fractions, variograms were calculated and modeled by a double, triple, or spherical model. Figure 11.2 and Figure 11.7 show the vario-

FIGURE 11.6 Localization of the 4887 topsoil texture data available within East Flanders, Belgium.

grams only for the sand fraction, and Table 11.1 contains their model parameters. The variograms of the other fractions are given in Van Cleemput.[10]

All variograms contain spatial structures extending over several kilometers, ranging from 3.24 km (medium sand area) to 10.4 km (alluvial deposits), illustrating the far-reaching spatial correlation of this textural fraction. Moreover, all variogams had a zero nugget, except for the NW polders, where the relative nugget effect was very small (0.7%).

11.3.4 INTERPOLATION

11.3.4.1 Ordinary Kriging

In its basic form, kriging calculates weights given to neighboring measurement points in order to predict the value of the investigated property at an unvisited location through the following general kriging equation:

$$Z^*(\mathbf{x}_0) - m(\mathbf{x}_0) = \sum_{\alpha=1}^{n(\mathbf{x}_0)} \lambda_\alpha \cdot \left[Z\left(\mathbf{x}_\alpha\right) - m\left(\mathbf{x}_\alpha\right) \right]$$

with λ_α, the unknown weights given to the $n(\mathbf{x}_0)$ measurements $z(\mathbf{x}_\alpha)$ at locations $\mathbf{x}$, used to interpolate Z at the unknown location $\mathbf{x}_0$: $Z^*(\mathbf{x}_0)$, $m(\mathbf{x}_\alpha)$ can represent anything between a spatial trend and a global stationary mean. How $m(\mathbf{x}_\alpha)$ is considered determines the version of kriging that is used. The most common

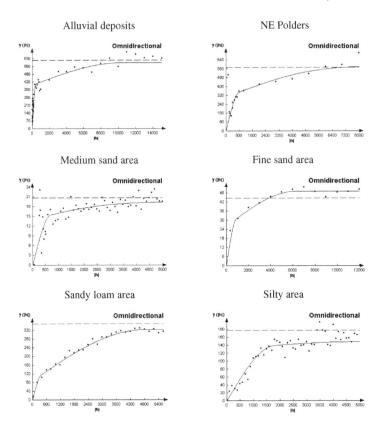

FIGURE 11.7 Omnidirectional variograms of six strata for the textural fraction sand (the variogram of the stratum NW Polder given in Figure 11.2).

TABLE 11.1
Variogram Parameters of the Double (Triple for the Polders) Spherical Models of the Sand Data of the Seven Strata

Stratum	C_0 (%²)	C_1 (%²)	a_1 (m)	C_2 (%²)	a_2 (m)	C_3 (%²)	a_3 (m)
NW polders	1.5	34	90	90	990	90	3500
NE polders	0	307	800	58	5120	214.5	8000
Medium sand area	0	25.8	780	9.8	3240		
Fine sand area	0	26.1	900	23	6240		
Sandy loam area	0	90	368	235	4800		
Silty area	0	129.6	1800	19.8	4800		
Alluvial deposits	0	396	400	200	10,400		

Note: See text for definitions of symbols.

version is ordinary kriging (OK), where $m(\mathbf{x}_\alpha)$ is considered to be locally stationary, i.e., constant within the neighborhood around $\mathbf{x}_0$. As a result, we can eliminate m from the equation, which is an advantage since we rarely know it. In order to be unbiased, the system also requires that the sum of the weights add to 1:

$$Z_{OK}^*(\mathbf{x}_0) = \sum_{\alpha=1}^{n(\mathbf{x}_0)} \lambda_\alpha Z(\mathbf{x}_\alpha) \qquad \text{with} \qquad \sum_{\alpha=1}^{n(\mathbf{x}_0)} \lambda_\alpha = 1$$

Solving this OK system requires knowledge of the variogram, but it has a unique solution obtained through matrix algebra and by introducing a Lagrange parameter. Ordinary kriging also allows a measure of the interpolation precision, the kriging variance. More details can be found in standard textbooks, like Goovaerts.[5]

11.3.4.2 Ordinary Kriging with Integrated Properties of Map Delineations

As mentioned before, the boundaries on soil maps represent abrupt discontinuities delineating internally homogeneous areas. However, soils rarely have abrupt boundaries; nevertheless, discontinuities may occur. Therefore, Boucneau et al.[11] presented criteria to identify four different types of map boundaries and developed modifications of OK in respect to the nature of these map boundaries. Their aim was to incorporate information provided by soil map boundaries into the interpolation of point data using OK.

In this chapter we identified two types of boundaries on the regional soil texture map (Figure 11.5): abrupt and gradual boundaries. Figure 11.8a represents the situation where the data are not stratified, so OK used all observation points within the neighborhood to interpolate at $\mathbf{x}_0$. In the case where a soil map is being used to stratify an area, the map boundaries could be considered to be abrupt (Figure 11.8b) or gradual (Figure 11.8c). In the first case, the boundary separates two strongly different soil populations. In this case, OK is not allowed to use data points located at the opposite side of the boundary, and so we used stratified OK. In the second situation, a gradual boundary between two types of soil occurs so that OK must be allowed to use all data, as well as the ones at the other side of the boundary. However, these two types of soil might still display a different internal structure of spatial variance, as modeled by their variograms. Consequently, when the area that is interpolated ($\mathbf{x}_0$) is located within area A, the variogram model of that soil type must be used, even for the data points located within area B, and the reverse. Therefore, in this case the boundary is not used to separate the observation points, but it is used to delineate the areas where a different variogram is used.

The regional soil texture map (Figure 11.5) contains eight boundaries, of which the boundaries separating two eolian sediments (for example, the ones

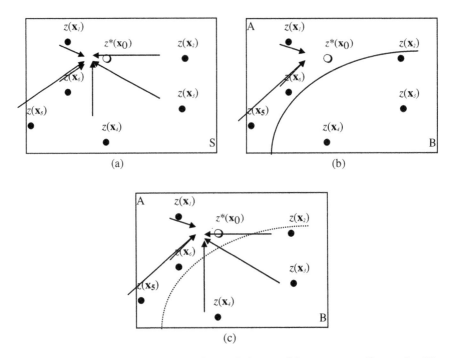

FIGURE 11.8 Illustration of the three interpolation conditions corresponding to the different types of map delineations: (a) ordinary kriging, (b) stratified ordinary kriging with an abrupt boundary, (c) ordinary kriging with a gradual boundary used for variogram stratification (arrows connect observations used for the interpolation with the location where Z has to be interpolated). (From G. Boucneau, M. van Meirvenne, O. Thas, and G. Hofman. *Eur. J. Soil Sci.,* 49: 213–229, 1998. With permission.)

between the medium sand area and the fine sand area, or between the sandy loam area and the silty area) were found to be gradual. So in the vicinity of these boundaries, OK was performed with a stratification of the variograms only (Figure 11.8c). The boundaries between alluvial and eolian sediments (for example, the ones between the polders and the medium sand area or between the river alluvia and the areas around them) were abrupt; therefore, in the vicinity of these boundaries OK was used with a stratification of both data and variograms (Figure 11.8b).

11.3.4.3 Compositional Ordinary Kriging

Ordinary kriging is an univariate interpolation method; thus, each textural fraction is considered independently from the other two. However, the sum of the three textural fractions must be 100 in order to be able to classify them with a textural triangle. This is not guaranteed by using OK. Therefore, we used compositional OK (COK), which considers all three fractions simultaneously by minimizing the sum of their prediction error variances and by taking the unbiasedness, nonnegativity, and constant sum constraints into account.[12] Assume that there are

K variables Z_k ($k = 1, \ldots, K$) that we want to interpolate simultaneously, then the interpolation of variable Z_k is obtained from its neighboring measurements by the COK estimator:[13]

$$Z^*_{COK,k}(\mathbf{x}_0) = \sum_{\alpha=1}^{n_k(\mathbf{x}_0)} \lambda_{\alpha,k} Z_k(\mathbf{x}_\alpha)$$

$$\text{with} \quad \sum_{\alpha=1}^{n_k(\mathbf{x}_0)} \lambda_{\alpha,k} = 1 \quad \forall\, k$$

$$\text{and} \quad \sum_{k=1}^{K} \sum_{\alpha=1}^{n_k(\mathbf{x}_0)} \lambda_{\alpha,k} Z_k(\mathbf{x}_\alpha) = 100$$

As with OK, solving the COK system requires knowledge of the variograms of all K variables and the introduction of K^2 Lagrange parameters. It also provides K kriging variances. As with OK, COK can be performed on a point basis or produce estimates for a given area or volume, termed block B.

11.3.4.4 Results

To evaluate the interpolation using COK compared to OK, applying the same stratification and using the same boundary types, we performed a cross-validation, i.e., estimating each measurement in turn from the remaining observations. Table 11.2 contains the results in terms of the root mean-square error (RMSE) summarized for all strata. The average predictive quality of COK is similar to OK despite the additional conditions.

Since the global predictive quality does not suggest preference of COK over OK, we used OK to predict the three textural fractions separately and checked

TABLE 11.2
Average Root Mean Square Estimation Error (RMSE) Obtained by a Cross-Validation Using OK Three Times (for Each Textural Fraction Independently) and COK for all Strata Combined (4887 Data Points)

Textural Fraction	OK RMSE (%)	COK RMSE (%)
Clay	4.88	4.90
Silt	8.19	8.17
Sand	10.33	10.32

TABLE 11.3
Sum of the Three Textural Fractions as Obtained
Independently from a Cross-Validation Using
Stratified OK

Stratum	n	m %	Minimum %	Maximum %
Alluvial deposits	372	100.01	91.85	107.40
NE polders	199	99.93	94.51	103.54
NW polders	365	99.96	95.56	104.56
Medium sand area	557	100.04	98.97	101.69
Fine sand area	1394	100.01	93.67	106.08
Sandy loam area	1444	100.02	88.57	106.63
Silty area	556	99.76	89.71	106.37

Note: n = number of data; *m* = mean.

if their sum was equal to 100 using again a cross-validation procedure. Table 11.3 shows the results for the seven strata separately, indicating the number of observation points per stratum, the overall mean of the sum of the three fractions, and the minimum and maximum sums encountered. Although the overall mean sum is close to 100 for all strata, the extremes sometimes depart as much as 11% from 100. Obviously, interpolating the three fractions independently is not a valid option if the results need to be combined for further analysis or classification into texture classes. The latter is done here.

We used stratified compositional ordinary block kriging to interpolate simultaneously the three average textural fractions of 47,454 blocks of 250 by 250 m (being a reasonable compromise between the number of blocks and the resolution), taking into account the nature of the boundaries used for stratification. Figure 11.9 shows the map of the sand fraction. Similar maps were obtained for clay and silt, while the sum of the three maps equals 100 for every pixel. Figure 11.9 shows that in some cases (e.g., between the polders and the neighboring sandy area), the boundaries represent major textural discontinuities, but in other situations, the within-stratum variation seems even larger than the variation between neighboring strata (e.g., the silty area).

The three textural maps represent the three basic layers of a GIS database about soil texture in this province. They could be used as input into regional modeling of soil processes (such as erosion) or into a pedotransfer function to obtain information about other soil properties (such as the soil moisture retention curve). We used them as input into a soil texture classification. In principle, any texture triangle could be used, but we preferred the USDA soil texture triangle[3] because it is the one that is most often internationally used (Figure 11.10). Such a map has never been produced before, since in Belgium a national soil texture triangle is mostly used. It is an indication for the textural variation of the study

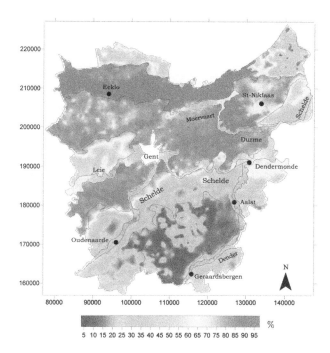

FIGURE 11.9 Topsoil sand content obtained by stratified compositional ordinary block kriging (block dimensions: 250×250 m) taking into account the nature of the boundaries used for stratification (lines). (See color version on accompanying CD.)

area that of the 12 textural classes of the USDA texture triangle, 11 occur on the map (only the class sandy clay is absent).

11.3.4.5 Sensitivity Analysis of Soil Texture Classification

Since every interpolation of each of the three textural fractions is uncertain, this uncertainty propagates through the classification. Therefore, one might be interested in the sensitivity of the classification for this uncertainty. Since the interpolation uncertainties vary over the map (depending on the sampling configuration and the variograms of the strata), the sensitivity of the texture classification has to be mapped as well.

The procedure we followed was based on the assumption that the global interpolation error follows a normal (Gaussian) distribution with two parameters: a mean value provided by the interpolated value and a variance equal to the kriging variance. Other, nonparametric techniques exist to obtain the local interpolation error numerically, like indicator kriging, Bayesian maximum entropy, or conditional simulations, but we did not want to overload this analysis by methods that are too advanced.

At each location, a Latin hypercube sampling[14] of 50 values of two of the three texture distributions was performed and combined. No correlation between

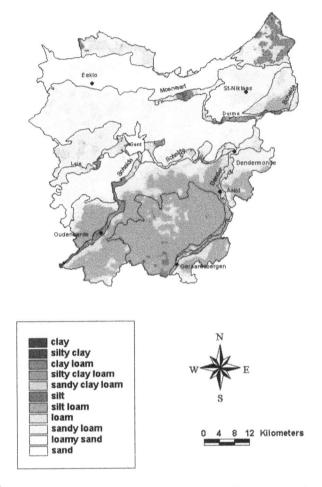

FIGURE 11.10 Soil texture classes according to the USDA texture triangle of East Flanders, Belgium. (See color version on accompanying CD.)

the three fractions was assumed, which is a conservative assumption, resulting possibly in some inflation of the uncertainty. The third textural fraction was obtained by subtracting the sum of the two sampled fractions from 100, yielding 2500 possible values of the three textural fractions.

This procedure was repeated for each of the three combinations of two textural fractions (clay–silt, silt–sand, and clay–sand). This yielded 7500 possible textural compositions at each location. These values were classified and the frequency of obtaining the same class as the predicted texture was taken as an indicator of the sensitivity of the classification (Figure 11.11). Thus, large frequencies indicate a stable classification, and the reverse.

One can observe that the lowest frequencies occur in the alluvial areas and the NE polders, both containing the most heterogeneous soils. Very stable clas-

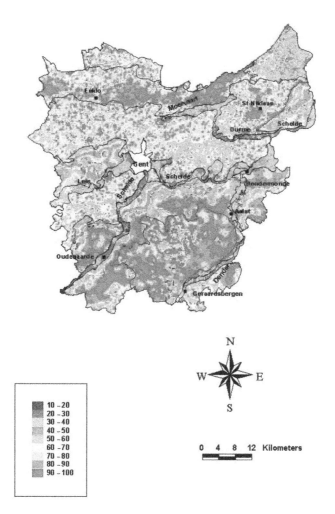

FIGURE 11.11 Frequency of obtaining the same texture class as predicted by COK (Figure 11.10) with the Latin hypercube sampling of the three textural distributions (using 7500 samples). Large frequencies indicate a stable classification. (See color version on accompanying CD.)

sifications were found in the sandy and silty areas. Lower frequencies were also found in areas with a texture near to a class boundary, causing linear patterns, often parallel to the stratum boundaries.

11.4 CONCLUSIONS

Mapping soil texture still receives a lot of attention in soil science since it is a key property guiding soil use and management. Soil texture refers to the distribution of grain size, but in practical soil survey, it is classified twice: first into texture fractions (mostly clay, silt, and sand) and second into texture classes (according to a texture triangle).

To avoid the second classification, we can use pedometrical techniques to build quantitative spatial data layers of textural fractions, which represent the basic layers about soil information in a soil information system. This contribution illustrated a sequence of steps that allows such layers to be constructed at different scales and shows how this variation can change from one scale to another. The regional mapping of soil texture accounted for two modifications:

1. The stratification was based on a general soil texture map delineating zones of different parent materials. However, the nature of the boundaries of this map was taken into account: abrupt boundaries were used to stratify data and variograms, and gradual boundaries stratified only the variograms, allowing much smoother transition zones between neighboring strata.
2. Compositional ordinary kriging was used to guarantee that the sum of the three textural fractions returned to 100. This condition was not met when the three fractions were interpolated independently.

The resulting data layers could serve many purposes; the one illustrated here was the classification according to a different texture triangle than the one used previously. Another outcome is the analysis of the quality of the classified map, represented by a sensitivity analysis based on the kriged values and their kriging variances.

It should be stressed that the data and background information used here originate mainly from the 1960s and 1970s. Only the methods of data analysis were developed more recently. This indicates the power of pedometrical techniques to retrieve new information from existing data, thus upgrading existing databases within a soil information system.

REFERENCES

1. MD Fredlund, DG Fredlund, GW Wilson. An equation to represent grain-size distribution. *Can Geotech* 37: 817–827, 2000.
2. CC Muggler, Th Pape, P Buurman. Laser grain-size determination in soil genetic studies. 2. Clay content, clay formation, and aggregation in some Brazilian Oxisols. *Soil Sci* 162: 219–228, 1997.
3. Food and Agriculture Organization of the United Nations (FAO). *Guidelines for Soil Profile Description*. FAO, 1977, p. 26.
4. PA Burrough, PFM van Gaans, R Hootsmans. Continuous classification in soil survey: spatial correlation, confusion and boundaries. *Geoderma* 77: 115–135, 1997.
5. P Goovaerts. *Geostatistics for Natural Resources Evaluation*. Oxford University Press, New York, 1997, p. 28.

6. M Van Meirvenne, G Hofman, J Van Hove, M Van Ruymbeke. A continuous spatial characterization of textural fractions and CaCO$_3$ content of the topsoil of the polder region of northwest East-Flanders, Belgium. *Soil Sci* 150: 710–716, 1990.

7. M Van Meirvenne, G Hofman. Spatial variability of soil nitrate nitrogen after potatoes and its change during winter. *Plant Soil* 40: 103–110, 1989.

8. A McBratney. Some considerations on methods for spatially aggregating and disaggregating soil information. *Nutrient Cycling Agroecosyst* 50: 51–62, 1998.

9. M Van Ruymbeke, L De Leenheer, F Appelmans, J Van Damme. La texture de la couche arable en flandre orientale. *Pedologie* 15: 255–340, 1965.

10. I Van Cleemput. Computercartografie van bodemtextuurfracties van Oost-Vlaanderen (in Dutch). M.Sc. thesis, Ghent University, Gent, Belgium, 2002.

11. G Boucneau, M Van Meirvenne, O Thas, G Hofman. Integrating properties of soil map delineations into ordinary kriging. *Eur J Soil Sci* 49: 213–229, 1998.

12. JJ De Gruijter, DJJ Walvoort, PFM van Gaans. Continuous soil maps: a fuzzy set approach to bridge the gap between aggregation levels of process and distribution models. *Geoderma* 77: 169–195, 1997.

13. DJJ Walvoort, JJ De Gruijter. Compositional kriging: a spatial interpolation method for compositional data. *Math Geol* 33: 951–966, 2001.

14. EJ Pebesma, GBM Heuvelink. Latin hypercube sampling of Gaussian random fields. *Technometrics* 41: 303–312, 1999.

12 Analysis of Complex Soil Variation Using Wavelets

R. Murray Lark

CONTENTS

ABSTRACT

We can use the wavelet transform to express the spatial variation of data in terms of a set of coefficients, each of which describes local variation at different spatial scales. It therefore requires no assumption that the variability is uniform in space. This is different from geostatistical methods that are commonly and fruitfully used for the analysis of spatial data on the soil. In this chapter the wavelet transform is introduced, focusing on the maximal overlap discrete wavelet transform as adapted for the analysis of relatively small datasets. It is possible to identify segments of a one-dimensional dataset within which the variance of the analyzed variable appears to be uniform at particular spatial scales by analysis of the wavelet coefficients. This may give insight into the complex causes of spatial variation in the soil-landscape. We can also analyze the covariation of two variables using their wavelet coefficients, to quantify how they are related at different spatial scales of generalization and to detect parts of the soil-landscape within which they may be related differently at different spatial scales. These ideas are illustrated with data on the organic carbon content and CO_2 emission rates of topsoils on a transect over farmland at Silsoe in England. The wavelet transform can be used to show how the variability of CO_2 emissions changes from one part of the landscape to another. It is also shown that, at one spatial scale, the organic carbon content of the soil and the rate of CO_2 emission are not uniformly related in space, but are strongly correlated in certain soils and uncorrelated elsewhere.

12.1 INTRODUCTION

The purpose of computing is insight, not numbers.

— R.W. Hamming[1]

Most quantitative methods used to study soils in the landscape are designed to solve practical problems. Geostatistical methods have proven particularly useful for this. We may apply kriging to predict the value of a soil property over a block or at an unsampled site, so that the farmer or engineer can decide how best to manage the soil to grow a crop or to build a road or to protect the water in an underlying aquifer.

When we use geostatistics we make assumptions. Specifically, we assume that our observation of soil property z at location $\mathbf{x}$, $z(\mathbf{x})$, is a realization of a random function, $Z(\mathbf{x})$, that is intrinsically stationary. This is a weak form of second-order stationarity and is met if two conditions hold. The first is that

$$E[Z(\mathbf{x}) - Z(\mathbf{x} + \mathbf{h})] = 0 \forall \mathbf{x},$$ (12.1)

where $\mathbf{h}$ is a separation in space, the *lag*. The second is that the variance of the differences,

$$2\gamma(\mathbf{h}) = E[\{Z(\mathbf{x}) - Z(\mathbf{x} + \mathbf{h})\}^2],$$ (12.2)

depends only on $\mathbf{h}$ and not on $\mathbf{x}$. The function $\gamma(\mathbf{h})$ is the semivariogram.

These assumptions also extend to the multivariate case, where we may use cokriging to estimate a soil property that is spatially correlated with a cheaply measured surrogate variable such as a remote sensor image.

Soil scientists, educated in the biological, chemical, and physical processes that together form the soil, may have reservations about treating soil properties as random processes. Does *random* not mean uncaused? Matheron[2] shows that the use of random models may be defended as a pragmatic but philosophically respectable response to complexity. Random models may one day be superseded in a particular area of science by mechanistic description. Indeed, a successful geostatistical analysis of the soil often combines a physical description of some components of the variation with a probabilistic description of other elements too complex, or perhaps chaotic,[3] to yield to physics. See, for example, the combination of kriging with a process model in Saito and Goovaerts's[4] analysis of the distribution of pollutants in soil.

There are other reasons why the soil scientist may have reservations about using geostatistics to analyze properties of the soil. These are centered on the plausibility of the intrinsic hypothesis stated above. Equation 12.1 implies stationarity of the mean of the process, which may be implausible when there is a strong trend. A trend that is explicable in terms of a mechanistic process may be eliminated if a process model is incorporated. Otherwise, the methods of universal kriging[5] or intrinsic random functions of order greater than zero[6] may be used. A less tractable problem is posed when our soil variable does not appear to be uniform in the variance. This is inconsistent with the second component of the intrinsic hypothesis, Equation 12.2.

It is not difficult to imagine a soil-landscape in which it is implausible to assume that soil properties are generated by a random process with uniform variance. The author learned much of his field soil science near Wytham in Oxfordshire. As you walk from the Thames to the top of Wytham Wood, alluvium with gravels and sands from braided streams gives way to terrace gravels overlaid by solifluction deposits from the Oxford Clay. The Oxford Clay itself dominates the hillslope, and a spring line marks the edge of the calcareous grit, a Corallian sandstone. The hill is capped by Coral Rag, an Upper Jurassic limestone. In parts of the hilltop, the relief is very marked with hillocks of Coral Rag surrounded by the Kimmeridge Clay, which overlies it in the local stratigraphy. It is clear that the variation of clay content of the soil, for example, will change along this notional transect. At the top of the hill, clay content will vary sharply according to the mesotopography at scales of around 10 m. Clay content of the soil on most of the hillslope will be large and fairly uniform, although marly deposits near the streamline, fragments of the calcareous grit, and drift of varying composition will impose short-range variation in the composition of the topsoil. Complex patterns of texture then emerge on the terrace gravels, and another pattern again on the floodplain. We may compute a semivariogram of clay content along this transect, but it is not clear that it characterizes a random function that we could plausibly assume to be realized anywhere on it.

The reader could take a similar imaginary walk along a familiar soil-landscape of her choice and probably note the implausibility of the intrinsic hypothesis in the case of at least a few soil properties. Cases are also found in the literature. Webster and de la Cuanalo[7] describe a 3-km transect across central England that they sampled at 10-m intervals. This crosses contrasting outcrops with more or less regular boundary spacing. Lark and Webster[8] and Webster[3] showed how the variance of clay content and pH computed within a moving window fluctuates along the transect in response to changes in the geology.

There are pragmatic solutions to such problems when our goal is geostatistical prediction. One approach, suitable when we have many data, is to compute local semivariograms within a moving window (e.g., Walter et al.[9]). This approach has drawbacks, as we shall see, particularly if we want our spatial analysis to give some insight into the variation of the soil at different spatial scales. What we want ideally is a method to analyze the spatial variation of soil that does not make assumptions of stationarity in the variance. This is the reason why there is growing interest in the wavelet transform as a statistical tool to analyze soil data and cognate variables.[8]

Wavelets are analyzing functions with mathematical properties that make them particularly suited to the study of complex spatial variation. In particular, a specific wavelet function responds only to the variation of a variable at a precise spatial scale in a local region. This makes the wavelet transform a powerful tool to analyze variation that is nonstationary in the variance. In this chapter the basic properties of the wavelet transform are considered and show how the transform can be used to analyze and elucidate the variation of soil properties and processes, through both univariate and multivariate analysis. Readers wanting more details are referred to papers by Lark and Webster[8,10] and to the articles of Kumar and Foufoula-Georgiou[11] and Mallat.[12]

12.2 THEORY

12.2.1 GENERAL PRINCIPLES

We start our discussion in one dimension. Let us represent our soil variable as a function of x, the location in space, $f(x)$. The wavelet transform is an integral transform. That is, it is a transform of a function by means of an integration with a second, a *basis function* $k(x)$, to generate a coefficient $K(\tau)$ so that in general,

$$K(\tau) = \int_{-\infty}^{+\infty} f(x)k(x \mid \tau)dx. \tag{12.3}$$

One familiar example of the integral transform is the Fourier transform. If $k(x \mid \tau)$ is a complex sinusoidal function of x of frequency τ, then $K(\tau)$ is a Fourier coefficient that characterizes the variation of $f(x)$ at frequency τ. The Fourier basis function is complex, and so the coefficient is also complex and can convey two pieces of information: the amplitude and the phase of the variation in $f(x)$ at frequency τ.

Since a sinusoidal function has uniform amplitude over the range $-\infty$ to $+\infty$, Fourier analysis of soil data will have similar problems as geostatistical analysis

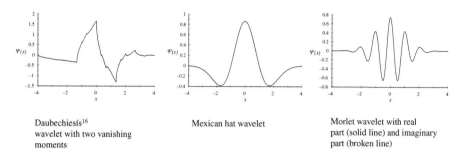

Daubechiesís[16] wavelet with two vanishing moments

Mexican hat wavelet

Morlet wavelet with real part (solid line) and imaginary part (broken line)

FIGURE 12.1 Some commonly used wavelet functions.

when a variable is not stationary in the variance; i.e., the apparent amplitude of variations at a particular frequency change in space. A Fourier transformation can always be computed, but the coefficients are unlikely to generate much insight into a soil variable unless it shows more or less uniform periodic variation. Thus, Webster[13] found the Fourier transform a useful tool to study soil variation in the special case of regularly patterned Gilgai terrain in Australia, but the method has not been widely used in pedometrics.

Wavelets do not have these drawbacks. A wavelet oscillates locally like a wave, but it also damps rapidly to zero on either side of its center. This means that the coefficient derived from an integral transform with a wavelet describes local variation within the interval where the wavelet takes nonzero values. We denote the basic wavelet function, or mother wavelet, by $\psi(x)$. Some examples of wavelet functions are given in Figure 12.1.

The mother wavelet is not an arbitrary function, but must satisfy three conditions:

1. Its mean is zero, i.e.,

$$\int_{-\infty}^{\infty} \psi(x)dx = 0. \tag{12.4}$$

2. Its squared norm is 1, i.e.,

$$\int |\psi(x)|^2 \, dx = 1. \tag{12.5}$$

3. It has a compact support; that is, it damps rapidly to zero and so operates very locally.

In order to generate a complete analysis of a set of data, the mother wavelet is transformed by two parameters. One of these is a scale parameter, λ. If the scale parameter is changed this either dilates or shrinks the wavelet; i.e., it expands

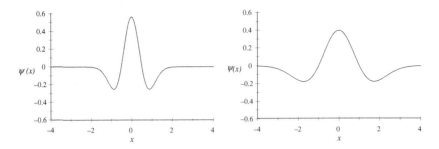

FIGURE 12.2 Two dilates of the Mexican hat wavelet function, achieved by changing the scale parameter λ.

or contracts the interval over which the function responds to variation in the data and so changes the scale at which the wavelet coefficient describes the data. Figure 12.2 shows two dilations of the Mexican hat wavelet, obtained with different values of the scale parameter. As well as dilating the wavelet we may *translate* it, i.e., move it over the transect in order to analyze the soil variation at different locations. The dilated and translated wavelet functions are denoted by $\psi_{\lambda,x}(w)$, where the parameters λ and x respectively define the scale and location represented by a particular wavelet function, and w is a displacement of position. The mother wavelet is dilated and translated by

$$\psi_{\lambda,x}(w) = \frac{1}{\sqrt{\lambda}}\,\psi\left(\frac{w-x}{\lambda}\right), \lambda > 0, x \in \Re, \tag{12.6}$$

where $\Re$ denotes the set of real numbers.

The integral transform of $f(x)$ with $\psi_{\lambda,x}$ thus generates a wavelet coefficient $Wf(\lambda, x)$ that describes the variation of $f(x)$ in the locality of x at a spatial scale determined by the parameter λ:

$$\begin{aligned}
Wf(\lambda, x) &= \int_{-\infty}^{\infty} f(w)\psi_{\lambda,x}(w)dw \\
&= \int_{-\infty}^{\infty} f(w)\frac{1}{\sqrt{\lambda}}\,\psi\left(\frac{w-x}{\lambda}\right)dw.
\end{aligned} \tag{12.7}$$

If we reduce the scale parameter then the transform returns a wavelet coefficient that describes fine-scale variation about the location x; increasing the parameter dilates the wavelet and returns a wavelet coefficient that describes variation around x at a coarser scale. This is the key advantage of the wavelet transform over methods that attempt to analyze local variation within a fixed window (e.g., Walter et al.[9]). In the wavelet transform the window is adapted to the spatial frequencies (equivalent to scale) that are analyzed, while in the other windowing methods all frequencies are analyzed within a more or less arbitrarily chosen window.

In Equation 12.7, scale and location are continuous. A function may be analyzed by a continuous wavelet transform, but data are always discontinuous at the limit, measured at discrete locations. Nonetheless, we may approximate to a continuous wavelet transform (CWT) of data. Lark and McBratney[14] show an example of the analysis of soil data with a continuous wavelet transform, and Si[15] gives another example. However, most wavelet applications use the discrete wavelet transform or developments of this (wavelet packets), and that is the focus in the present chapter.

12.2.2 THE DISCRETE WAVELET TRANSFORM

In the discrete wavelet transform we discretize the scale parameter and the location by which the mother wavelet is dilated and translated. The scale parameter is increased dyadically so that

$$\lambda = \lambda_0^j, j = 1, 2, \dots . \tag{12.8}$$

By convention, λ_0 is set to 2. The finest-scale parameter is therefore twice the basic interval between discrete data values. The location is incremented in scale-dependent steps:

$$x = n x_0 \lambda_0^j, n = 1, 2, \dots . \tag{12.9}$$

where x_0 is conventionally set to 1, the unit interval between discrete samples. On the basis of these conventions, the discrete wavelet function is given by

$$\psi_{j,n}(x) = \frac{1}{\sqrt{\lambda_0^j}} \psi \left(\frac{x - n x_0 \lambda_0^j}{\lambda_0^j} \right)$$

$$= \lambda_0^{-j/2} \psi(\lambda_0^{-j} x - n x_0). \tag{12.10}$$

Under the conventions that $\lambda_0 = 2$ and $x_0 = 1$, Equation 12.10 becomes

$$\psi_{j,n}(x) = \frac{1}{\sqrt{2^j}} \psi \left(\frac{x - n 2^j}{2^j} \right)$$

$$= 2^{-j/2} \psi(2^{-j} x - n). \tag{12.11}$$

Thus, the discrete wavelet coefficient for the nth translation of the jth dilation of the mother wavelet is $D_{m,n}$, an inner product:

$$D_{j,n} = \left\langle f, \psi_{j,n} \right\rangle = 2^{-j/2} \int f(x)\psi(2^{-j}x - n)\mathrm{d}x. \qquad (12.12)$$

If the mother wavelet function is appropriately chosen[16] then the set of dilates and translates under the discrete wavelet transform (DWT) constitute a complete orthonormal basis for all functions, $f(x)$, of finite variance. This means that we may approximate any data sequence of finite variance to any level of precision from the DWT coefficients by multiplying them by the corresponding wavelet and summing the products. Thus,

$$f(x) = \sum_{j=-\infty}^{\infty} \sum_{n=-\infty}^{\infty} \left\langle f, \psi_{j,n} \right\rangle \psi_{j,n}(x). \qquad (12.13)$$

The first summation is over scales, and the second is over translations. In essence, if we were to exclude all wavelet coefficients from the approximation with a scale parameter less than some threshold value, then the result would be a smoothed approximation to the original data since components represented at finer scales than the threshold have been filtered out. This is discussed below in the context of multiresolution analysis.

In practice, the DWT is applied with the pyramid algorithm.[17] The DWT can be thought of as a convolution of the data $f(x)$ with the scaled wavelet $\psi_{j,0}(x)$, the output of the convolution then being subsampled at interval $n_0 2^j$. The pyramid algorithm achieves this efficiently. The details are not considered here, but in effect the filter elements are unchanged while the data are downsampled by a factor of 2 in successive steps, so that the data are shrunk relative to the wavelet and coefficients of successively large scale are extracted.

One obvious problem in the practical implementation of the DWT is the infinite range of the translations in Equation 12.13. In practice this means that when the data are convolved with a wavelet function, we have a problem near the start or end of the sequence where the filter overlaps the end of the data. One solution is to wrap around, in effect to treat the data as if they are in a circle. This is appropriate for some wavelet applications, but not for the analysis of spatial data. After all, we are interested in wavelets because they allow localized analysis of variation. It would be perverse to use wavelet coefficients that are partly derived from local data and partly from data at the other end of the transect. Another solution is to use adapted wavelet filters that only operate over a finite interval. The adapted wavelet coefficients proposed by Cohen and colleagues[18] have been applied to the analysis of soil variation,[10] and Figure 12.3 shows the adapted version of the Daubechies[16] wavelet in Figure 12.1.

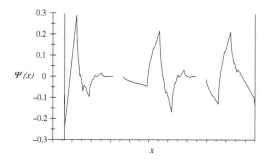

FIGURE 12.3 Adapted versions of Daubechies's[16] wavelet with two vanishing moments, after Cohen et al. (1993). Shown are the adapted filter for the left-hand side of the transect, the center (unadapted), and the right-hand adapted filter.

12.2.3 MULTIRESOLUTION ANALYSIS

Equations 12.12 and 12.13 show how we can conduct a wavelet transform of data, and then invert this to reconstruct the data. If, after a forward transform, we set to zero all wavelet coefficients with a scale parameter equal to 2^1, then on inverse transformation we would generate a smooth approximation to the data in which the components of variation at this spatial scale have been filtered out. If the filtered data are transformed, then the coefficients corresponding to scale parameter 2^2 are set to zero before another inverse transformation, and we would create a version of the data that is smoother again.

At each dilation j we may compute a detail component, the difference between the smooth version of the data at this scale parameter and the less smooth version at scale parameter 2^{j-1}. We may obtain these detail components by transforming the original data, setting to zero all the coefficients other than those corresponding to the scale of interest, and then back-transforming. Because of the orthogonality of the DWT, the resulting components are additive components of the original data, and we call this partition of the data into scale-specific components a multiresolution analysis. An example of a multiresolution analysis of soil data is given in the case study.

A more rigorous account of multiresolution analysis is given by Mallat[12] and Lark and Webster.[8] One point worth noting is that we can think of each step in the scale parameter of the multiresolution analysis as a projection of the original data onto a scale-specific subspace of the set of all vectors of real numbers. This subspace is nested within the subspaces for all smaller-scale parameters. Within a particular subspace, say for scale parameter 2^j, our representation of the data is the direct sum of two orthogonal subspaces; the component in one of these is the detail representation of the data at scale 2^{j+1}, and the other is the smooth version of the data at this scale. The basis of the first subspace is the wavelet function dilated by $\lambda = 2^j$. The basis of the second subspace is a so-called scaling function, a smoothing function of norm 1, which is a unique complement to any wavelet.

12.3 STATISTICAL DEVELOPMENTS

12.3.1 Wavelet Variance and Change Detection

The DWT decomposes a set of data into additive components of different spatial scales. The variance of the data is similarly partitioned into scale-specific components. The variance of N data on variable u at scale parameter 2^j with n_j DWT coefficients is estimated by

$$\hat{\sigma}^2_{u,j} = \frac{1}{2^j n_j} \sum_{n=1}^{n_j} D^2_{j,n} , \qquad (12.14)$$

Percival[19] calls this the sample wavelet variance.

Percival and Guttorp[20] defined a maximal overlap discrete wavelet transform (MODWT). This is similar to the DWT, except that locations are incremented in steps of 1 at all scales. This means that, in effect, the MODWT coefficients are obtained by a convolution of the data with a dilated wavelet function, but while the pyramid algorithm for the DWT subsamples the filtered values in steps of 2^j, all the values from the convolution are retained in the MODWT. Percival and Guttorp[20] showed that it is more efficient to estimate the wavelet variance from MODWT coefficients, since these retain information that is discarded by the DWT. However, the MODWT coefficients are not orthogonal, and so there are fewer effective degrees of freedom than there are coefficients.[20]

Lark and Webster[10] proposed an adapted MODWT (AMODWT), using the adapted wavelet filters of Cohen et al.[18] in order to return MODWT coefficients near the ends of a transect. As they showed, this introduces some bias into wavelet variance components, especially at the beginning of a transect at scale parameters 2^1 and 2^2. Relatively little information is lost if these affected coefficients are deleted.

If $\breve{d}_{j,n}$ is the nth AMODWT coefficient for scale parameter 2^j, then the AMODWT wavelet variance for this scale, at which there are $\breve{n}_j$ coefficients, is

$$\breve{\sigma}^2_{u,j} = \frac{1}{2^j \breve{n}_j} \sum_{n=1}^{\breve{n}_j} \breve{d}^2_{j,n}. \qquad (12.15)$$

Confidence limits may be calculated for this wavelet variance (see Lark and Webster[10]).

The wavelet variance is a useful tool to describe scale-specific variation, but since it is a measure of variability pooled over all locations (by the summation in Equation 12.15), it is directly comparable to the power spectrum or the semivariogram. That is, its interpretation requires some stationarity assumption. However, our interest in wavelets is motivated by the observation that soil variability is not itself uniform. Each wavelet coefficient is a local scale-specific measure of variability, and we should use this information to characterize how the variability of a soil property changes in space. Whitcher[21] showed that changes in variance at a particular scale can be detected from DWT coefficients. For n_j coefficients we define

$$S_k \equiv \frac{\sum_{m=1}^{k} D_{j,m}^2}{\sum_{m=1}^{n_j} D_{j,m}^2}, k = 1, 2, \ldots, n_j - 1. \tag{12.16}$$

If a variable has uniform variance, then S_k is expected to increase linearly with k. If the variance changes at a particular location, then there will be a break of slope in the plot of S_k against k. Breaks of slope may be detected by defining

$$B \equiv \max\left[\left\{\max_{1 \leq k \leq n_j - 1}\left(\frac{k}{n_j - 1} - S_k\right)\right\}, \left\{\max_{1 \leq k \leq n_j - 1}\left(S_k - \frac{k-1}{n_j - 1}\right)\right\}\right]. \tag{12.17}$$

When we compute B from a set of DWT coefficients, then location k, which gives rise through Equation 12.17 to the value of B, is the location such that the variances on either side are most different. It is a candidate change point. Whitcher[21] used Monte-Carlo methods to compute a sample distribution of B under the null hypothesis that the wavelet coefficients have uniform variance. This entailed the assumption that the DWT coefficients are independent random variables. This assumption is reasonable for DWT coefficients of certain classes of processes, but it does not hold for MODWT or AMODWT coefficients.

In the case study below, a similar approach is used, but AMODWT coefficients were used throughout. The B statistic was computed from all n AMODWT coefficients for scale parameter 2^j in Equation 12.17. The AMODWT coefficients cannot be assumed to be independent, so I obtained a sample distribution for the B statistic by Monte-Carlo simulation. The B statistic was computed for each of 5000 sets of n AMODWT coefficients of simulated data transects with the same semivariogram as the basic data. The simulations were generated by sequential Gaussian simulation with the routine SGSIM in the GSLIB library.[23] If the computed B statistic for the data exceeded the 95th percentile of the Monte-Carlo values, then the null hypothesis that the real data have uniform variance over the sequence of n value that is of interest, and at the scale parameter that has been specified, was rejected. The location of the change in variance is l, where the value of S_k at $k = l$ (see Equation 12.16) gives rise to the value of the B statistic in Equation 12.17.

This method is first applied to all $\breve{n}_j$ locations at a particular scale parameter. If a significant change in variance was detected at location l, then the procedure was repeated for the segments $1, 2, \dots l$ and $l+1,\ l+2, \dots, \tilde{n}_j$. Once a particular segment was judged to be uniform according to the B statistic, then no further change points were sought. I therefore proceed in the search for change points until the transect is divided into segments that are all judged uniform.

12.3.2 WAVELET COVARIANCE AND CHANGE DETECTION

Lark and Webster[10] proposed, after Whitcher and colleagues,[22] that a scale-specific component of the covariance of properties u and v can be estimated from AMODWT coefficients; this is the AMODWT covariance given by

$$\breve{C}_{u,v,j} = \frac{1}{2^j \tilde{n}_j} \sum_{n=1}^{\tilde{n}_j} \breve{d}_{j,n}^{u} \breve{d}_{j,n}^{v} \tag{12.18}$$

where $\breve{d}_{j,n}^{u}$ and $\breve{d}_{j,n}^{v}$ are respectively the nth AMODWT coefficients of variables u and v for scale parameter 2^j.

If we standardize the wavelet covariance for a particular scale by the wavelet variances of u and v, we obtain a scale-specific wavelet correlation, $\breve{\rho}_j$:

$$\breve{\rho}_j = \frac{\breve{C}_{u,v,j}}{\sigma_{u,j}\sigma_{v,j}} \ . \tag{12.19}$$

This expresses the correlation of u and v specifically for scale parameter 2^j. In order to obtain confidence limits for this estimate, it is transformed by Fisher's z transformation:

$$h\!\left(\breve{\rho}_j\right) = \tanh^{-1}\!\left(\breve{\rho}_j\right). \tag{12.20}$$

The transformed estimate of the correlation is distributed approximately normally with a variance of $1/(n-3)$, where the correlation is derived from n independent observations. Now the wavelet correlation $\breve{\rho}_j$ is not derived from $\tilde{n}_j$ independent observations. The ordinary discrete wavelet functions (all dilations and translations) are orthogonal, and the resulting wavelet coefficients are uncorrelated for several classes of processes, but these DWT coefficients are a subsampled set of the AMODWT coefficients, every 2^j th. Whitcher and colleagues[22] therefore suggested that n be set to $N/2^j$ when calculating sample variances for wavelet correlations (where N is the total number of data). Lark and Webster[10] calculated effective degrees of freedom for the AMODWT wavelet correlation by Monte-Carlo analysis. The effective degrees of freedom were more than $N/2^j$, especially for small j. This shows that Whitcher et al.'s[22] approach is conservative.

The wavelet correlation measures the overall scale-specific correlation between two variables. In this way, it is comparable to the codispersion coefficient from geostatistics[24] or the cross-spectrum in spectral analysis. Like these measures, it is an overall statistic obtained from all the local information contained in the wavelet coefficients. As noted above in the discussion of wavelet variance, the motivation for wavelet analysis of soil data is to exploit the localized nature of the information in wavelet coefficients, so we cannot be satisfied with the wavelet correlation as a statistical descriptor of spatial covariation. Lark and Webster[10] showed how changes in the covariance of two variables, at a particular scale parameter, may be detected by examining the cumulative sum of products of wavelet coefficients. The wavelet correlations on either side of the candidate change point may then be compared statistically.

This method uses a normalized cumulative sum of products statistic computed from the AMODWT coefficients for variables u and v for each scale parameter. This statistic is directly analogous to the cumulative sum of squares, S_k, defined in Equation 12.16:

$$S_k^{u,v} = \frac{\sum_{m=1}^{k} d_{j,m}^u d_{j,m}^v}{\sum_{m=1}^{\tilde{n}_j} d_{j,m}^u d_{j,m}^v}, k = 1,\ldots,\tilde{n}_j - 1. \tag{12.21}$$

Under a null hypothesis that the covariance over all locations is uniform, $S_k^{u,v}$ is expected to increase linearly with k. As in the detection of changes in the variance, breaks of slope in the plot of $S_k^{u,v}$ against k are found by identifying the location k that gives rise to the value $B^{u,v}$:

$$B^{u,v} \equiv \max\left[\left\{\max_{1 \leq k \leq n_j - 1}\left(\frac{k}{n_j - 1} - S_k^{u,v}\right)\right\}, \left\{\max_{1 \leq k \leq n_j - 1}\left(S_k^{u,v} - \frac{k-1}{n_j - 1}\right)\right\}\right]. \tag{12.22}$$

This k is a candidate change point. The wavelet correlations are then computed on each side of the change point, and the absolute difference between them is compared with critical values of the distribution of this statistic, for random variables of uniform correlation, obtained by a Monte-Carlo simulation. If the correlations on either side of the change point are found to differ significantly ($p < 0.05$), then this procedure is repeated for each of the two segments defined by the change point; otherwise, the search for change points is stopped. If this procedure is iterated, the transect is divided into segments within which the wavelet correlation (at a specified scale) is uniform.

12.4 CASE STUDY

12.4.1 THE DATA

The data presented in this chapter were collected as an adjunct to a study on nitrous oxide emissions from the soil.[25] This study is to be published elsewhere, but here we examine data on CO_2 emissions measured at the same time, since these usefully illustrate how the wavelet transform can give insight into complex spatial variability of soil properties and processes.

The gas flux measurements were made on soil cores that had been collected in the field and refrigerated. This was because I wanted to use a range of wavelet analysis tools on the data, and this requires that the number of data is an integer power of two. It was decided to collect data from 256 locations; this precludes direct measurement of gas fluxes in the field, since to obtain measurements from 256 sites by *in situ* measurements without confounding spatial variation with variation over a substantial time interval is not feasible.

The 256 sites were sampled on a linear transect at 4-m intervals. The transect crossed farmed fields and some waste ground at Silsoe Research Institute in England. The transect spans four fields: Cashmore field (positions 1 to 50) Banqueting field (positions 57 to 76), Obelisk East field (positions 80 to 152), and Bypass field (positions 154 to 256). Intervening positions were in grassed waste ground. Cashmore field had been drilled with a winter barley crop 14 days before the soil was sampled, and Bypass and Obelisk East fields with winter wheat 12 days before. Banqueting field was in wheat stubble.

Cashmore field is over the Lower Greensand to the north and Gault Clay to the south (both Cretaceous formations). The geological boundary is between positions 30 and 40 on the transect. The Gault Clay underlies all the sample sites south of Cashmore field. The measurements in this study were made on topsoil samples, the properties of which are determined by superficial deposits. Deposits of glacial drift of variable texture are important,[26] derived mostly from Jurassic clays and Gault Clay, but also influenced by the Lower Greensand. At all locations over the Lower Greensand there are superficial deposits. The Gault Clay occurs at the surface in some parts of this landscape, but at most locations drift overlies it. Around the lowest-lying position on the transect is a narrow ditch, spanned by sample positions 50 and 51. The topsoil here is derived from alluvium that varies in texture from clay to fine loam. Peat lenses are also found in the soil in this part of the landscape.

This variability in the solid and drift geology influences the properties of the surface soil. The following description uses the textural classes of the Soil Survey of England and Wales.[27] The lightest-textured soils, sandy loam, occur at the top of the transect in drift over the Lower Greensand. Sandy clay loams and sandy clays are formed in drift over the Gault Clay in Cashmore field, and clay loams and clays occur near the ditch at position 51. South of the ditch the transect crosses two low ridges, the break of slope between these occurring near the boundary between Bypass and Obelisk East fields. The soils of the northernmost

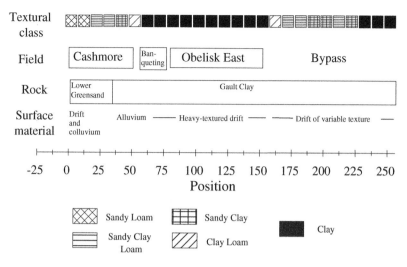

FIGURE 12.4 Summary of variation along the transect used in the case study. Shown are (top) textural classes (27) for the topsoil at 40-m intervals (the key is at the bottom of the figure), field boundaries and names, solid geology, and a general description of superficial deposits. The numbers on the abscissa are (dimensionless) sample numbers in order, 1 to 256. These are at sites separated by 4-m intervals.

ridge (Obelisk East and Banqueting fields) are all clays in the topsoil, while the soils of the second ridge are more variable in texture; sandy clay loams, sandy clays, clay loams, and clay occur, the latter at the end of the transect. The key features of the transect are summarized in Figure 12.4.

All the soil samples were collected within a period of 7 hours. At each site four cores of 44 mm diameter for 0 to 150 mm depth were collected with a gouge auger. The cores at each site were collected from as close together as possible. The cores were taken with minimum delay to a 4°C cold room and were kept at this temperature until they were analyzed.

One core from each location was selected randomly, and its fresh weight and length were recorded. Cores were preincubated at 15°C for 17 to 24 hours in a 1-l Kilner jar, with the jar lids positioned but without a rubber gasket or clamp. This prevented desiccation, but allowed some gas transfer from the jar. The jars were then flushed with laboratory air and resealed, but this time with a rubber gasket and clamp in position so that they were gas tight. An initial sample (20 ml) of the gas headspace was collected and injected into an evacuated vaco tube. The jars were incubated at 15°C for 24 hours, then two further 20-ml samples from the headspace were collected and each injected into vaco tubes. Within a few days, the gas samples in the vaco tubes were collected in a 20-ml syringe and injected into an Ai93 gas chromatograph, which analyzed CO_2 by a thermal conductivity detector (TCD). Standards for CO_2 were also injected to calibrate the instrument. From the change in concentrations of CO_2 in the headspace, the rate of gas emission was determined. The moisture content of the core was then

TABLE 12.1
Summary Statistics of the Data on Soil Variables

	CO_2 Emission µg kg^{-1} day^{-1}	CO_2 Emission ln µg kg^{-1} day^{-1}	Total Organic Carbon %	Total Organic Carbon ln %
Mean	5315.1	8.37	2.49	0.88
Median	4085.5	8.32	2.47	0.90
Standard deviation	5261.4	0.58	0.72	0.26
Skewness	5.07	0.90	2.16	0.19

determined by oven drying a sample to constant weight at 105°C. The rate of emission was then expressed per unit weight of dry soil.

Other analyses were done on the soil. For the present discussion, we are interested in soil organic carbon content. This was determined for an air-dried subsample passed through a 2-mm sieve. The determination was done by combustion in a Leco analyzer. This instrument uses a version of the Dumas digestion method described by Tabatabai and Bremner.[28] A correction was made for the carbonate content, which was determined by Williams's[29] water-filled calcimeter method.

Table 12.1 shows summary statistics of the CO_2 emission rates and the total carbon contents and of the same variables transformed to natural logarithms. Both variables were strongly skewed before transformation, but log transformation reduced the coefficient of skew to less than 1. The remaining analyses were done with the log-transformed data.

Figure 12.5 and Figure 12.6 show the log-transformed CO_2 emission rate and soil carbon content plotted against position on the transect. There are similar

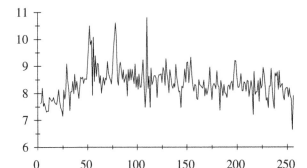

FIGURE 12.5 Measured CO_2 emissions along the transect, transformed to natural logarithms. The numbers on the abscissa are (dimensionless) sample numbers in order, 1 to 256. These are at sites separated by 4-m intervals.

Organic C/ln (%)

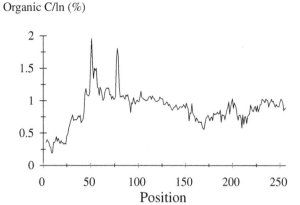

FIGURE 12.6 Measured soil total carbon content along the transect, transformed to natural logarithms. The numbers on the abscissa are (dimensionless) sample numbers in order, 1 to 256. These are at sites separated by 4-m intervals.

trends in the data and two peaks at corresponding positions (over the alluvium near the ditch at position 50 and near position 75 in the stubble on Banqueting field). There is another peak in CO_2 emission in the heavy-textured drift over Obelisk East field. The overall correlation coefficient between the two log-transformed variables is 0.65. A wavelet analysis can elucidate the trends and short-range variability, and more information will emerge about local and scale-dependent relationships between these variables.

12.4.2 MULTIRESOLUTION ANALYSIS

A multiresolution analysis was performed on the CO_2 emission rates with Daubechies's[16] wavelet with two vanishing moments, as adapted to the finite interval by Cohen et al. (1993) (Figure 12.3). I used the shift-averaging method proposed by Lark and Webster[8] after Coifman and Donoho.[30] The wavelet transform is conducted in a moving window of length $2^7 \times 4$ m. Detail components are computed within the window for scale parameter $\lambda = 2^1 \times 4$ m, ..., $2^5 \times 4$ m. The window is then shifted by one position and the procedure is repeated. The resulting series of detail components are then averaged. This procedure was first proposed by Coifman and Donoho[30] to avoid artifacts in the multiresolution analysis due to aliasing of features on a transect with the tiling of the transect into sections of 2^j units.

Figure 12.7 shows the detail components of log CO_2 emission rates computed in this way for the scales 8 to 128 m. Figure 12.8 shows the smooth representation of the data at scale 128 m (detail components for all scales up to 128 m filtered out), and with each detail component added in turn up to scale 16 m (if the 8-m detail component is added, this simply reconstructs the original data).

The smooth representation at scale 128 m shows a simple trend in the CO_2 emissions. There is a sharp decline at the top of the transect to the lowest values

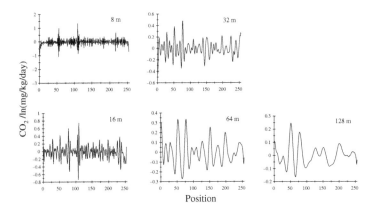

FIGURE 12.7 Detail components from multiresolution analysis of CO_2 emissions. The numbers on the abscissa are (dimensionless) sample numbers in order, 1 to 256. These are at sites separated by 4-m intervals.

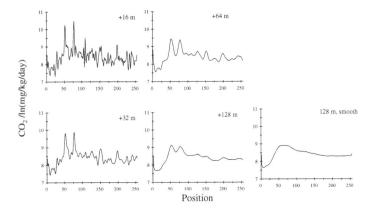

FIGURE 12.8 Smooth representations from multiresolution analysis of CO_2 emissions. The numbers on the abscissa are (dimensionless) sample numbers in order, 1 to 256. These are at sites separated by 4-m intervals.

near the center of Cashmore field, then an increase to largest values over the alluvium and heavy-textured drift from near positions 50 to 75. There is a small decline beyond position 75, although CO_2 emission rates for the cores from the Gault Clay are all larger than the rates over the Lower Greensand.

It is notable that peaks in CO_2 emission near positions 50 and 75 appear at scale parameter 128 m (detail) (see Figure 12.7 and Figure 12.8). Further variation occurs here at finer scales, but this may obscure the presence of structure at relatively coarse scale at these locations. By contrast, the strong peak at position 110, the largest emission rate on the transect, only appears at scale parameters 16 and 8 m. This implies that the first two peaks are related to broader-scale soil variations (increases in soil wetness and organic content, which promote CO_2

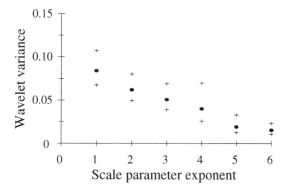

FIGURE 12.9 Sample AMODWT wavelet variances of CO_2 emissions with 95% confidence intervals. The numbers on the abscissa are (dimensionless) sample numbers in order, 1 to 256. These are at sites separated by 4-m intervals.

production) over alluvium and stubble over clay, while the latter represents a strongly localized hot spot of soil respiration. The identification of such features is particularly facilitated by the wavelet-based multiresolution analysis.

12.4.3 UNIFORMITY OF THE VARIANCE

Figure 12.9 shows the sample AMODWT wavelet variances computed with Equation 12.15, and with 95% confidence limits. These show a decline in the variance with increasing scale parameter; i.e., most of the variation in CO_2 emissions is associated with spatial variations at the finest scale. This interpretation is straightforward insofar as the CO_2 emissions can be treated as realizations of a process stationary in the variance at all scales.

When we examine the detail components in Figure 12.7, it seems clear that the variance of these is not uniform. At scale parameter 32 m, for example, the variation seems much larger at positions up to around 75 than on the rest of the transect. One way to examine this nonuniformity of the variance is to compute the local components of variance at each scale. The component at position n for scale parameter 2^j is calculated by

$$\frac{\tilde{d}^2_{j,n}}{2^j}.$$

(12.23)

These components can then be plotted. This is done for log CO_2 emissions in Figure 12.10. This is a layered plot; i.e., the thickness of the lowest layer at any position is equal to the local variance component at that scale, obtained with Equation 12.23. Layers representing finer scales are then stacked on top of this, so that the overall height of the graph at any position is the sum at this position of components for all scales.

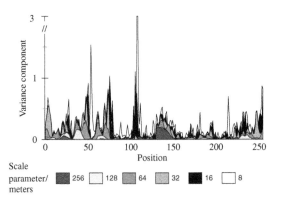

FIGURE 12.10 Local components of wavelet variance by scale parameter.

This graph shows, for example, the very large variation at the finer scales around the hot spot at position 110 (note the truncated scale). It also shows a general reduction in the importance of coarser-scale components of variance at positions after around 75.

Graphs such as Figure 12.10 are useful tools for visualizing changes in the variance of a variable at different spatial scales. However, we need a method to identify changes in variance that appear to be significant — that is, to test the change against a null hypothesis of an underlying stationary process. This was done with the methods described in Section 12.3.1.

For each scale parameter I computed the normalized cumulative sum of squared AMODWT coefficients, S_k, described in Equation 12.16. A plot of this is shown in Figure 12.11 (for a scale parameter of 32 m). The B statistic (Equation 12.17) identifies position 77 as the one where the plot deviates most from the straight-line expectation under the null hypothesis that the underlying process has uniform variance.

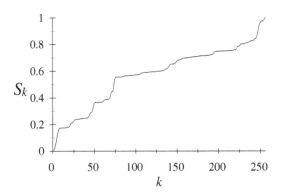

FIGURE 12.11 Plot of normalized cumulative sum of squared AMODWT coefficients of CO_2 emissions against k (position on the transect) for a scale parameter of 32 m.

TABLE 12.2
Detecting Changes in Variance in
Log-Transformed CO_2 Emission Rates

	Positions	Wavelet Variance	*B*
		Order of Division 1	
Candidate Change Point: Position 77			
Segment 1	1–76	0.095	
Segment 2	76–256	0.032	0.26*
		Order of Division 2	
Candidate Change Point: Position 8			
Segment 1	1–8	0.28	
Segment 2	9–76	0.073	0.22 NS
Candidate Change Point: Position 222			
Segment 1	77–221	0.018	
Segment 2	222–256	0.09	0.35 **
		Order of Division 3	
Candidate Change Point: Position 135			
Segment 1	77–134	0.011	
Segment 2	135–221	0.023	0.16 NS
Candidate Change Point: Position 246			
Segment 1	222–245	0.048	
Segment 2	246–256	0.180	0.34 NS

Note: These results are for a scale parameter of 32 m. The *B* statistic was compared to percentiles of a distribution obtained by Monte-Carlo simulation. The evidence against a null hypothesis that the variance is uniform is presented at four levels: $p > 0.05$ (NS), $p < 0.05$ (*), $p < 0.01$ (**), and $p < 0.001$ (***).

Table 12.2 shows the steps in identifying change points in the variance for a scale parameter of 32 m with the procedure described above. The overall conclusion is that there are three segments with different variance: from the start of the transect to position 76, where the variance is largest; from position 77 to 222, where the variance is small; and then from 223 to the end, where the variance is larger again. This latter change point corresponds to the position on the transect where topsoils in textural class clay reappear after lighter-textured drift (see Figure 12.4).

Figure 12.12 shows the segments of different variance for the three scales (8, 32, and 64 m) at which significant changes were detected. This analysis of the variance of CO_2 emissions into scale-dependent localized components shows two patterns of variation in the process. At the finest scale, there are two small regions of elevated variance. One of these (with the largest variance) is associated with the hot spot identified by the multiresolution analysis. The first is over the alluvium where short-range variation in CO_2 production due to the presence of discrete peat lenses is expected. At the coarser scales, the variance is largest over

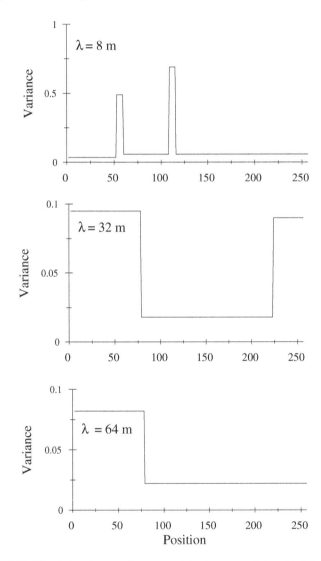

FIGURE 12.12 Wavelet variances of regions of the transect within which these are uniform for the three scale parameters where significant change points were found.

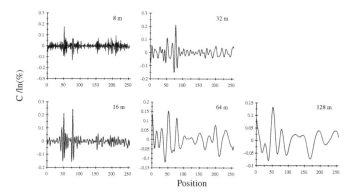

FIGURE 12.13 Detail components from multiresolution analysis of soil carbon content. The numbers on the abscissa are (dimensionless) sample numbers in order, 1 to 256. These are at sites separated by 4-m intervals.

the first 76 positions, then declines, increasing again near the end of the transect at scale 32 m. These changes in variation probably arise from patterns of variation in parent material and recent cultivations. To elucidate the factors driving changes in the soil process requires analysis of the joint variation of CO_2 emission rates with soil properties, and this is now illustrated with the data on soil carbon content.

12.4.4 COVARIATION WITH SOIL CARBON CONTENT

The (log-transformed) data on soil carbon content are shown in Figure 12.6. It was noted above that the two transformed variables have a correlation coefficient of 0.65. A positive correlation between organic carbon content and CO_2 production is expected because soil carbon is the basic substrate for microbial respiration and is a basic driver of microbial activity in the soil. Figure 12.13 shows the detail components in the multiresolution analysis of soil carbon content. There are some common features with the analysis on CO_2 emission, notably the peak near position 50 that appears at scale parameters 16, 64, and 128 m.

The top graph in Figure 12.14 shows the wavelet variances of soil carbon content. Unlike CO_2 emission, the variance increases with the scale parameter, so while there are short-range variations in carbon content, longer-range variations are more important overall. The middle graph shows the AMODWT wavelet covariances of soil carbon content and CO_2 emissions, computed with Equation 12.18. These also increase with the scale parameter and are all positive. The wavelet correlations are shown in the lower graph. These increase with scale parameter to around 0.8 or larger at the three coarsest scales. The correlation at the finest scale is not significantly different from zero.

The wavelet correlation analysis suggests that broad-scale variations in CO_2 emissions may be driven by corresponding trends in soil carbon content, but that short-range variations are not. However, this assumes stationarity in the spatial covariation of the variables. Therefore, the method described in Section 12.3.2

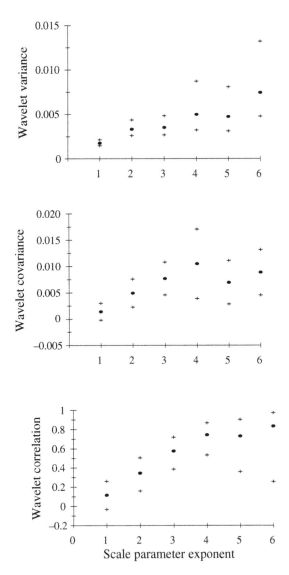

FIGURE 12.14 AMODWT wavelet variances for soil carbon content (top), and wavelet covariances (middle) and wavelet correlations (bottom) for soil carbon content and CO_2 emission.

was used to look for changes in the correlation at each scale. The results are shown in Figure 12.15. This shows regions of the transect within which the wavelet correlation of the two variables is uniform and significantly different from zero.

This analysis shows that overall wavelet correlations may sometimes obscure interesting features of spatial covariation. At a scale parameter of 16 m the overall wavelet correlation is small, although significantly different from zero. The local

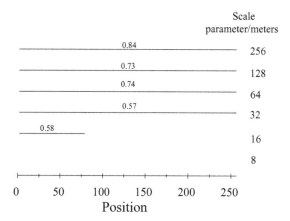

FIGURE 12.15 Regions of uniform and significant wavelet correlations between soil carbon content and CO_2 emission. The numbers on the abscissa are (dimensionless) sample numbers in order, 1 to 256. These are at sites separated by 4-m intervals.

analysis shows that the variables are significantly correlated at this scale from the beginning of the transect to just beyond position 75 (i.e., over the Lower Greensand and alluvial soils and just onto the clay drift). In the soils on the rest of the transect, formed in drift over the Gault Clay, the variables are not correlated at this scale. The implication of this is that variations in CO_2 emissions at this spatial scale may reflect variations in carbon content of the soil (e.g., differences due to within-field variations in primary production and inputs from crop residues), but this is not the case under all soil conditions. The multiresolution analysis in Figure 12.7 shows that there are fluctuations at this spatial scale (16 m) in the CO_2 emissions from the soil formed in drift over the Gault Clay, but the local analysis of the wavelet correlations indicates that some factor other than the organic content of these soils causes these variations.

In summary, how does wavelet analysis allow the soil scientist to extract more information on the covariation of these variables than is contained in the overall correlation coefficient? At the coarsest four scales the correlation of the variables is significant, and spatially uniform across the transect. At scales of 64 m and coarser the correlations are larger than 0.7. The broad trends in soil carbon content across the whole of this landscape appear, as expected, to drive the variation at comparable scale in CO_2 emission from the soil; the relationships between these trends are similar over contrasting soil conditions and are more strongly associated than the ordinary correlation coefficient would suggest. By contrast, the wavelet correlation at the finest scale (8 m) is small and not significant, and this is true at all locations in the transect. This indicates that some factor other than the soil carbon drives the finest-scale variations in CO_2 emission, and this is the case over all the soil conditions in the transect. The occurrence of strong peaks in CO_2 emission at single sites, in contrast with neighboring sites, cannot be explained simply from the presence of a local peak in soil organic carbon. The wavelet correlations at 16 m are small overall (although significantly positive). At this scale we find evidence that variations in soil carbon content may determine variations in CO_2 emissions at the same scale, but this only pertains in some parts of the landscape, and not in the soils formed in drift over the Gault Clay.

12.5 CONCLUSIONS

It has been shown that wavelet analysis is a powerful mathematical technique to examine variation and covariation of soil properties and processes at different spatial scales, free of any assumptions of stationarity in the variance. The case study illustrates the potential of the methods. In a relatively simple soil-landscape the spatial variation of an important soil process, CO_2 emission, and its covariation with a soil property has been shown to be complex, scale dependent, and local (i.e., nonstationary). The wavelet analysis gives more insight into the variation and covariation of soil properties and processes than do statistics that ignore spatial scale (e.g., the simple correlation) or those that assume variation is uniform in space. It is clear that the relationship between CO_2 emission and soil carbon content is not spatially uniform at all spatial scales and that this reflects differences in soil conditions along the transect.

There is more to be said about existing wavelet techniques than can be achieved within the constraints of an introductory discussion. This discussion has focused on the analysis of data on transects, but the DWT is readily generalized to two dimensions. In a two-dimensional DWT the detail is obtained with three wavelet functions: one that responds to variation along the columns of the data array, one that responds to the rows, and one that responds to variation on the diagonals. The adaptation to the finite interval is a harder problem in two dimensions, and the methods for inference about change in variance and covariance that are described above do not generalize simply from the one-dimensional case. Nonetheless, it is clear that the two-dimensional and maybe higher-dimensional wavelet transforms are likely to prove as useful tools for soil scientists in the future, and an example is given by Lark and Webster.[31]

Wavelet packet analysis is also likely to emerge as a useful tool to study spatial variability. In effect, this is a way to adapt the DWT method to obtain a finer-resolution analysis in the scale domain. This is likely to prove a more fruitful technology for this purpose than continuous wavelet transforms.

There is also interest in the application wavelets to simulate processes with nonstationary behavior. This has been done in signal analysis,[32] and there is an example in soil science.[25]

The principal drawbacks of wavelet analysis in soil science are probably due to the availability of data in suitable numbers and spatial configurations. In order to make fullest use of the wavelet transform, we require data sampled on regular grids or transects, and we need a lot of data. In the case study we had 256 sample points, and this constrained the data collection to laboratory rather than *in situ* measurements of gaseous emissions. Adequate datasets, sampled regularly, are relatively rare in soil science.

Two observations may be made here. First, it is likely that wavelets will come into their own for the analysis of soil data collected with sensors rather than with an auger or spade. These data may be very dense and collected more or less regularly. Lark and colleagues[33] showed how multiresolution analysis of data on the electrical conductivity of the soil obtained with a Wenner array could give insight into its spatial variability and improve the predictions of soil properties made from the data. Other sensor data such as remote sensor data might usefully be analyzed with wavelet transforms. The analysis will give insight into the spatial variability of the measured variables that may be treated as proxy information on the variability of certain soil properties, and used to plan how best to sample these.

Second, there are some developments in the wavelet analysis of data that are not sampled regularly.[34] These may prove useful in the analysis of soil data, although it should be noted that the full partition of variation over discrete scales still requires regular sampling.

Geostatistics serves pedometricians well, but geostatistical tools have their purposes and their limitations. They are not well suited to aid the interpretation of complex soil variation across landscapes where the assumption of an underlying intrinsically stationary process is clearly implausible. In these conditions the wavelet transform may come into its own, particularly if we remember the words of R.W. Hamming[1] that preface this chapter.

ACKNOWLEDGMENTS

The developments presented in this chapter were made in work funded by the U.K. Biotechnology and Biological Sciences Research Council (BBSRC) through its core grant to Silsoe Research Institute and through Grant MAF12269. This latter grant also funded the collection of the data reported here. I am grateful to collaborators in that funded project at Rothamsted Research for their permission to use these data, and to Síle O'Flaherty, Colin Webster, and Steven Francis, who did the laboratory work.

REFERENCES

1. RW Hamming. *Numerical Methods for Scientists and Engineers.* McGraw-Hill, New York, 1962.
2. G Matheron. *Estimating and Choosing.* Springer-Verlag, Berlin, 1989.
3. R Webster. Is soil variation random? *Geoderma* 97: 149–163, 2000.
4. H Saito, P Goovaerts. Accounting for source location and transport direction into geostatistical prediction of contaminants. *Environ Sci Technol* 35: 4823–4829, 2001.
5. G Matheron. *Le krigeage universel*, Cahiers du Centre de Morphologie Mathématique 1. Ecole des Mines de Paris, Fontainebleau, 1969.
6. G Matheron. The intrinsic random functions and their applications. *Adv Appl Probab* 5: 439–468, 1973.

7. R Webster, HE de la Cuanalo. Soil transect correlograms of north Oxfordshire and their interpretation. *J Soil Sci* 26: 176–194, 1975.

8. RM Lark, R Webster. Analysis and elucidation of soil variation using wavelets. *Eur J Soil Sci* 50: 185–206, 1999.

9. C Walter, AB McBratney, A Douaoui, B Minasny. Spatial prediction of topsoil salinity in the Chelif valley, Algeria, using ordinary kriging with local variograms versus a whole-area variogram. *Aust J Soil Res* 39: 259–272, 2001.

10. RM Lark, R Webster. Changes in variance and correlation of soil properties with scale and location: analysis using an adapted maximal overlap discrete wavelet transform. *Eur J Soil Sci* 52: 547–562, 2001.

11. P Kumar, E Foufoula-Georgiou. Wavelet analysis in geophysics: an introduction. In *Wavelets in Geophysics*, E Foufoula-Georgiou, P Kumar, Eds. Academic Press, New York, 1994, pp. 1–43.

12. SG Mallat. A theory for multiresolution signal decomposition: the wavelet representation. *IEEE Trans Pattern Anal Mach Intell* 11: 674–693, 1989.

13. R Webster. Spectral analysis of gilgai soil. *Aust J Soil Res* 15: 191–204, 1977.

14. RM Lark, AB McBratney. Wavelet analysis. In *Methods of Soil Analysis*, Part 4, *Physical Methods*, JH Dane, C Topp, Eds. Soil Science Society of America, Madison, WI, 2002, pp. 184–195.

15. B Si. Scale and location dependent soil hydraulic properties in a hummocky landscape: a wavelet approach. In *Scaling Methods in Soil Physics*, Y Pachepsky, D Radcliffe, H Magdi Selim, Eds. CRC Press LLC, Boca Raton, FL, 2003, pp. 169–187.

16. I Daubechies. Orthonormal bases of compactly supported wavelets. *Commun Pure Appl Math* 41: 909–996, 1988.

17. WH Press, SA Teukolsky, WT Vetterling, BP Flannery. *Numerical Recipes (Fortran)*, 2nd ed. Cambridge University Press, Cambridge, U.K., 1992.

18. A Cohen, I Daubechies, P Vial. Wavelets on the interval and fast wavelet transforms. *Appl Comput Harmonic Anal* 1: 54–81, 1993.

19. DP Percival. On estimation of the wavelet variance. *Biometrika* 82: 619–631, 1995.

20. DB Percival, P Guttorp. Long-memory processes, the Allan variance and wavelets. In *Wavelets in Geophysics*, E Foufoula-Georgiou, P Kumar, Eds. Academic Press, New York, 1994, pp. 325–344.

21. BJ Whitcher. Assessing Nonstationary Time Series Using Wavelets. Ph.D. thesis, University of Washington, Seattle. 1998. Available at www.cgd.ucar.edu/staff whitcher/papers/thesis.ps.gz.

22. BJ Whitcher, P Guttorp, DB Percival. Wavelet analysis of covariance with application to atmospheric time series. *J Geophys Res Atmos* 105: 14941–14962, 2000.

23. CV Deutsch, AG Journel. *GSLIB Geostatistical Software and User's Guide*. Oxford University Press, New York, 1992.

24. P Goovaerts, R Webster. Scale-dependent correlation between topsoil copper and cobalt concentrations in Scotland. *Eur J Soil Sci* 45: 79–95, 1994.

25. RM Lark, AE Milne, TM Addiscott, KWT Goulding, CP Webster, S O'Flaherty. Analysing spatially intermittent variation of nitrous oxide emissions from soil with wavelets and the implications for sampling. *Eur J Soil Sci* 55, 601–610, 2004.

26. DW King. *Soils of the Luton and Bedford District*, Special Survey 1. Soil Survey of England and Wales, Harpenden, U.K., 1969.

27. JM Hodgson, Ed. *Soil Survey Field Handbook*, 2nd ed., Soil Survey Technical Monograph 5. Rothamsted Experimental Station, Harpenden, U.K., 1976.

28. MA Tabatabai, JM Bremner. Automated instruments for determinations of total carbon, nitrogen, and sulfur in soils by combustion techniques. In *Soil Analysis: Modern Instrumental Techniques*, 2nd ed., KA Smith, Ed. Marcel Dekker, New York, 1991, pp. 261–286.

29. DE Williams. A rapid manometric method for the determination of carbonate in soil. *Soil Sci Soc Am Proc* 25: 248–250, 1949.

30. RR Coifman, DL Donoho. Translation-invariant denoising. In *Wavelets and Statistics*, Lecture Notes in Statistics 103, A Antoniadis, G Oppenheim, Eds. Springer-Verlag, New York, 1995, pp. 125–150.

31. RM Lark, R Webster. Analysing soil variation in two dimensions with the discrete wavelet transform. *Eur J Soil Sci* 55, 777–797, 2004.

32. J Iyama, H Kuwamura. Application of wavelets to analysis and simulation of earthquake motions. *Earthquake Eng Struct Dyn* 28: 255–272, 1999.

33. RM Lark, SR Kaffka, DL Corwin. Multiresolution analysis of soil electrical conductivity data using wavelets. *J Hydrol* 272: 276–290, 2003.

34. I Daubechies, I Guskov, P Schröder, W Sweldens. Wavelets on irregular point sets. *Phil Trans R Soc London Ser A* 357: 2397–2413, 1999.

13 Three-Dimensional Reconstruction and Scientific Visualization of Soil-Landscapes

Sabine Grunwald

CONTENTS

ABSTRACT

An approach to reconstruct and visualize virtual soil-landscape models and space–time simulations based on ontological modeling comprising the physical, logic/representation, implementation, and cognitive universes were presented. The modeling process was disaggregated into conceptualization, reconstruction, and scientific visualization. The ontological modeling concept was employed to produce three- and four-dimensional virtual models integrating soil, land use, and topographic datasets for soil-landscapes in southern Wisconsin and northeastern Florida. Spatially and temporally explicit modeling of robust soil properties (e.g., bulk density) as well as dynamic soil properties (e.g., water level) was demonstrated. Ordinary kriging and cokriging were used to reconstruct the soil-landscapes, and Virtual Reality Modeling Language to render virtual objects. To engage users, interactive functions were programmed in Java and External Authoring Interfaces. Scientific visualization combined with quantitative reconstruction techniques has the potential to translate real soil-landscapes and ecosystem processes into a transparent format to enhance our understanding of real-world phenomena and complex environmental systems. Web-based virtual soil-landscape models and space–time simulation models facilitate the exploration, analysis, synthesis, and presentation of georeferenced environmental datasets. Combining multimedia elements (e.g., Internet, interactivity, three-dimensional scientific visualization) can produce insight that would not arise from use of the elements alone.

13.1 INTRODUCTION

Scientists have focused on two contrasting concepts to study soil-landscapes and ecosystem processes, which are both equally important. The reductionist approach promotes ever more detailed studies of distributions, soil classes, events, and processes, followed by their interpretation. The other approach develops and enunciates an integrative, unifying point of view encompassing and integrating previous observations and results. Both concepts have been employed for quantitative spatially explicit modeling of soil-forming factors evolving through time.

Different kinds of models have been used to translate real soil-landscapes and ecosystem processes into virtual environments. Hoosbeek and Bryant[1] provided an overview of pedogenetic models using the following criteria: (1) the relative degree of computation (qualitative vs. quantitative models), (2) complexity of the model structure (functional vs. mechanistic models), and (3) level of organization (soil region, pedon to molecular scale). Modeling is about choosing the appropriate metaphor or analogy with which to better understand a phenomenon, e.g., the spatial distribution of soils and their behavior and relationship to other environmental factors. In this sense, we create media about phenomena to bridge the gap between what we do not know and what we are trying to comprehend. Media such as slides, maps, animations, three-dimensional virtual worlds, and digital libraries are models. Although one might talk of absolutes such as reality and truth, all we have at our disposal are models, which mediate the world for us. Transcending real into virtual soil-landscapes depends on the (1) space domain, (2) time domain, (3) spatial and temporal scale, (4) ecosystem conditions, (5) spatial and temporal variability, (6) interrelationships between environmental factors and soil properties, and (7) causal linkages and behavior of the system. The modeling process itself can be disaggregated into (1) conceptualization, i.e., defining the model structure and design; (2) reconstruction, i.e., describing and quantifying underlying conditions and behavior using mathematics; and (3) scientific visualization (SciVis), i.e., abstracting real soil-landscapes into a format that we can comprehend and that helps us to understand the complexity of soil-landscapes.

Real soil-landscapes are complex, consisting of an inextricable mix of systematic and random patterns of environmental variables (e.g., soils, topography, land use) varying continuously in the space–time continuum. Soils and parent material show gradual variations in the horizontal and vertical planes, forming three-dimensional bodies that are commonly anisotropic. There is no real beginning and endpoint in real soil-landscapes because environmental conditions are dynamically changing through water flow and biogeochemical processes. In addition, human activities confound naturally occurring patterns and processes.

Transforming real into virtual soil-landscapes is based on model predictions and estimations. Estimations use sample data to make an inference about a population, whereas predictions refer to a statement about the future or reasoning about the future. Methods used in science for the derivation of predictions of unknown facts from known facts include (modified after Bunge[2]):

1. Logical inference that includes deduction, induction, and abduction, the latter one referring to the generation of hypotheses to explain observations
2. Structural laws that help predict new properties from the known properties of material or formal structures
3. Phenomenological laws that predict phenomena on the basis of known constant associations
4. Functional laws that infer functional properties of a system from knowledge of the functional role of the parts and their interconnections
5. Statistical laws that help derive collective properties of classes of events from an analysis of such classes
6. Mechanical laws that extrapolate future (or past) states on the basis of known current states and relations (e.g., Newtonian laws)

Recently, much attention has been given to transform what Egenhofer and Mark[3] called the naïve geography, a body of knowledge that captures the way people reason about geographic space and time, into complex geographic ontologies and semantics.[4] Since ontology is the means to discover the structures and generalities in reality that (metaphysical) realism predicts, ontology is a crucial part of the science world. An ontology is a formal, explicit specification of a shared conceptualization. *Conceptualization* refers to an abstract model of phenomena in the world by having identified the relevant concepts of those phenomena. *Explicit* means that the type of concepts used, and the constraints on their use, are explicitly defined. *Formal* refers to the fact that the ontology should be machine readable. *Shared* reflects that ontology should capture consensual knowledge accepted by the communities. Since most ontological schemes use spatial and temporal concepts, GIScience often serves as an ontological precursor to the design or discovery of phenomena in scientific investigations.[5] Fonseca et al.[6] proposed an ontology-driven geographic information system (ODGIS) that acts as a system integrator independently of the model. The idea is to build a next-generation GIS that entails a systematic collection and specification of geographic entities, their properties, and relations. Smith[7] suggests a terminological distinction between referent or reality-based ontology (*R-ontology*) and elicited or epistemological ontology *(E-ontology)*. R-ontology is a theory about how the whole universe is organized, and corresponds to the philosopher's point of view. E-ontology, on the other hand, fits the purposes of software engineers and information scientistis, and is defined as a theory about how a given individual, group, language, or science conceptualizes a given domain.

Smyth[8] and Davis[9] suggest domain-specific ontologies that are based on the following ontological elements:

{content/entities + time + geometry + physics + logic}

These ontological components, in turn, can be modeled computationally as objects, spatial representation, temporal representation, numerical models, and knowledge base and inference systems, respectively. Applied to soil-landscape modeling, *content* can be modeled, e.g., as soil attributes, soil classes, or other environmental attributes. Typically, hierarchical trees can be built that represent an ontological component, e.g., a catena composed of soil types, soil attributes that distinguish soil types, soil horizons that define soil types, soil attributes that define horizons, etc. This is in analogy to a hierarchical model that allows the identity of individual trees to be subsumed by a forest; it makes sense to see either the individual trees or the forest. Similarly, we can create soil-landscape snapshot models that represent either the spatial distribution of soil horizons or one specific soil attribute (e.g., soil phosphorus content). There may be several *time* and *geometrical* representations of the same underlying entity. For example, the *geometry* to represent soil characteristics (e.g., silt content) can be a two-dimensional polygon, three-dimensional polyhedron, pixel (two-dimensional raster cell), or voxel (three-dimensional cell). According to Davis,[9] *physics* refers to the rules of interaction between or among entities typically expressed as mathematical formulas. A *logic* allows new facts to be deduced about a (soil-landscape) world for a given configuration.

To understand the role of ontologies in geographic data modeling, Gomes and Velho[10] suggested the four-universes paradigm for modeling a digital (virtual) representation. The four universes are:

1. The *physical universe*, which comprises the objects and phenomena of the real world that will be modeled and transcended into a virtual world.
2. The *mathematical universe* or *logical universe*, a formal definition of these objects and phenomena.
3. The *representation universe*, which uses metaphoric (symbolic) or mirror real-world descriptions of the elements defined in the mathematical universe. For example, a soil profile might be represented as a dot on a map (symbolic representation) attached with a label/legend, or set of multiple polyhedrons that represents soil attributes, soil horizons, and the shape of the soil profile.
4. The *implementation universe*, which is used to map the elements from the representation universe into data structures implemented in a computer language.

Fonseca et al.[6] added the *cognitive universe*, which captures what people perceive about the physical universe. Figure 13.1 characterizes the different universes for a soil-landscape in southern Wisconsin using soil horizons and soil characteristics (e.g., bulk density), crisp objects and voxels, geostatistical methods to reconstruct the soil and terrain features mathematically, and Virtual Reality Modeling Language (VRML) to visualize the soil-landscape models (Figure 13.1).

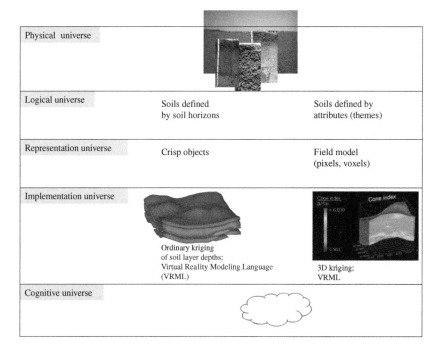

FIGURE 13.1 Schematics of universes used to model a soil-landscape in southern Wisconsin. (See color version on accompanying CD.)

13.2 SPACE AND TIME CONCEPTS

Conceptions of space and time have profoundly influenced our notion of how we observe, describe, and perceive soil-landscapes. Generally, it is necessary to divide geographical space into discrete spatial units, and the resulting tessellation is taken as a reasonable approximation of reality at the level of resolution under consideration. Peuquet,[11] Goodchild et al.,[12] and Burrough and McDonnell[13] propose two contrasting spatial discretization methods: (1) crisp irregular two-dimensionl polygons or three-dimensional polyhedrons and (2) regular-shaped pixels or voxels. In essence, both methods discretize Euclidean geographic space that has a constant zero curvature; i.e., it is a plane and has one and only one parallel to a line passing through a given point. Conceptually there is a slight difference, but computationally a large difference, between both discretization methods. This can be explained by the fact that vector and raster data types are encoded differently, where the former one encodes nodes and vertices of geographic features as well as the topology explicitly, whereas the latter one is based on a much simpler matrix representation. Many different spatial algorithms have been developed and customized to accommodate either vector or raster data in a GIS environment.

Frank[14] and Raper et al.[5] provide an overview of time models. Newtonian time is focused on a succession of phenomena along a linear time coordinate, providing the simplest time concept characterized by causal inertness. Time is viewed as a neutral framework against which independently unfolding events are projected, sorted, and measured. Newton argued that time is absolute, implying that the universe has a single universal clock capable of determining that two occurrences are simultaneous. The present moment, which is changing constantly, forms the center point. Backcasting and forecasting models predict past and future events (e.g., formation of Spodosols, land use change) with exponentially increasing prediction errors from the present moment. Other soil-landscape models are snapshot models that are limited to describe current environmental conditions. Frank[14] proposed alternative time concepts, such as cyclic time, branching time, transaction time, and event-driven time. Other time concepts remain in the phenomenological realm, such Husserl's human subjective time, Minkowski's light cone of time, and Einstein's relative time.

Almost all existing soil-landscape simulation models are based on Newtonian time that is well suited to represent physical and chemical processes. However, the modeling of anthropogenic-driven events (e.g., land use change, sometimes irrational decisions by land stewards) or chaotic climatic events modeled within a Newtonian framework poses problems. Characteristics of real time include that events are nonrepeatable, sometimes structured and at other times chaotic. Real time is relative and depends on the context of previous events and expected future. More generally, the timing of an event changes its nature to the extent that the unique context of other events within which it occurs affects its role in the determination of subsequent events. No two instants can be the same, each one relating to a different set of preceding and succeeding moments and their remembered or anticipated contents.

Space–time domains curved or warped by the presence of mass and energy within them revolutionized our thinking, unifying space and time in the general theory of relativity. Such epistemological space–time concepts have been adopted to represent reality but they are difficult to implement. Much simpler space–time models have been adopted for soil-landscape modeling. Hayes[15] proposed the concept of histories, reasoning that no two histories may overlap in space–time and that a history may be projected onto a point in space–time. Roshannejad and Kainz[16] suggested the logical data model shown in Figure 5.8. The geographic coordinates are described by x, y, and z coordinates representing easting, northing, and depth from the soil surface, respectively. Time is considered the fourth dimension, relegating attribute values to the fifth dimension. Deterministic event models describe the change of ecosystem states in a chain-like fashion according to mathematical algorithms. The result is a spatio-temporal thematic object triplet describing data evolution. Hence, data evolution is described by a sequence of records $\{(s, t, e)\}$, where s is a spatial coordinate, t is a time stamp, and e is an ecosystem function (= behavioral function, e.g., mineralization, leaching of nitrate).

The linkage between space and time conceptualization is through the *process* to be described. Conventionally, simulating ecological processes correspond to a sequence of events with an orderly structure, where one event occurs after the other (deterministic view). Each process that is in focus provides its own context and conceptual frame for the cognition of space and time. Change must be seen as a composite of processes (and interactions between those processes) that occur on a wide band of timescales in the atmospheric, biological, geological, and human domains. Chrisman[17] and others suggested treating time as an axis, a dimension of measurement similar to the spatial case — simply spatializing time.

McSweeney et al.[18] pointed out that soil-landscapes are three-dimensional systems and should be represented using geographic information technology. Burrough and McDonnell[13] argue that the term *three-dimensional* is usually (and properly) reserved for situations in which an attribute varies continuously through a three-dimensional spatial frame of reference, e.g., soil system, whereas land use, land cover, and topography are surfaces and can be represented as two-dimensional geographic features in three-dimensional view. Timeless two-dimensional space concepts will continue to be useful to represent soil-landscapes. Yet recently there have been major advances in computational capabilities, geostatistical analyses, and SciVis, resulting in striking multidimensional soil-landscape representations. Such models will revolutionize how we comprehend soil-landscape systems.

Some soil-landscape studies are limited to one specific period of data collection,[19] others to two dimensions,[20,21] and few are dynamic, addressing changes over time.[22] The development of three- and four-dimensional models has been hampered in the past by the large amount of input data required and the complexities of soil-forming and ecosystem processes. Models that are highly specialized describe the spatial distribution of one property explicitly, such as digital terrain models,[23] land cover layers,[24] or soil maps.[21] There are fewer models that infuse above- and belowground properties to reconstruct soil-landscapes and environmental systems in three-dimensional geographic space.[25–27] Numerous studies present techniques for three-dimensional visualization.[28–30] However, they do not extend their approach below the soil surface to address a three-dimensional ecosystem. Even fewer studies attempt four-dimensional space–time modeling of soil and landscape properties.[31] Since soil-landscapes are truly three-dimensional, undergoing continuous evolution of their components (e.g., due to soil formation, water table dynamics), adequate computerized techniques are needed to reconstruct and visualize these multidimensional and multicategorical environmental systems.

Currently, no universal spatio-temporal modeling and information system exists, but there are a variety of prototypes. Abraham and Roddick[32] presented a comprehensive review on conceptualizations of spatio-temporal databases. Koeppel and Ahlmer[33] distinguish between attribute-oriented spatio-temporal databases that track changes in information about spatial entities, while topology-oriented spatio-temporal databases track changes in positional information about features and their spatial relationships. Whigham[34] proposed a dual-ordered hierarchical structure where time and events are represented in their own hierarchies, placed on a spatial background reference. Hermosilla[35] argues for a temporal GIS with reasoning capabilities based on artificial intelligence. Peuquet and Duan[36] suggest an Event-Based Spatio-Temporal Data Model (ESTDM) focusing on events that are represented along a temporal vector in chain-like fashion. Yuan[37] suggests a three-domain model representing semantics, space, and time separately and providing links between them to describe geographic processes and phenomena. Many of these tools are focused on database management rather than reconstruction and scientific visualization.

13.3   RECONSTRUCTION AND SCIENTIFIC VISUALIZATION

State-of-the-art reconstruction techniques are described throughout this textbook. An overview is given in Chapter 1. Mathematical and statistical methods used to reconstruct soil-landscapes have been described by Goovaerts,[38] Chilès and Delfiner,[39] Stein,[40] McBratney et al.,[41] Webster and Oliver,[42] and Berthouex and Brown.[43] The strengths of soil-landscape modeling lie in hypothesis testing, understanding causal linkages between environmental factors, and their interrelationships within a spatial and temporal explicit context.

Scientific visualization can be implemented using programming languages such as Java, C++, or VRML. The last one is a three-dimensional open-source graphics language suitable for stand-alone or browser-based interactive viewing on the Internet and is used to render the face geometry of soil-landscape models.[44,45] Within the VRML-capable browser, the user can move around VRML worlds in three dimensions, scale and rotate objects, and view updates in real time. Capabilities of VRML include three-dimensional interactive animations, three-dimensional worlds (scenes) comprising several different three-dimensional objects, scaling of objects, material properties and texture mapping (e.g., draping of photographs or bitmap art over the face of a three-dimensional object), setting different viewpoints, and use of light sources. In short, VRML provides the technology that integrated two- and three-dimensional objects, text, and multimedia into a coherent modeling framework. When these media types are combined with scripting languages, an entirely new genre of interactive applications becomes possible. The key elements of the VRML are nodes that describe the shapes, colors, lights, viewpoints, how to position and orient shapes, animation timers, interpolators, etc., and their properties in a virtual world. Fields define attribute characteristics of a node, and every value is of a specified field type.[45] Object-oriented languages such as VRML, C++, and Java support the concept of data abstraction and modularity in program design.

Troy and Czapar[46] employed VRML to evaluate conservation practices. They generated a three-dimensional environment for the Lake Springfield watershed in order to visualize environmental factors and to direct the planning, installation, and maintenance of conservation practices. The authors found VRML useful in the promotion of properly placed best-management practices. Miller et al.[47] used VRML to model rural environments and document land use changes. They emphasized that the virtual reality environment has the potential to aid in communication, decision making, and scenario testing. Lovett et al.[48] used a VRML-based approach for sustainable agricultural management exemplified by a virtual landscape. Grunwald et al.[25] developed similar soil-landscape models implemented in VRML at four different spatial scales for sites in southern Wisconsin. Models were implemented using polyhedrons to represent soil layers that were integrated with a digital elevation model (DEM) to describe soil-landscapes. Voxels were used to create three-dimensional soil property models for bulk density, soil texture, and penetration resistance.

The Internet, geographic information technology, and SciVis provide new education and information delivery capabilities. Numerous studies have shown that SciVis is effective for enhancing rote memorization and higher-order cognitive skills.[49–51] Stibbard[52] found that information is absorbed best when using more than one human sense; i.e., 10% of the information is taken in by reading, 30% by reading and visuals, 50% by reading, visuals, and sound, and 80% by reading, visuals, sound, and interaction. Koussoulakou and Kraak[49] tested the usefulness of different SciVis methods, including static maps, series of static maps, and animated maps, and found significantly better response times for animated maps. Barraclough and Guymer[50] reported that advanced visualization techniques served to better communicate spatial information between people in different fields, such as scientists, administrators, educators, students, and the general public. Just as maps can visually enhance the spatial understanding of phenomena, interactive spatio-temporal applications can enhance our understanding of complex environmental systems and the underlying transport processes driving soil and water quality. According to Fisher and Unwin,[53] visual interfaces maximize our natural perception abilities, improve our comprehension of huge amounts of data, allow the perception of emergent properties that were not anticipated, and facilitate understanding of both large-scale and small-scale geographic features of ecosystems. According to Gordon and Pea,[54] SciVis improves learning because it supports the thought process and methodologies practiced by scientists. These include learning from observation, developing hypotheses to explain observations, and testing of hypotheses with datasets, thereby iteratively developing more detailed and sophisticated analyses. Raper[55] and Morris et al.[56] present innovative management of multivariate and multidimensional datasets and display environmental data in a three-dimensional format. The development of immersive and desktop virtual reality techniques has been instrumental to develop virtual soil-landscapes and environments. VRML models enhanced with Java and External Authoring Interfaces provide capabilities to display real soil-landscapes in three- and four-dimensional digital formats.[57] Characteristics of virtual reality include (1) immersion, (2) navigation (freedom for the user to explore), and (3) interaction.

13.4 APPLICATIONS

The following applications describe the ontological components used to implement different virtual soil-landscape models.

13.4.1 SOIL-LANDSCAPE MODEL 1

13.4.1.1 Physical Universe

The study site, a 2.73-ha field, was located on the University of Wisconsin–Madison Agricultural Research Station, West Madison. Shallow reworked loess cover was found on the eroded soils on shoulder and backslope positions, whereas thick reworked loess deposits were found on footslope and toeslope positions. Soils were mapped as fine-loamy, mixed, mesic Typic Argiudolls. The reworked loess was underlain by sandy loam glacial till. Land use was a corn (*Zea mays*)–alfalfa (*Medicago sativa*) rotation. Climate is temperate humid. A constant-rate profile cone penetrometer (PCP), described in detail in Grunwald,[58] was used to collect cone index measurements up to a soil depth of 1.30 m at 273 locations on a 10 × 10 m grid. Bulk density (BD) measurements were collected along soil profiles in 10-cm-depth increments at 77 locations spatially distributed throughout the site. The BD sampling design targeted locations that showed heterogeneous terrain patterns. Cone index measurements were dense ($n = 273$), whereas bulk density measurements were sparse ($n = 77$). A Trimble 4600 LS differential global positioning system (dGPS) (Trimble Inc., Sunnyvale, CA) with base station and beacon differential correction was used to georeference sampling locations and collect elevation data using a dense kinematic mapping technique.

13.4.1.2 Logical and Representation Universe

The goal was to map the spatial distribution of BD across the site in southern Wisconsin adopting a voxel model. The soil and terrain data were fused to reconstruct a coherent three-dimensional model representing terrain and soil patterns.

13.4.1.3 Implementation Universe

Three-dimensional collocated cokriging was used to predict BD at unsampled locations using cone index data as the secondary variable. Cokriging is an extension of autokriging. It takes into account additional correlated information in the subsidiary variables[59] according to Equation 13.1. The dataset was partially heterotopic. A heterotopic situation can be characterized by a variable of interest known at a few points and an auxiliary variable known everywhere in the domain (or at least at all nodes of a given estimation grid and at the data locations of the variable of interest):

$$\hat{Z}(x_0) = w_0 S(x_0) + \sum_{\alpha=1}^{n} (w_Z^{\alpha} Z(x_\alpha) + w_S^{\alpha} S(x_\alpha)) \qquad (13.1)$$

where

$\hat{Z}$ = estimated value
w = weights
$Z(x_0)$ = primary random variable Z at location x_0
$S(x_0)$ = secondary random variable S at location x_0
x_α = points used to estimate x_0; $\alpha = 1, \ldots, n$

The three-dimensional bulk density voxel model was fused with elevation data to produce a three-dimensional soil-terrain model. Virtual Reality Modeling Language was used to render face geometry of voxels. Each voxel represents one estimated BD value. Models in Figure 13.2 show the spatial distribution of BD across the study site. Cross-validation showed a mean squared estimation error of 0.05. A scientific visualization technique called slicing was employed to show the variation of BD across the site (Figure 13.3).

13.4.1.4 Cognitive Universe

Soil-landscape models are Web based and provide interactivity functions to engage users (e.g., zoom, rotate). Three-dimensional models stimulate the geographic abstraction skills. Bulk density values are viewed in concert with terrain properties, providing insight into soil-topographic relationships. Grunwald et al.[59] found that in glaciated landscapes in southern Wisconsin BD is closely related to soil materials such as glacial till and reworked loess. Commonly, large BD values (≥ 1.6 Mg m^{-3}) are associated with sandy loam glacial till, and medium BD values (≥ 1.3 and < 1.6 Mg m^{-3}) are found in loess material. The presented

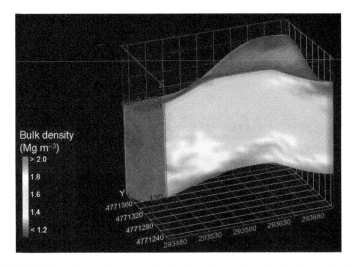

FIGURE 13.2 Three-dimensional model showing the spatial distribution of bulk density values. (See color version on accompanying CD.)

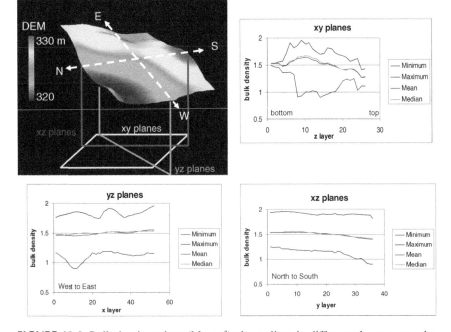

FIGURE 13.3 Bulk density values (Mg m⁻³) along slices in different planes across the Wisconsin study site. (See color version on accompanying CD.)

models provide the foundation for sustainable land use management, optimizing crop growth while minimizing adverse effects on the environment.

13.4.2 Soil-Landscape Model 2

13.4.2.1 Physical Universe

The study area comprised a 42-ha site in northeastern Florida with hydric and nonhydric soils. About one third of the site was covered by cypress (*Taxodium distichum*) and about two thirds by slash pine (*Pinus elliottii*). In 1994, three silvicultural treatments were administered. While the southwest block was left as a control (uncut), the southeast block was clear-cut. On the northwest block only the forest on the hydric soils was cut; that on the nonhydric soils was left untouched. Morphological and taxonomic soil data were collected at 123 locations on a 100 × 100 m grid. Topography was characterized by laser level and ranged from 27 to 31 m. A detailed description of the study can be found in Bliss and Comerford.[60]

13.4.2.2 Logical and Representation Universe

The objective was to generate a model that displays the spatial distribution of soil horizons across the study site. The crisp object model was adopted to represent horizons focusing on soil genetic aspects of this site. The soil and terrain data were integrated to reconstruct a coherent three-dimensional model representing terrain and soil patterns.

13.4.2.3 Implementation Universe

Two-dimensional ordinary kriging[42] in the horizontal plane and linear interpolation in the vertical plane were used to create three-dimensional face geometry of soil layers. The output product was a stratigraphic model representing soil horizons as polyhedrons or volume objects (Figure 13.4a). The *IndexedFaceSet* VRML class was employed to render polyhedrons. A point-arc geographic data model was used to create *IndexFaceSets*. The appearance of volume objects was coded using the RGB (red-green-blue) color classification system. A model showing representative soil pedons across the site used VRML texture mapping, draping photographs of soil profiles over the face of cylinder objects (Figure 13.4b). Cylinders were linked to quantitative datasets using Java Script embedded into VRML. Each soil profile has a sensor node, which senses the computer mouse. Once the user selects a soil profile, an event is sent to the Java Script through the Java External Authoring Interface and attribute data are displayed in a textbox adjacent to the three-dimensional soil model. Geographic objects in VRML were created using the Environmental Visualization System software (EVS-PRO, CTech Development Corporation, Huntington Beach, CA).

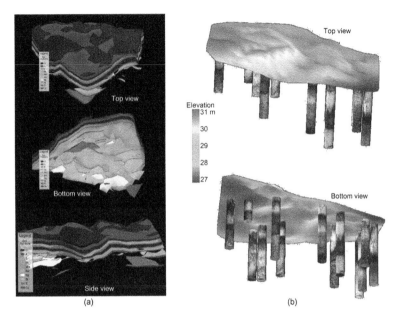

FIGURE 13.4 (a) Three-dimensional soil horizon model. (b) Soil profile model. (See color version on accompanying CD.)

13.4.2.4 Cognitive Universe

The model was implemented using desktop virtual reality and is accessible at http://3Dmodel.ifas.ufl.edu. Employing SciVis facilitated multiple views of the soil-landscape, which stimulates a greater understanding and insight of the flatwood system. It is this synthesis of geographic datasets that distinguishes the virtual environment from conventional instructional media (e.g., two-dimensional GIS maps).

13.4.3 SPACE–TIME SIMULATION OF WATER TABLE DYNAMIC

13.4.3.1 Physical Universe

The water table was monitored biweekly at 123 wells on the flatwood site described under Section 13.4.2 from April 1992 to March 1998. The wells were 1 to 1.4 m deep and positioned on a 100×100 m grid. Daily precipitation data were collected for the same period and aggregated to biweekly time increments.

13.4.3.2 Logical and Representation Universe

The objective was to generate models that show the spatial and temporal distributions of water table dynamics across the site. The terrain model was fused with the water model. The first representation model was based on voxels (space–time

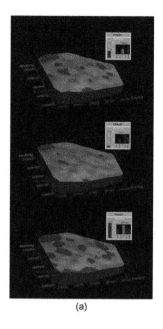 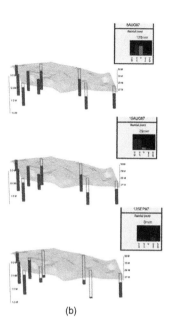

(a)   (b)

FIGURE 13.5 (a) Space–time models of inundation. (b) Space–time models show water table depth and rainfall observed for different periods. (See color version on accompanying CD.)

inundation model), and the second representation model was based on objects (space–time model of water table depth).

13.4.3.3 Implementation Universe

13.4.3.3.1 Space–Time Inundation Model (Figure 13.5a)

Hundreds of semivariograms for water table depth had to be generated, each representing one specific period between April 1992 and March 1998. Water table levels were interpolated using two-dimensional ordinary kriging. The water table surface was sliced with the digital elevation model to distinguish inundated from noninundated areas. The *CoordinateInterpolator* VRML node was used to produce a smooth display of water table depths between observation periods. The *IndexedFaceSet* VRML class was employed to render the extent of the study site.

13.4.3.3.2 Space–Time Model of Water Table Depth (Figure 13.5b)

Each well was rendered using empty cubes and positioned at observed geographic locations. After triggering the simulation, a Java program reads the measured water table levels, defined by their respective x, y, and z coordinates, at each well for a specific period. The *CoordinateInterpolator* VRML node was used to interpolate water levels across the flat-wood site. A graphical user interface was added as front end to enable users to select the wells and period they want to display. The cubes are filled with water on-the-fly according to respective observed water levels constrained to the subset of user-defined parameters. The simulation can be repeated for different well locations or periods. Java Server Pages from Sun Microsystems was used for developing this dynamic Web simulation.

13.4.3.4 Cognitive Universe

The space–time models are available at http://3Dmodel.ifas.ufl.edu. An interactive framework was used to engage users. Users can trigger an event by constraining the border conditions (e.g., time period, geographic domain) of simulations. Adaptive selective simulation stimulates experimental learning through the observation of ecosystem processes using a sequence of events: trigger an event, observe ecosystem process, interpret, assimilate. Causal linkages between terrain properties, land use, soils, precipitation, and water table dynamics within a spatially and temporally explicit context can be described and visualized. Employing SciVis facilitated multiple views of the content world, which stimulates a greater understanding and insight of the flatwood system. Desktop virtual reality, when combined with other forms of digital media, may offer great potential for a cognitive approach to research and education.

13.5 FINAL REMARKS

The use of software components extracted from ontologies is a way to share knowledge and integrate different kinds of information. Ontological component modeling is structured and facilitates transparent documentation of the modeling process. Ontologies are useful at the database, reconstruction, and visualization levels.

SciVis combined with quantitative reconstruction techniques has the potential to translate real soil-landscapes and ecosystem processes into a transparent format to enhance our understanding of real-world phenomena (e.g., water dynamics, nitrate leaching) and complex environmental systems. Limitations of the presented approach are largely those due to the availability of soil and other environmental datasets used to reconstruct models and the size of the study area. Soil-landscape models that extend over large areas are challenging to visualize because soils are typically mapped only to a depth up to 1 to 2 m. Therefore, it is necessary to use exaggeration factors to visualize soil profiles or properties across a landscape that extends over hundreds of square kilometers. Web-based virtual soil-landscape models and space–time simulation models facilitate the exploration, analysis, synthesis, and presentation of georeferenced environmental datasets. Shiffer[61] argues that users gain an improved understanding by viewing information from several different graphical perspectives. Krygier[62] notes that combining multimedia elements (e.g., Internet, interactivity, three-dimensional SciVis) can produce insight that would not arise from use of the elements alone. Virtual soil-landscape models are beneficial in disseminating georeferenced soil and landscape data to educators, researchers, government agencies, and the general public. In the realm of education, Freudenschuh and Hellevik[63] pointed out that students should be encouraged to become active participants, rather than passive learners, by appealing to their multisensory learning ability with interactive media. Training and integration of quantitative modeling and programming skills into curricula is a prerequisite to produce the next generation of interactive, virtual soil-landscape models.

ACKNOWLEDGMENTS

I thank the National Council for Air and Stream Improvement (NCASI) and the U.S. Forest Service for funds that allowed the data collection at the Florida site conducted by C.M. Comerford and C.M. Bliss. Thanks to G.W. Hurt, who provided photographs of soil profiles, and to A. Mangeot and V. Ramasundaram, who assisted with programming. I also thank K. McSweeney and B. Lowery for providing funding for the Wisconsin project and P. Barak for his stimulating ideas to move from two-dimensional to interactive three- and four-dimensional virtual worlds.

This research was supported by the Florida Agricultural Experiment Station and approved for publication as Journal Series R-10602.

REFERENCES

1. MR Hoosbeek, RB Bryant. Developing and adapting soil process submodels for use in the pedodynamic orthod model. In *Quantitative Modeling of Soil Forming Processes*, SSSA Special Publication 39, RB Bryant, RW Arnold, Eds. SSSA, Madison, WI, 1994, pp. 111–128.
2. M Bunge. *Causality*. Harvard University Press, Cambridge, MA, 1959.
3. M Egenhofer, D Mark. Naïve geography. In *Spatial Information Theory: A Theoretical Basis for GIS*, Lecture Notes in Computer Science 988, A Frank, W Kuhn, Eds. Springer-Verlag, Berlin, 1995, pp. 1–15.
4. GJ Hunter. Understanding semantics and ontologies: they are quite simple really — if you know what I mean! *Trans GIS* 6: 83–87, 2002.
5. J Raper, T McCarthy, N Williams. Georeferenced four-dimensional virtual environments: principles and applications. *Comput Environ Urban Syst* 22: 529–539, 1998.
6. FT Fonseca, MJ Egenhofer, R Agouris, G Câmara. Using ontologies for integrated geographic information systems. *Trans GIS* 6: 231–257, 2002.
7. B Smith. An introduction to ontology. In *The Ontology of Fields*, D Peuquet, B Smith, B Brogaard, Eds. National Center for Geographic Information and Analysis, Santa Barbara, CA, 1998, pp. 10–14.
8. CS Smyth. A representational framework for geographic modeling. In *Spatial and Temporal Reasoning in Geographic Information Systems*, MJ Egenhofer, RG Golledge, Eds. Oxford University Press, New York, 1998, pp. 191–213.
9. E Davis. *Representations of Commonsense Knowledge*. Morgan Kaufmann, San Mateo, CA, 1990.

10. J Gomes, L Velho. Abstraction paradigms for computer graphics. *Visual Comput* 11: 227–239, 1995.
11. D Peuquet. Presentations of geographic space: towards a conceptual synthesis. *Ann Assoc Am Geogr* 78: 375–394, 1988.
12. M Goodchild, G Sun, S Yang. Development and test of an error model for categorical data. *Int J Geogr Inf Syst* 6: 87–104, 1992.
13. PA Burrough, RA McDonnell. *Principles of Geographical Information Systems.* Oxford University Press, New York, 1998.
14. AU Frank. Different types of "times" in GIS. In *Spatial and Temporal Reasoning in Geographic Information Systems*, MJ Egenhofer, RG Golledge, Eds. Oxford University Press, New York, 1998.
15. PJ Hayes. The second naïve manifesto. In *Formal Theories of a Commonsense World*, J Hobbes, R Moore, Eds. American Association for Artificial Intelligence, Menlo Park, CA, 1985, pp. 1–36.
16. AA Roshannejad, W Kainz. Handling Identities in Spatio-Temporal Databases. Paper presented at Proceedings of the ASCM/ASPRS Annual Convention and Exposition, Bethesda, MD. Technical Papers, 1995.
17. NR Chrisman. Beyond the snapshot: changing the approach to change, error, and process. In *Spatial and Temporal Reasoning in Geographic Information Systems*, MJ Egenhofer, RG Golledge, Eds. Oxford University Press, New York, 1998, pp. 85–93.
18. K McSweeney, PE Gessler, BK Slater, RD Hammer, GW Peterson, JC Bell. Towards a new framework for modeling the soil-landscape continuum. In *Factors in Soil Formation*, SSSA Special Publication 33. SSSA, Madison, WI, 1994.
19. K Verheyen, D Adriaens, M Hermy, S Deckers. High-resolution continuous soil classification using morphological soil profile descriptions. *Geoderma* 101: 31–48, 2001.
20. DJ Pennock, DF Acton. Hydrological and sedimentological influences on Boroll catenas, central Saskatchewan. *Soil Sci Soc Am J* 53: 904–910, 1989.
21. LJ Osher, SW Buol. Relationship of soil properties to parent material and landscape position in eastern Madre de Dios, Peru. *Geoderma* 83: 143–166, 1998.
22. B Minasny, AB McBratney. A rudimentary mechanistic model for soil production and landscape development. *Geoderma* 90: 3–21, 1999.
23. JA Thompson, JC Bell, CA Butler. Digital elevation model resolution: effects on terrain attribute calculation and quantitative soil-landscape modeling. *Geoderma* 100: 67–89, 2001.
24. C Domenikiotis, NR Dalezio, A Loukas, M Karteris. Agreement assessment of NOAA/AVHRR NDVI with Landsat TM NDVI for mapping burned forested areas. *Int J Remote Sensing* 23: 4235–4247, 2002.
25. S Grunwald, P Barak, K McSweeney, B Lowery. Soil landscape models at different scales portrayed in Virtual Reality Modeling Language. *J Soil Sci* 165: 598–615, 2000.
26. NM Sirakov, FH Muge. A system for reconstructing and visualizing three-dimensional objects. *Comput Geosci* 27: 59–69, 2001.
27. R Marschallinger. Three-dimensional reconstruction and visualization of geological materials with IDL: examples and source code. *Comput Geosci* 27: 419–426, 2001.
28. J Döllner, K Hinrichs. An object-oriented approach for integrating 3D visualization systems and GIS. *Comput Geosci* 26: 67–76, 2000.

29. M Kreuseler. Visualization of geographically related multidimensional data in virtual scenes. *Comput Geosci* 26: 101–108, 2000.
30. JM Lees. Geotouch: software for three- and four-dimensional GIS in the earth sciences. *Comput Geosci* 26: 751–761, 2000.
31. MFP Bierkens. Spatio-temporal modeling of the soil water balance using a stochastic model and soil profile descriptions. *Geoderma* 103: 27–50, 2001.
32. T Abraham, JF Roddick. Survey of spatio-temporal databases. *Geoinformatica* 3: 61–99, 1999.
33. IJ Koeppel, SD Ahlmer. Integrating the Dimension of Time into AM/FM Systems. In Proceedings of the AM/FM XVI International Annual Conference, Aurora, CO, 1993.
34. PA Whigham. Hierarchies of space and time. In *Spatial Information Theory: A Theoretical Basis for GIS*, Lecture Notes in Computer Science 716. Springer, New York, 1993.
35. LH Hermosilla. A Unified Approach for Developing a Temporal GIS with Database and Reasoning Capabilities. Paper presented at Proceedings of EGIS 94, The 5th European Conference and Exhibition on Geographical Information Systems, Paris, 1994.
36. DJ Peuquet, N Duan. An Event-Based Spatio-Temporal Data Model (ESTDM) for temporal analysis of geographical data. *Int J Geogr Inf Syst* 9: 7–24, 1995.
37. M Yuan. Modeling semantics, spatial and temporal information in a GIS. In *Progress in Trans-Atlantic Geographic Information Research*, M Craglia, H Couleclis, Eds. Taylor & Francis, New York, 1997.
38. P Goovaerts. *Geostatistics for Natural Resources Evaluation*. Oxford University Press, New York, 1997.
39. JP Chilès, P Delfiner. *Geostatistics: Modeling Spatial Uncertainty*. John Wiley & Sons, New York, 1999.
40. ML Stein. *Interpolation of Spatial Data: Some Theory for Kriging*, Springer-Verlag, Berlin, 1999.
41. AB McBratney, IOA Odeh, TFA Bishop, MS Dunbar, TM Shatar. An overview of pedometric techniques for use in soil survey. *Geoderma* 97: 293–327, 2000.
42. R Webster, MA Oliver. *Geostatistics for Environmental Scientists*. John Wiley & Sons, Chichester, England, 2001.
43. PM Berthouex, LC Brown. *Statistics for Environmental Engineers*. Lewis Publ., New York, 2002.
44. L Lemay, J Couch, K Murdock. *3D graphics and VRML 2*. Sams.net Publ., Indianapolis, 1999.
45. Web3D Consortium. 2003. Available at http://www.web3d.org/.
46. JL Troy, GF Czapar. Evaluating conservation practices using GIS and Virtual Reality Modeling Language. *J Nat Resour Life Sci Educ* 31: 38–43, 2002.
47. DR Miller, RA Dunham, W Chen. The application of VR modelling in assessing potential visual impacts of rural development. In *Virtual Reality in Geography*, P Fisher, D Unwin, Eds. Taylor & Francis, New York, 2002, pp. 131–143.
48. A Lovett, R Kennaway, G Sünnenberg, D Cobb, P Dolman, T O'Riordan, D Arnold. Visualizing sustainable agricultural landscapes. In *Virtual Reality in Geography*, P Fisher, D Unwin, Eds. Taylor & Francis, New York, 2002, pp. 102–130.
49. A Koussoulakou, MJ Kraak. Spatio-temporal maps and cartographic communication. *Cartogr J* 29: 101–108, 1992.

50. A Barraclough, I Guymer. Virtual reality: a role in environmental engineering education? *Water Sci Technol* 38: 303–310, 1998.
51. JD Hays, S Phirman, B Blumenthal, K Kastens, W Menke. Earth science instruction with digital data. *J Comput Geosci* 26: 657–668, 2000.
52. A Stibbard. Warwick University Forum, No. 6, 1997.
53. P Fisher, D Unwin. *Virtual Reality in Geography.* Taylor & Francis, New York, 2002.
54. DN Gordon, RD Pea. Prospects for scientific visualization as an educational technology. *J Learn Sci* 4: 249–279, 1995.
55. J Raper. *Multidimensional Geographic Information Science.* Taylor & Francis, New York, 2000.
56. K Morris, D Hill, A Moore. Mapping the environment through three-dimensional space and time. *J Comput Environ Urban Syst* 24: 435–450, 2000.
57. V Ramasundaram, S Grunwald, A Mangeot, NB Comerford, CM Bliss. Development of an environmental virtual field laboratory. *J Comput Educ* 45: 21–34, 2005.
58. S Grunwald, B Lowery, DJ Rooney, K McSweeney. Profile cone penetrometer data used to distinguish between soil materials. *Soil Tillage Res* 62: 27–40, 2001.
59. H Wackernagel. *Multivariate Geostatistics.* Springer, Berlin, 2003.
60. CM Bliss, NB Comerford. Forest harvesting influence on water table dynamics in a Florida flatwood landscape. *Soil Sci Soc Am J* 66: 1344–1349, 2002.
61. MJ Shiffer. Towards a collaborative planning system. *Environ Plann B* 19: 709–722, 1992.
62. JB Krygier. Cartographic multimedia and praxis in human geography and the social sciences. In *Multimedia Cartography,* W Cartwright, M Peterson, G Gartner, Eds. Springer, Berlin, 1999, pp. 245–256.
63. SM Freudenschuh, W Hellevik. Multimedia technology in cartography and geographic education. In *Multimedia Cartography,* W Cartwright, M Peterson, G Gartner, Eds. Springer, Berlin, 1999, pp. 271–280.

14 On Spatial Lattice Modeling of Soil Properties

Jun Zhu, Richard P. Wolkowski, Wei Yue, and Ruifeng Xu

CONTENTS

ABSTRACT

Kriging methods are popular for mapping soil properties by interpolating a finite number of samples in a field. Although kriging is often known as the best unbiased linear prediction in a geostatistical model, the idea of kriging can also be implemented in a spatial lattice model that decomposes an observation into signal plus noise. In this chapter, we review a type of spatial lattice model, namely, a multiresolution tree-structured spatial linear model. Because of the multiresolution tree structure, kriging can be conducted using a fast change-of-resolution

Kalman filter algorithm. We apply the methodology to obtain the best unbiased linear prediction of soil properties in a field in Wisconsin, while accounting for field conditions using a linear regression. Comparison is made among linear regression, a traditional spatial linear model for lattice data, and the multiresolution tree-structured spatial linear model. The result shows that the multiresolution tree-structured spatial linear model does not always fit the data better, but does offer a fast alternative for mapping soil properties.

14.1 INTRODUCTION

Observations of a soil property in a field, such as the depth of the A horizon, soil texture in the A and B horizons, and spatial distribution of key nutrients, can be modeled as realizations from a spatial random process $\{Z(\mathbf{s})\}$, where for a given location $\mathbf{s}$, $Z(\mathbf{s})$ is a random variable subject to a probability distribution, and the location $\mathbf{s}$ ranges continuously over a spatial domain D. To capture different sources of variation, the spatial random process is often decomposed into a global variation, $\{m(\mathbf{s})\}$, which consists of a deterministic mean structure, a local variation, $\{u(\mathbf{s})\}$, which consists of a spatially correlated zero-mean random process, and a measurement error, $\{e(\mathbf{s})\}$, which consists of a zero-mean white noise process independent of the local variation:

$$Z(\mathbf{s}) = m(\mathbf{s}) + u(\mathbf{s}) + e(\mathbf{s}) \tag{14.1}$$

(see, e.g., Section 3.1 in Cressie[4]). For example, the global variation may be modeled by a linear regression on a set of explanatory variables, while the local variation may be modeled by a stationary Gaussian process with an exponential variogram. Let

$$Y(\mathbf{s}) = m(\mathbf{s}) + u(\mathbf{s}) \tag{14.2}$$

denote the signal process that consists of both the global and the local variation. Then the decomposition of $Z(\mathbf{s})$ can be rewritten in the form of signal plus noise:

$$Z(\mathbf{s}) = Y(\mathbf{s}) + e(\mathbf{s}) \tag{14.3}$$

There are two primary objectives in spatial random process modeling. One objective is to estimate the global variation $\{m(\mathbf{s})\}$, usually in relation to potential explanatory variables. Thus, the aim of modeling the local variation $\{u(\mathbf{s})\}$ is to capture any remaining spatial dependence after the global variation, $\{m(\mathbf{s})\}$, has been accounted for. Although sometimes viewed as a nuisance, modeling the local variation is important for ensuring a valid statistical inference of the global variation.

The other objective is to predict the signal process, based on knowledge of the global variation, the spatial dependence structure in the local variation, and the measurement error. Given a set of observations, $\{Z(\mathbf{s}_1), \ldots, Z(\mathbf{s}_n)\}$, observed at sampling locations, $s_1, ..., s_n$, the best linear unbiased predictor (BLUP) of $Y(s_0)$, at a given location s_0, which minimizes the mean squared prediction error (MSPE), can be obtained using kriging (see, e.g., Odeh et al.,[12,13] Knotters et al.,[9] Bourennane et al.,[2] Carre and Girard[3]). In the absence of measurement error, prediction of $Y(s_0)$ is the same as prediction of $Z(s_0)$, since $Z(s_0) = Y(s_0) = m(s_0) + u(s_0)$. If an observation is available at s_0, then the BLUP of $Y(s_0)$ is the observation $Z(s_0)$ itself. If an observation is not available at s_0, then the BLUP of $Y(s_0)$ is a weighted sum of the observations where the optimal weights are obtained from a set of kriging estimation equations. Depending on the assumptions made about the global variation $\{m(s)\}$, some commonly used krigings include simple kriging, where $m(s)$ is assumed to be known; ordinary kriging, where $m(s)$ is unknown but is assumed to be a constant; and universal kriging, where $m(s)$ consists of a trend or linear regression with unknown regression coefficients. On the other hand, in the presence of measurement error, prediction of $Y(s_0)$ is also a weighted sum of the observations with optimal weights obtained from a similar set of kriging estimation equations. In fact, the BLUP of $Y(s_0)$ at an unsampled location is the same as that of $Z(s_0)$, since the measurement error at the unsampled location does not affect how the underlying signal is predicted. However, depending on the level of measurement error, the BLUP of $Y(s_0)$ at a sampled location does not need to be the same as the observation $Z(s_0)$. In the extreme case where there is no local variation but only measurement error, $Y(s_0) = m(s_0)$ is fixed or deterministic. Suppose further that $m(s_0)$ is a fixed constant, then the BLUP of $Y(s_0)$ is the average of the observations, regardless of the location s_0. That is, the optimal weights in the BLUP of $Y(s_0)$ are equal, as there is no spatial dependence among the signal variables. More details of these issues can be found in Section 3.2.1 of Cressie.[4]

In the spatial statistics literature, an alternative to the aforementioned geostatistical approach is lattice data modeling, where a spatial random process takes place on a regular or irregular lattice, either due to the discrete nature of data locations or due to the aggregation of an underlying random process with continuous spatial index. For instance, the soil property of interest in a spatial region D could be modeled by a random process $Y(s)$. Partition D into a set of N cells $\{D_1, ..., D_N\}$ and define

$$y_k = \frac{1}{|D_k|} \int_{D_k} Y(s) \, ds \qquad (14.4)$$

as an average value of the random process aggregated over the locations in the k-th cell D_k, where $k = 1, ..., N$. Instead of modeling $\{Y(s)\}$ directly, a lattice modeling approach could focus on the aggregated process $\{y_k\}$. The value in each grid cell is usually modeled as spatially correlated to its neighboring values, according to a certain neighborhood structure. Neighborhood structures may be defined in terms of proximity, direction, and contiguousness of grid cells so that meaningful spatial correlations can be established among neighbors. For example, a neighborhood structure could define two grid cells to be neighbors of each other if they share a side in a particular compass direction (see, e.g., Upton and Fingleton[14]).

Despite the differences in a lattice data model specification from a geostatistical model, some of the modeling strategies are similar. In particular, the spatial random process y_k is often decomposed into a global variation and a local variation:

$$y_k = m_k + u_k \tag{14.5}$$

where $m_k = \dfrac{1}{|D_k|} \displaystyle\int_{D_k} m(\mathbf{s}) \, d\mathbf{s}$ and $u_k = \dfrac{1}{|D_k|} \displaystyle\int_{D_k} u(\mathbf{s}) \, d\mathbf{s}$ are the global variation and local variation processes, respectively, at the aggregated scale. An observation within a cell, denoted by z_k, is modeled by

$$z_k = y_k + e_k \tag{14.6}$$

where e_k is a measurement error process. Parameter estimation of the global variation can be obtained while accounting for the spatial dependence in the local variation. Prediction of y_k can be obtained by minimizing the MSPE using kriging, as in the geostatistical modeling approach. For a cell that has an observation, the BLUP of y_k is the same as the observed value z_k if there is no measurement error, but does not need to be the same as the observed value z_k if there is measurement error included in the model. Unlike in the geostatistical modeling approach, kriging in a lattice model approach often involves predicting the underlying signal process on the lattice, based on observations with measurement errors.

In this chapter, we consider a multiresolution tree-structured spatial linear model (MTSLM) developed by Zhu and Yue,[15] and our main purpose is to demonstrate the use of an MTSLM in mapping soil properties. The MTSLM is a generalization of the multiresolution tree-structured model developed by Huang et al.[6] The main features of an MTSLM include a linear regression in the global variation, a spatial dependence structure in the local variation derived from a multiresolution tree structure, and a measurement error process. For statistical inference, model parameters are estimated by maximum likelihood using an expectation-maximization (EM) algorithm, while the signal process is predicted by a change-of-resolution Kalman filter algorithm. The linear regression in the global variation allows for modeling a soil property in relation to explanatory variables. Moreover, because of the multiresolution tree structure, the prediction algorithm can be implemented in one pass and is very fast.

The remainder of the chapter is organized as follows. In Section 14.2, we describe the MTSLM and the related statistical inference. In Section 14.3, we apply the MTSLM to map soil properties in a Wisconsin field. Comparison is made between the MTSLM and the traditional linear models. Discussion of the results is given in Section 14.4, and a conclusion is given in Section 14.5.

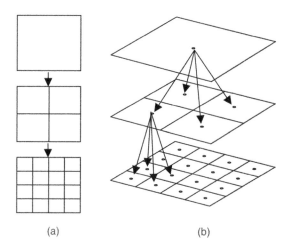

(a)   (b)

FIGURE 14.1 (a) The spatial region is partitioned into four cells, each partitioned into four subcells. (b) The multiresolution tree structure has three resolutions, one root, and four child nodes for each node that is a parent node.

14.2 MULTIRESOLUTION TREE-STRUCTURED SPATIAL LINEAR MODEL

To construct a tree structure, the spatial region D of interest is partitioned in a nested fashion as follows. On the first (coarsest) resolution, we partition the spatial region D into a set of N_1 cells $\{D_{1,k}\}$, where $D_{1,k}$ denotes the k-th cell on the first resolution; $k = 1, ..., N_1$. We further partition each cell $D_{1,k}$ on the first resolution into $n_{1,k}$ subcells on the second resolution, resulting in a set of $N_2 = \sum_{k=1}^{N_1} n_{1,k}$

cells $\{D_{2,k}\}$, where $D_{2,k}$ denotes the kth cell on the second resolution; $k=1, ..., N_2$. The partitioning is continued until a desired fine-resolution J with a set of

$N_J = \sum_{k=1}^{N_{J-1}} n_{J-1,k}$ cells $\{D_{J,k}\}$, where $D_{J,k}$ denotes the kth cell on the J-th (finest)

resolution; $k = 1, ..., N_J$. For simplicity, we assume that the partition of each cell is homogeneous, such that it gives equal-size subcells with $|D_{j,1}| = |D_{j,2}| = ... = |D_{j,Nj}|$ for a given j, except for the first (coarsest) resolution. See Figure 14.1a for an illustration of partitioning a cell into subcells. Given these partitions, we define

$y_{j,k} = \frac{1}{|D_{j,k}|} \int_{D_{j,k}} Y(\mathbf{s}) \, d\mathbf{s}$ to be the average value of the signal process $Y(\mathbf{s})$ in

$D_{j,k}$, aggregated over the locations in cell $D_{j,k}$, where $k=1, ..., N_j, j = 1, ..., J$. In this section, we model $\{y_{j,k}\}$ by a multiresolution tree-structured spatial linear model (MTSLM), estimate the model parameters by maximum likelihood, and predict $\{y_{j,k}\}$ by a change-of-resolution Kalman filter algorithm with a known level of uncertainty, based on the observations $\{z_{j,k}\}$.

14.2.1 THE GLOBAL VARIATION

The MTSLM assumes that the signal process $\{y_{j,k}\}$ is decomposed into a global variation and a local variation:

$$y_{j,k} = x^T_{j,k}\boldsymbol{\beta} + u_{j,k} \tag{14.7}$$

Here the global variation is in the form of a linear regression $x^T_{j,k}\boldsymbol{\beta}$, where $x_{j,k}$ is a vector of explanatory variables associated with the k-th cell on the j-th resolution and $\boldsymbol{\beta}$ is a vector of regression coefficients, both in p dimensions. Further, the local variation $\{u_{j,k}\}$ has a zero mean and a spatial dependence structure that is determined by a multiresolution tree structure. In a field study to be shown in Section 14.3, $y_{j,k}$ would be a property such as soil phosphorus (P), while the explanatory variables would be topography such as elevation.

14.2.2 THE LOCAL VARIATION

A multiresolution tree structure consists of a set of nodes interconnected by a set of edges directed from the coarser resolutions to the finer resolutions. Let the (j,k)-th node denote the k-th node on the j-th resolution located in the center of the cell $D_{j,k}$. The *parent* node of the (j,k)-th node, $pa(j,k)$, is the node on the $(j-1)$-th resolution, which has an edge directed from $pa(j,k)$ to (j,k), whereas a *child* node of the (j,k)-th node, $ch(j,k)$, is the node on the $(j+1)$-th resolution, which has an edge directed from (j,k) to $ch(j,k)$. We assume that the spatial resolution is the coarsest on the first resolution $j = 1$ and the finest on the last resolution $j = J$. A node on the coarsest resolution does not have a parent and is called a *root* node. A node on the finest resolution does not have any child and is called a *leaf* node. Figure 14.1b shows a schematic plot of a three-resolution one-root quad tree, where a root node on the coarsest resolution has four child nodes on the middle resolution, and similarly, each node on the middle resolution has four child nodes on the finest resolution. Here *quad* refers to four child nodes per node. Hence, the total number of resolutions is $J = 3$ and the numbers of nodes on each resolution are $N_1 = 1$, $N_2 = 4$, and $N_3 = 16$.

Associating the (j,k)-th node with the k-th cell on the j-th resolution $D_{j,k}$, we model $\{u_{j,k}\}$ by a multiresolution tree structure:

$$u_{j,k} = u_{pa(j,k)} + \omega_{j,k}, \ \omega_{j,k} \sim N(0,\sigma_j^2), \ k = 1, \Lambda, N_j, j = 2, \Lambda, J,$$

$$u_{1,k} \sim N(0,\sigma_1^2), \ k = 1, \Lambda, N_1 \tag{14.8}$$

where $u_{j,k}$ is the local variation in the cell $D_{j,k}$ associated with the (j,k)-th node. The mathematical expectation of $u_{j,k}$ is assumed to be zero. The term $\omega_{j,k}$ corresponds to a random independent departure of $u_{j,k}$ on the j-th resolution from $u_{pa(j,k)}$ on the $(j-1)$-th resolution and is assumed to have a Gaussian distribution with mean zero and noise variance σ_j^2. Thus, the value $u_{j,k}$ is influenced by both the value of its parent $u_{pa(j,k)}$ and some independent random fluctuation $\omega_{j,k}$. A smaller noise variance σ_j^2 of the model would force $u_{j,k}$ to resemble its parent value $u_{pa(j,k)}$ more than a larger σ_j^2.

To further match the physical conditions, it is necessary that an average of all the children's values be equal to their parent's value, satisfying a so-called mass-balance property. That is,

$$\frac{1}{n_{j,k}} \sum_{allch(j,k)} y_{ch(j,k)} = y_{j,k} \tag{14.9}$$

where $n_{j,k}$ is the number of child nodes of the (j,k)-th node. The MTSLM assumes two conditions under which the mass-balance property holds:

$$\frac{1}{n_{j,k}} \sum_{allch(j,k)} u_{ch(j,k)} = u_{j,k} \quad \text{and} \quad \frac{1}{n_{j,k}} \sum_{allch(j,k)} x_{ch(j,k)} = x_{j,k} \tag{14.10}$$

The spatial covariance structure of the multiresolution model is induced by the parent–child relationship, such that the strength of the spatial structure is determined by how parent nodes are shared among the child nodes of the finest resolution. Given two nodes (j,k) and (j,k') on the j-th resolution, denote the first common ancestor on the j_{an}-th resolution as $an(j,k,k')$. If the two nodes do not have a common ancestor, then they are descendants of two different root nodes and we let $j_{an} = 0$. If the two nodes are the same node, then $j_{an} = j$. The covariance between $u_{j,k}$ and $u_{j,k'}$ is:

$$\text{cov}(u_{j,k}, u_{j,k'}) = \begin{cases} \sigma_1^2 + \cdots + \sigma_j^2; \, j_{an} = j \\ \sigma_1^2 + \cdots + \sigma_{jan}^2 - \dfrac{\sigma_{jan+1}^2}{n_{jan} - 1}; \, j_{an} = 1,...,j-1 \\ 0; \, j_{an} = 0 \end{cases} \tag{14.11}$$

Note that the spatial covariance (or spatial correlation) between two nodes on the j-th resolution decreases as j_{an} of the first common ancestor becomes smaller. Moreover, the spatial covariance function $cov(u_{j,k}, u_{j,k'})$ is invariant to the geographical locations of the nodes (j,k) and (j',k') as long as the first common ancestor $an(j,k,k')$ is on the same j_{an}-th resolution. In essence, the spatial dependence structures are a result of the multiresolution tree structure, and the strength of the spatial dependence is determined by the relative magnitude of the noise variances $\{\sigma_J^2, \sigma_{J-1}^2, \ldots, \sigma_1^2\}$.

Compared with a variogram model in geostatistics, the spatial dependence structure here is simpler, primarily because the spatial domain of interest is divided into cells and the spatial dependence is induced by the multiresolution tree structure. However, one benefit of using a simpler model is in the computational efficiency. See also Zhu et al.[16] for examples that compare the multiresolution models with kriging methods. Because of the multiresolution tree structure, a one-pass change-of-resolution algorithm can be used to carry out spatial prediction. We note that the change-of-resolution algorithm is similar to other tools, such as the Bayesian multiscale models in Kolaczyk[10] and the wavelet characterization in Epinat et al.[5]

14.2.3 THE MEASUREMENT ERROR

Measurements about the soil property $\{y_{J,k}\}$ are obtained on the J-th resolution in a spatial region D. Within the k-th cell on the J-th resolution $D_{J,k}$, an observation, $z_{J,k}$, is modeled by

$$z_{J,k} = y_{J,k} + e_{J,k},\ e_{J,k} \sim N(0, \phi_J^2),\ k = 1, \Lambda, N_J \tag{14.12}$$

Here the term $e_{J,k}$ is the measurement error associated with the observation $z_{J,k}$, which captures the random independent departure of the observed value $z_{J,k}$ from the signal value $y_{J,k}$. The measurement error $e_{J,k}$ is assumed to have an independent Gaussian distribution with mean zero and variance ϕ_J^2. In soil property mapping, the observation $z_{J,k}$ could be the k-th soil P measured in the cell $D_{J,k}$, whereas the signal $y_{J,k}$ would then be the actual soil P in the cell $D_{J,k}$ and is the unknown quantity that we would like to predict.

14.2.4 STATISTICAL INFERENCE

According to Equations 14.7 to 14.9 and 14.12, there are three types of unknown parameters: the regression coefficients β, the measurement error variance ϕ_J^2, and the noise variances of the model σ_j^2, $j = 1, \ldots, J$. The measurement error variance will be prespecified based on an estimate of the nugget effect. Given the estimated ϕ_J^2, the regression coefficients β and the noise variances σ_j^2 are estimated by maximum likelihood using an expectation-maximization (EM) algorithm. The EM algorithm also gives the variance-covariance matrix of the parameter estimates, which could be used for hypothesis testing and confidence intervals of the true parameter values.

For prediction of the signal process, let $\mathbf{Z}$ denote the vector of all the data. Under the Gaussian assumption, the best unbiased predictor of $y_{J,k}$, which minimizes the mean squared prediction error (MSPE), is $E(y_{J,k}|\mathbf{Z})$. For the given multiresolution tree-structured model, the estimation equations in the matrix form are

$$E(y_{J,k}|Z) = x^T_{j,k}\beta + E(u_{J,k}|Z) \qquad (14.13)$$

where $E(u_{J,k}|Z) = cov(y_{J,k}, Z) var(Z)^{-1}(Z - X\beta)$ and X is the matrix containing all the $x_{j,k}$.

In $E(y_{J,k}|Z)$, we first substitute the true regression coefficients β by the maximum likelihood estimate β_{MLE}. Then we compute $E(u_{J,k}|Z)$, where the model parameters are replaced by the maximum likelihood estimates. Under a kriging setup, the computation of $E(u_{J,k}|Z)$ would involve an inversion of the matrix $var(Z)$. However, because of the multiresolution tree structure, the matrix inversion can be replaced by a change-of-resolution computing algorithm, involving a fine-to-coarse-resolution filtering step, followed by a coarse-to-fine-resolution smoothing step. In the filtering step, the algorithm moves from fine resolutions to coarse resolutions, recursively computing the values $u_{j,k}$, based on the data on the relevant finer resolutions. Once the coarsest resolution is reached, the algorithm goes back from coarse resolutions to fine resolutions, recursively computing the values $u_{j,k}$ on each resolution based on all the data. In the final step of the recursion, the prediction of $u_{J,k}$ on the finest resolution and the prediction error variance are calculated. Moreover, as a by-product, the change-of-resolution algorithm handles missing data automatically by replacing any unobserved (missing) datum on a node with its most plausible estimator $E(u_{j,k}|Z)$. The change-of-resolution computing algorithm is an extension from the Kalman filter algorithm for time series data (see, e.g., Meinhold and Singpurwalla[11]).

The main features of the MTSLM can be summarized as follows. The change-of-resolution algorithm gives prediction of $\{y_{J,k}\}$ on the desired finest resolution. The computation is fast and hence feasible for practical use. Moreover, the inclusion of a linear regression enables modeling of relationships among variables while accounting for spatial dependence in the residuals. Hence, the model can be used for two important purposes in practice: regression on explanatory variables and prediction of the signal process.

14.2.5 MODEL SELECTION AND EVALUATION

Sound statistical inference, including parameter estimation and signal prediction, hinges on the careful selection of an underlying model. To use the MTSLM described above for a given dataset, various choices of the model need to be made. The linear regression mean structure depends on which explanatory variables are incorporated in the model. In addition, the spatial dependence is specified via the tree structure and depends on several factors, such as the number of resolutions, the partition on the coarsest resolution, and the number of children for each parent node. On one hand, we favor models that fit the data well according to a certain criterion, which is the maximum likelihood here. On the other hand, more complex models (e.g., with more explanatory variables) may increase the likelihood without substantially enhancing the explanatory or predictive power of the model. Hence, we also favor models that are parsimonious. In this regard, Akaike's information criterion (AIC), defined as minus twice the log-likelihood plus twice the number of parameters, is suitable for model selection.[1] A smaller AIC value is associated with a larger likelihood and a more parsimonious model by penalizing unnecessarily complex models. We shall use AIC as a criterion for selecting the best models. Our strategy is to first select the best multiresolution tree structure, given the full regression model that includes all the potential explanatory variables. Then given the selected multiresolution tree structure, we select the best set of explanatory variables for a reduced regression model.

To further evaluate the MTSLM quality, we compare model fitting and model selection using a simple linear regression (SLR) model, spatial linear model (SLM), and MTSLM. The three models all assume a linear regression and a Gaussian distribution in the error, but differ in the dependence structure of the error. The SLR model assumes that the errors are independent, and hence does not allow for spatial dependence. Both SLM and MTSLM allow for spatial dependence in the error, but differ in the type of spatial dependence. We consider SLM with a first-order neighborhood, such that the neighbors of a given cell consist of the adjacent cells to the north, west, south, and east. Other choices of the neighborhood structures are diagonal, such that the neighbors consist of the cells in the northeast, northwest, southwest, and southeast, or second order, which combines the neighbors from the first order and the diagonal neighborhood. Moreover, we consider simultaneous autoregressive (SAR) models, which specify the spatial dependence using a joint distribution. Other commonly used model specifications are conditional autoregressive (CAR) models and moving average (MA) models, but they do not fit our data as well as the SAR model and thus are not used. For details of the SAR, CAR, and MA models, see Chapters 6 and 7 of Cressie[4] and Chapter 5 of Kaluzny et al.[7]

14.3 MATERIALS AND METHODS

A study of soil properties was conducted in a 57-ha field near Arena, WI (43 × 15'N 89×93'W). The soils in the field are Sparta loamy fine sand (Enthic Hapludoll) and Dakota sandy loam (Typic Argiudoll). These soils are positioned on a gently rolling landscape, well drained, and developed under prairie vegetation on stream terraces in the Lower Wisconsin River Valley. The U.S. Department of Agriculture's Natural Resource Conservation Service (USDA-NRCS) Soil Survey for Iowa County, Wisconsin (USDA-SCS, 196), shows that the field was historically subdivided and managed as various small fields. In recent years, the field has been managed as a single unit in a center-pivot irrigated vegetable crop rotation (mainly potato (*Solanum tuberosum* L.), sweet corn (*Zea mays* L.), and green beans (*Phaseolis vulgaris* L.)). Standard agronomic practices include a chisel plow/disk tillage system and recommended pesticide and fertilization management.[2]

Soil samples were taken from the field in October 2000 using a uniform systematic grid pattern on a 0.40-ha spacing. Eight to 10 soil cores were composited from the top 20 cm of the soil within a 1-m radius of a point of known latitude and longitude (Figure 14.2). The soil samples were analyzed by the Soil

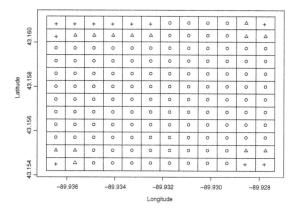

FIGURE 14.2 The sampling locations (marked by o and Δ) according to a uniform systematic sampling scheme in a field near Arena, WI. The samples marked by Δ are not used in the parameter estimation, but are used in prediction. The cells marked by + are not part of the field.

Testing and Plant Analysis Laboratory at the University of Wisconsin for pH, phosphorus (P) in ppm, potassium (K) in ppm, calcium (Ca) in ppm, and magnesium (Mg) in ppm.[8] Hence, the observations consist of the soil property measurements and the corresponding spatial locations on the grid. In addition, various factors, such as topography, soil types, weather conditions, and fertilization, would affect the spatial distribution of these soil nutrients. Here we consider elevation (in m) as a possible explanatory variable.

The MTSLM was applied to analyze the data for the following purposes: (1) examining the relation, if any, among the soil property variables, (2) assessing the potential influence of topography on the soil property variables, and (3) mapping the soil property based on these relations as well as using the topographic information. For a given soil property variable (pH, P, K, Ca, or Mg), all other soil property variables and elevation were used as the explanatory variables in the regression. In this study, there were a total of 133 observations. We focused on a 12 × 12 grid of cells over the field and used a three-resolution nine-root quad tree (i.e., $N_1 = 9$, $N_2 = 36$, $N_3 = 144$) as the multiresolution tree structure. Among the 144 cells, 11 cells along the edges (marked by + in Figure 14.2) were not part of the field and were treated as missing nodes in the multiresolution tree structure. Given the multiresolution tree structure, we fitted all possible MTSLMs to the data using maximum likelihood and selected the best subset of explanatory variables using AIC.

Statistical inference was performed based on the best model for each soil variable. The measurement error variance was estimated by the nugget effect in fitting an exponential variogram model to the empirical variogram. Maximum likelihood estimates of the regression coefficients and the noise variances, along with their corresponding standard errors, were computed using the EM algorithm. Given the estimated parameters, the change-of-resolution Kalman filter algorithm was applied to the data for obtaining the predicted values of $\{y_{j,k}\}$, as shown in Equation 14.13, and the corresponding prediction standard errors.

For comparison, data analysis was performed using simple linear regression (SLR) where the residuals were assumed to have independent and identical Gaussian distribution. Furthermore, a spatial linear model (SLM) was considered, for which the regression residuals were assumed to be spatially dependent using a first-order neighborhood structure. In both SLR and SLM, model selection was performed in a similar manner as in the MTSLM using AIC. Parameter estimation and model selection results were compared among SLR, SLM, and MTSLM. Because of the mass-balance property, 13 of the 133 observations on the edge of the field did not play a role in the parameter estimation. For a fair comparison, we performed SLR and SLM analysis on the remaining 120 observations (Figure 14.2).

For model checking, Moran's I test was performed to determine whether the residuals after fitting an SLR, SLM, or MTSLM were spatially correlated. The quality of the model was assessed by R-square for the SLR model and by AIC for the SLM and MTSLM.

14.4 RESULTS AND DISCUSSION

14.4.1 Exploratory Data Analysis

The summary statistics are given in Table 14.1, including the minimum, first quartile, median, mean, third quartile, and maximum of the data. Table 14.1 and the histograms (not shown) of the data suggest that the distribution of the data is fairly symmetric, except that the distribution of Mg is somewhat skewed. The scatter plots of all pairs of the variables are shown in Figure 14.3. There is indication of a positive relation among K, Ca, and Mg, with the sample correlation coefficients 0.71, 0.57, and 0.73 for K and Ca, K and Mg, and Ca and Mg, respectively. Although not as strong, there is some indication of a negative relation between elevation and P, K, Ca, and Mg, but a positive relation between elevation and pH. That is, as the elevation increases, the values of P, K, Ca, and Mg decrease,

TABLE 14.1
Summary Statistics of the Data

	PH (ppm)	P (ppm)	K (ppm)	Ca (ppm)	Mg (ppm)	Elevation (m)
Minimum	6.0	33	40	120	50	191.00
1st quartile	6.5	63	113	855	90	191.48
Median	6.7	79	159	1055	130	191.71
Mean	6.718	83.31	157.0	1106	146.5	191.69
3rd quartile	7.0	100	199	1292	193	191.90
Maximum	7.4	175	321	2210	350	192.29

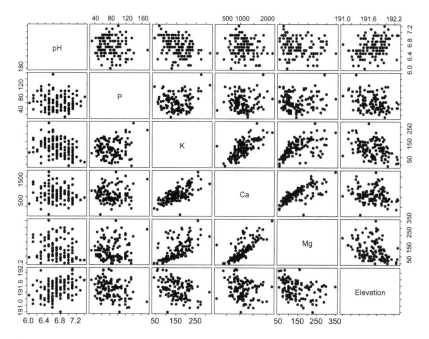

FIGURE 14.3 The pair-wise scatter plots of the variables pH, P, K, Ca, Mg, and elevation.

although the reverse is the case with the pH. The sample correlation coefficients between the elevation and the soil property variables are 0.38, −0.26, −0.37, −0.21, and −0.31 for pH, P, K, Ca, and Mg, respectively.

14.4.2 MODEL SELECTION AND PARAMETER ESTIMATION

For each soil property variable, parameter estimation results are shown in Table 14.2 based on the full MTSLM with all the explanatory variables included and in Table 14.3 based on the reduced MTSLM with only the best subset of explanatory variables included. Reported are the estimates of the regression coefficients, the noise variances, and the measurement error variances. Moreover, Moran's I values and the corresponding p values for testing the spatial independence among the residuals, as well as the AIC values, are presented in Table 14.2 and Table 14.3.

First we focus our discussion on the results from the best (reduced) model (Table 14.3). In the model for pH, K has a negative relation with pH, Ca and elevation have a positive relation with pH, and P and Mg do not have a significant influence on pH. Elevation appears to have the most significant influence on pH. In the model for P, the only variable that has an influence on P is elevation. As the elevation increases, the level of P decreases. In the model for K, the best model is the full model. That is, all the variables have an influence on K, except that the effect of Mg is not significant. The relation is positive between K and P, Ca, while the relation is negative between K and pH, elevation. In the model for Ca, K, and Mg have the strongest positive relation with Ca, as seen from the

TABLE 14.2
Parameter Estimation in the Full Multiresolution Tree-Structured Spatial Linear Model (MTSLM)

Response	pH		P		K		Ca		Mg	
	Estimate	Standard Error	Estimate	Standard Error	Estimate	Standard Error	Estimate	Standard Error	Estimate	Standard Error
Intercept	-71.70	9.51	5386.48	1626.84	5938.49	1295.34	-2024.71	5507.49	3400.71	1190.80
pH			3.596	4.444	-33.16	6.12	155.01	58.08	-24.85	9.20
P	0.0002	0.0009			0.10	0.07	-0.25	0.71	-0.05	0.16
K	-0.0015	0.0006	0.0572	0.0161			2.80	0.39	-0.04	0.10
Ca	0.0003	0.0001	-0.0082	0.0191	0.10	0.01			0.14	0.01
Mg	-0.0008	0.0006	0.0140	0.1178	0.02	0.11	3.00	0.35		
Elevation	0.41	0.05	-27.80	8.08	-29.61	6.12	6.44	25.68	-16.84	4.37
ϕ_1^2	0.014	0.003	11.44	19.69	79.53	24.94	7095.80	1099.29	414.79	45.99
ϕ_2^2	0.013	0.006	77.14	24.65	215.38	61.74	8694.47	4053.48	567.84	223.93
ϕ_3^2	0.013	0.006	47.03	70.17	909.00	141.02	4818.35	3220.58	253.94	131.66
ϕ_4^2	0.03		525.47		21.44		22,289.00		847.85	
Moran's I correlation	0.07		-0.03		0.25		0.04		-0.03	
p value	0.22		0.79		0.00		0.47		0.72	
AIC	284.49		1215.88		14,265.80		1825.47		1564.32	

TABLE 14.3

Parameter Estimation in the Best Multiresolution Tree-Structured Spatial Linear Model (MTSLM)

Response	pH Estimate	pH Standard Error	P Estimate	P Standard Error	K Estimate	K Standard Error	Ca Estimate	Ca Standard Error	Mg Estimate	Mg Standard Error
Intercept	−73.75	3.71	5688.90	1887.63	5938.49	1295.34	−17,752.30	4574.05	5214.96	958.38
PH					−33.16	6.12				
P					0.10	0.07				
K	−0.0015	0.0006			0.10	0.01	2.72	0.38		
Ca	0.0002	0.0001							0.13	0.01
Mg					0.02	0.11	2.96	0.35		
Elevation	0.42	0.02	−29.24	9.85	−29.61	6.12	93.89	23.88	−27.20	5.05
ϕ_1^2	0.011	0.001	13.77	26.51	79.53	24.94	5067.40	874.81	347.02	51.13
ϕ_2^2	0.013	0.006	76.33	54.44	215.38	61.74	8933.74	4110.15	556.30	229.04
ϕ_3^2	0.015	0.006	50.03	66.57	909.00	141.02	6472.64	3468.00	300.08	139.46
ϕ_3^3	0.03		525.47		21.44		22,289.00		847.85	
Moran's I correlation	0.10		−0.02		0.25		0.05		0.00	
p value	0.11		0.89		0.00		0.34		0.94	
AIC	274.08		1209.97		14,265.80		1819.86		1548.09	

pair-wise scatter plots (Figure 14.3). Although the scatter plot between Ca and elevation indicates a negative relation, the regression coefficient of elevation is positive. We suspect that the positive relation is unstable due to possible multi-collinearity among the explanatory variables. Finally, in the model for Mg, Ca has a positive relation, elevation has a negative relation, and the other variables pH, P, and K have no significant relation, all with Mg. Note that elevation is the one variable that has a significant influence on all the soil property variables: pH, P, K, Ca, and Mg.

The full model (Table 14.2), on the other hand, has similar parameter estimates as those in the best model, while including all the other nonsignificant variables in the model. As a result, although the log-likelihood of the full model is higher than that of the best model, the AIC values of the best model are lower. Furthermore, for both the best models and the full models, Moran's I tests have large p values (>0.10) for pH, P, Ca, and Mg, suggesting that there is no spatial correlation left in the residuals after the spatial dependence has been accounted for by the multiresolution tree structure. The exception is the model for K, where Moran's I test statistic is 0.25 and the p value is less than 0.01. The multiresolution tree structure does not adequately describe the spatial dependence structure in the K data.

14.4.3 Model Comparisons

The parameter estimation results of fitting the MTSLM are compared with those of the SLR model. For each soil property variable, parameter estimation results are shown in Table 14.4 based on the full SLR model with all the explanatory variables included and in Table 14.5 based on the reduced SLR model with only the best subset of explanatory variables included. Reported are the estimates of the regression coefficients and the error variances (σ^2). Moreover, Moran's I values and the corresponding p values for testing the spatial independence among the residuals, as well as the R-square values, are included in Table 14.4 and Table 14.5.

For modeling P, Ca, and Mg, the best SLR model has selected the same set of explanatory variables as the MTSLM by AIC. For modeling pH, the best SLR model does not include Ca, whereas for modeling K, the best SLR model does not include P and Mg, while the best MTSLM does. Moran's I tests for spatial independence have very small p values (<0.01) in the models for pH, K, Ca, and Mg, and a p value of 0.09 for P. That is, there is strong evidence of spatial correlation among the residuals after fitting the SLR model. As a consequence, the statistical inferences, including parameter estimation and model selection based on the SLR models, are not correct. In contrast, Moran's I tests suggest that the spatial dependence in the residuals has indeed been accounted for by the MTSLM (except the model for K).

Similarly, the parameter estimation results of fitting the MTSLM are compared with those of SAR SLM. For each soil property variable, parameter estimation results are shown in Table 14.6 based on the full SLM with all the explanatory variables included and in Table 14.7 based on the reduced SLM with

TABLE 14.4
Parameter Estimation in the Full Simple Linear Regression (SLR) Model

Response	pH		P		K		Ca		Mg	
	Estimate	Standard Error	Estimate	Standard Error	Estimate	Standard Error	Estimate	Standard Error	Estimate	Standard Error
Intercept	−55.35	20.46	4984.27	1972.52	5994.21	2703.50	−23,626.38	15,899.33	7341.79	3488.50
pH			4.9603	8.9817	−32.78	11.86	88.71	70.74	−5.35	15.77
P	0.0005	0.0010			0.10	0.13	−0.46	0.74	0.04	0.16
K	−0.0019	0.0007	0.0544	0.0686			3.24	0.45	0.03	0.12
Ca	0.0002	0.0001	−0.0074	0.0118	0.10	0.01			0.13	0.02
Mg	−0.0002	0.0006	0.0118	0.0534	0.02	0.07	2.64	0.34		
Elevation	0.32	0.11	−25.75	10.36	−29.91	14.22	121.44	83.47	−38.13	18.31
ϕ^2	0.079		695.10		1289.80		43,594.00		2138.80	
Moran's I correlation	0.32		0.08		0.25		0.28		0.32	
p value	0.00		0.17		0.00		0.00		0.00	
R-square	0.20		0.08		0.59		0.68		0.57	

TABLE 14.5
Parameter Estimation in the Best Simple Linear Regression (SLR) Model

Response	pH Estimate	pH Standard Error	P Estimate	P Standard Error	K Estimate	K Standard Error	Ca Estimate	Ca Standard Error	Mg Estimate	Mg Standard Error
Intercept	−56.21	19.32	5177.95	1706.79	6673.64	2541.29	−31,006.25	14,880.02	8165.87	3069.72
pH					−32.57	11.78				
P	−0.0013	0.0005								
K							3.11	0.44		
Ca					0.10	0.01			0.13	0.01
Mg							2.66	0.34		
Elevation	0.33	0.10	−26.58	8.90	−33.42	13.39	162.95	77.50	−42.60	16.00
ϕ^2	0.075		676.40		1275.30		43,555.00		2089.00	
Moran's I correlation	0.30		0.09		0.26		0.26		0.33	
p value	0.00		0.13		0.00		0.00		0.00	
R-square	0.19		0.07		0.59		0.68		0.56	

TABLE 14.6
Parameter Estimation in the Full Spatial Linear Model (SLM) with a Simultaneous Autoregression and a First-Order Neighborhood

Response	pH		P		K		Ca		Mg	
	Estimate	Standard Error	Estimate	Standard Error	Estimate	Standard Error	Estimate	Standard Error	Estimate	Standard Error
Intercept	-51.64	26.22	4931.32	2170.02	3645.84	3457.07	4255.30	20,498.49	7369.40	4371.14
pH			3.2342	9.3530	-38.65	12.15	175.42	70.49	-16.80	15.80
P	-0.0003	0.0008			0.07	0.12	0.37	0.66	-0.04	0.14
K	-0.0021	0.0007	0.0528	0.0706			3.17	0.44	-0.04	0.12
Ca	0.0003	0.0001	-0.0038	0.0122	0.10	0.01			0.13	0.02
Mg	-0.0006	0.0005	0.0022	0.0556	-0.02	0.0749	2.77	0.34		
Elevation	0.31	0.14	-25.43	11.39	-17.43	18.11	-27.48	107.28	-37.81	22.89
r	0.16		0.04		0.14		0.15		0.14	
φ	0.056		680.75		1049.57		33,320.68		1622.39	
Moran's I correlation	-0.01		0.00		0.02		-0.02		-0.01	
p value	0.99		0.96		0.65		0.82		0.93	
AIC	245.20		1364.00		1423.73		1841.27		1477.30	

TABLE 14.7

Parameter Estimation in the Best Spatial Linear Model (SLM) with a Simultaneous Autoregression and a First-Order Neighborhood

Response	pH		P		K		Ca		Mg	
	Estimate	Standard Error	Estimate	Standard Error	Estimate	Standard Error	Estimate	Standard Error	Estimate	Standard Error
Intercept	−57.70	25.21	4781.80	2063.60	329.53	79.47	−939.42	480.96	8148.76	4101.77
pH					−41.96	11.57	168.01	68.39		
P										
K	−0.0021	0.0007	0.0032	0.049			3.21	0.43		
Ca	0.0002	0.0001			0.10	0.01			0.13	0.01
Mg							2.79	0.33		
Elevation	0.34	0.13	−24.54	10.75					−42.49	21.38
r	0.15		0.05		0.15		0.15		0.14	
ϕ^2	0.056		663.47		1020.8		33,146.14		1603.78	
Moran's I correlation	0.00		−0.00		0.03		−0.02		−0.01	
p value	0.94		0.95		0.60		0.81		1.00	
AIC	242.58		1358.20		1419.16		1837.66		1472.50	

only the best subset of explanatory variables included. The estimates of the regression coefficients, the spatial correlation coefficients among first-order neighbors (ρ), and the error variances (σ^2) are presented. Also presented in Table 14.6 and Table 14.7 are Moran's I values and the corresponding p values for testing the spatial independence among the residuals, as well as the AIC values.

For modeling pH and Mg, the best SLM has selected the same set of explanatory variables as the MTSLM. For modeling P, the best SLM has selected K and elevation as the explanatory variables, but the best MTSLM has selected only elevation. However, the AIC value in the MTSLM is lower than that in the SLM, indicating that the MTSLM is a better fit to the data. Indeed, the variable K is not significant in the best SLM. For modeling K, on the other hand, the SLM has a much lower AIC value than the MTSLM. The best SLM has selected pH and Ca as the influential explanatory variables, whereas the MTSLM has kept all the variables and is not able to identify the most important variables. Finally, for modeling Ca, the best SLM has selected pH, K, and Mg, whereas the best MTSLM has selected K, Mg, and elevation. Since the AIC value of the MTSLM is lower than that in the SLM, the MTSLM is again a better fit to the data.

Note that, for all the variables, Moran's I tests for spatial independence have very large p values (>0.10). That is, there is no evidence of spatial correlation among the residuals after fitting the SLM. Thus, both the SLM and the MTSLM have accounted for the spatial dependence structure adequately. Nonetheless, there is no clear winner between the SLM and the MTSLM, according to the AIC values. For modeling pH, K, and Mg, the SLM has a lower AIC and is better than the MTSLM, whereas for modeling P and Ca, the MTSLM has a lower AIC and is better than the SLM. Although in general the model selection and parameter estimate results are close between the two models, the models for K are exceptional. While both the SLR model and the MTSLM do not fit K well, the SLM is much better. We have not found a plausible reason as to why the MTSLM could not account for the spatial dependence as well as the SLM does.

14.4.4 PREDICTION OF SOIL PROPERTIES

Based on the fitted best MTSLM, the change-of-resolution computing algorithm was implemented to predict each soil property variable in the field. The observed and the predicted signal values of pH, P, K, Ca, and Mg are shown in the image plots using Splus command image (Figure 14.4).[7] In general, the predicted signals are smoother than the observed values, as expected. The amount of shrinkage in the predicted values seems to depend on the magnitude of the measurement error variances, or more importantly, the ratio of the measurement error variance to the total variance. If the measurement error variance is large, then there would be more shrinkage from the observed values to the predicted signal values, and vice versa.

For pH, the higher values are in the western, northeastern, and southern parts of the field. The spatial pattern of the signal values is close to that of the observed values, but has a narrower range (shown as more pale hues), because the mea-

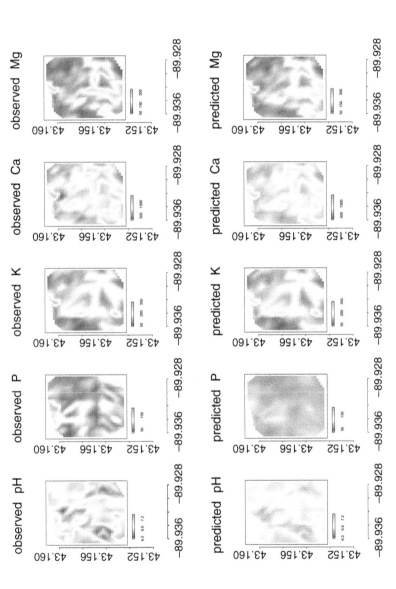

FIGURE 14.4 The observed (top panel) and predicted (bottom panel) values of pH, P, K, Ca, and Mg in the field. (See color version on accompanying CD.)

surement error variance is fairly large. In contrast, the shrinkage in the signal values is substantial for P, because the ratio of the measurement error to the total variance is the highest with P. For K and Ca, the higher values tend to occur in the middle of the field. The spatial pattern of the signal values is close to that of the observed values, because the measurement error variance is relatively small. Finally, for Mg, there are a few patches of higher values in the north and the south, and the spatial patterns of the signal values and the observed values are again close.

The corresponding prediction standard errors associated with pH, P, K, Ca, and Mg are 0.12 ppm, 9.71 ppm, 4.59 ppm, 88.08 ppm, and 18.80 ppm at the 120 locations (marked by o in Figure 14.2), and are 0.19 ppm, 11.56 ppm, 34.69 ppm, 138.67 ppm, and 34.13 ppm at the 13 locations where the observations were not used in parameter estimation (marked by Δ in Figure 14.2).

14.5 CONCLUSIONS

In this chapter, we have reviewed a type of spatial lattice model, namely, a multiresolution tree-structured spatial linear model. Because of the multiresolution tree structure, kriging can be obtained using a fast change-of-resolution Kalman filter algorithm. We have applied the methodology to obtain the best unbiased linear prediction of soil properties in a Wisconsin field, while accounting for field conditions using a linear regression. Comparison has been made among linear regression, a traditional spatial linear model for lattice data, and the multiresolution tree-structured spatial linear model. The multiresolution tree-structured spatial linear model does not always fit the data better, but it does offer a fast alternative for accounting for measurement error and mapping the signal processes (here, the soil properties).

It is worth noting that the linearity in the spatial linear models refers to linearity in the model parameters. Complicated nonlinear relations among variables can be captured by spatial linear models, as in the case of linear regression models. For more complicated relations that demand nonlinear regression models, extension of the multiresolution tree-structured spatial linear model might not be so straightforward, as computational efficiency of the change-of-resolution Kalman filter algorithm relies on the linear structure of the model.

ACKNOWLEDGMENTS

Funding has been provided for this research from the USDA Cooperative State Research, Education and Extension Service (CSREES) Hatch project WIS04676. Data used in this study were organized with help from Christine Molling. We thank the reviewers and the editors for constructive comments that improved the presentation of the chapter.

REFERENCES

1. H Akaike. Information theory and extension of the maximum likelihood principle. In *2nd International Symposium on Information Theory*, BN Petrov, F Csaki, Eds. Akademiai Kiado, Budapest, 1973, pp. 267–281.
2. H Bourennane, D King, A Couturier. Comparison of kriging with external drift and simple linear regression for predicting soil horizon thickness with different sample densities. *Geoderma* 97: 255–271, 2000.
3. F Carre, MC Girard. Quantitative mapping of soil types based on regression kriging of taxonomic distances with landform and land cover attributes. *Geoderma* 110: 241–263, 2002.
4. N Cressie. *Statistics for Spatial Data*, revised ed. Wiley, New York, 1993.
5. V Epinat, A Stein, SM De Jong, J Bouma. A wavelet characterization of high-resolution NDVI patterns for precision agriculture. *Int J Appl Earth Obs Geoinf* 3: 121–132, 2001.
6. HC Huang, N Cressie, J Gabrosek. Fast, resolution-consistent spatial prediction of global processes from satellite data. *J Comput Graphical Stat* 11: 1–26, 2002.
7. SP Kaluzny, SC Vega, TP Cardoso, AA Shelly. *S+SpatialStats User's Manual*. MathSoft, Seattle, 1997.
8. KA Kelling, LG Bundy, SM Combs, JB Peters. *Soil Test Recommendations for Field, Vegetable, and Fruit Crops*, Publication A2809. University of Wisconsin Extensions, Madison, WI, 1998.
9. M Knotters, DJ Brus, JH Oude Voshaar. A comparison of kriging, co-kriging and kriging combined with regression for spatial interpolation of horizon depth with censored observations. *Geoderma* 67: 227–246, 1995.
10. ED Kolaczyk. Bayesian multi-scale models for Poisson processes. *J Am Stat Assoc* 94: 920–933, 1999.
11. J Meinhold, ND Singpurwalla. Understanding the Kalman filter. *Am Stat* 37: 123–127, 1983.
12. IOA Odeh, AB McBratney, DJ Chittleborough. Spatial prediction of soil properties from landform attributes derived from a digital elevation model. *Geoderma* 63: 197–214, 1994.
13. IOA Odeh, AB McBratney, DJ Chittleborough. Further results on prediction of soil properties from terrain attributes: heterotopic cokriging and regression-kriging. *Geoderma* 67: 215–225, 1995.
14. G Upton, B Fingleton. *Spatial Data Analysis by Example: Point Pattern and Quantitative Data*, Vol. 1. Wiley, New York, 1995.
15. J Zhu, W Yue. A multiresolution tree-structured spatial linear model. *J Comput Graphical Stat*, 14: 168–184, 2005.
16. J Zhu, CLS Morgan, JM Norman, W Yue, B Lowery. Combined mapping of soil properties using a multi-scale tree-structured spatial model. *Geoderma* 118: 321–334, 2004.

15 Multiscale Soil-Landscape Process Modeling

Jeroen M. Schoorl and Antonie Veldkamp

CONTENTS

ABSTRACT

This chapter deals with the aspects of scale in soil-landscape modeling and presents a thorough review of the importance and effect of scale, including space and time resolution and the extent of space and time. The introduction covers the importance of the landscape in soil science and modeling throughout the past centuries, presents the aspects of soil-landscape modeling and scale issues, and briefly discusses sustainability and human influence, introduced as the fifth dimension. The general objective of this chapter is to illustrate the role of soils and geomorphological processes in the multiscale soil-landscape context. To illustrate and investigate this multiscale role, a simple model is used, with only a few parameters, which can be applied at different scales. Modeling results are given for the effects of changing spatial extension and resolution of a digital elevation model (DEM) and changing temporal extension (i.e., number of time steps) for these different DEMs' resolutions and extensions. The results are discussed in the context of scale problems occurring in hydrological and geomorphological modeling approaches, such as (1) emerging properties, (2) spatial heterogeneity of processes, (3) nonlinear behavior of process rates in time, (4) threshold dependency, (5) varying dominant processes, and (6) differing responses to disturbances. The example presented in this chapter, although with only a limited set of variables, indicates a spatial and temporal scale resolution- and extension-dependent response for different DEMs to the same set of input parameters, thus illustrating the multiscale character of the landscape and the existence of many of the well-known scale problems within the soil-landscape context.

15.1 GENERAL INTRODUCTION

Soils are an essential part of and basically controlled by landscape. The central role of the multiscale landscape in this chapter is based on the consideration that the landscape is the main driving factor behind many processes at different temporal and spatial levels in geoenvironmental sciences. At the same time, landscape can be considered the consequence of geological evolution and the result of geomorphological processes. Therefore, landscape can be defined in terms of genesis (how formed, processes), geomorphology (its present form, shape), lithology/ soil (its composition), land cover (surface characteristics), land use (its use, human function), and even land management (human factor). Consequently, these interrelated issues have to be considered within the multiscale soil-landscape modeling context.

15.2 THE HISTORICAL IMPORTANCE OF
THE LANDSCAPE IN SOIL SCIENCE
AND MODELING

The role of the landscape in soil science was prominent in the early years of the 19th century when the first systematic geomorphological and geological descriptions and observations of the land surface were made. It was during this period that the first geological maps appeared.[1] Meanwhile, in Russia the first soil maps were produced by soil scientists who discovered the relation between soil and climate.[2,3] During these early years, mainly geologists, chemists, and agronomists were investigating the soil. Soil science as an individual discipline established itself in the Netherlands from the beginning of the 20th century.[4]

Soil science initially focused on soil taxonomy and mapping issues for many years. All around the world major efforts were directed toward classifying soils and mapping of land surfaces at different scales. During the first half of the 20th century the landscape still played an important role because of the widely applied physiographic mapping techniques, linking soil units to geomorphological features.[5] In addition, during this period concepts like the *soil catena* and *chronosequence* were developed, placing the soils in their logical landscape context, i.e., descriptions of properties in relation to surrounding topographical, hydrological, and geomorphological processes.[6]

These soil survey efforts have resulted in the division of the Earth's surface into classified entities and colored polygons, which culminated in the Food and Agriculture Organization (FAO) global soil map.[7] In the early years, different classification systems were developed all over the world. Later, the numbers of systems were limited, and some standards are now more widely accepted.[7,8] However, many countries still have their own national classification systems.

In the second half of the 20th century landscape started to lose its visibility in soil science, mainly because of the introduction of descriptive morphometric properties in the map legends, such as texture, structure, pedogenesis, etc.[5] The interest in using these chemical and physical soil properties in the soil surveys was a consequence of the more detailed mapping units that became standard in those days (<1:50,000) and the shift of interest from mapping units to taxonomy. Consequently, the large-scale landscape perspective and cooperation between soil science and geomorphology decreased.[9-11]

Driven by the need for data in the taxonomy-oriented scientific community, the soil pedon obtained a central position, where pedon is defined as the smallest three-dimensional volume that can be called a soil, large enough to permit the study of all horizons.[12] Soil properties were treated at this pedon scale only, and therefore, soil science (pedology) started to study the soil pedon inch by inch. The main focus was directed toward pedon dynamics mainly in a two-dimensional way (top-down) — first mainly in vertical fluxes, much later also some horizontal and lateral inputs. Consequently, soil science was focused on profile dynamics and the landscape around the soil pedon was reduced to a variable set of boundary conditions.

However, we have to be aware of the consequences of using soil maps with generalized and sometimes limiting properties. In this sense, we can consider the dualistic position of geostatistics. On the one hand, it is useful and powerful in determining and even predicting spatial variation, dependency, and variability between and within soil classes, optimizing sample strategies, predicting soil attributes, and so on.[13-15] On the other hand, it gives an apparent accuracy for land units that may need a complete new classification at the first place that properly reflects landscape processes.[16] At the same time, it appears that sometimes geostatistics provides an elaborate statistical analysis for recognizing soil properties or geomorphological features that can be distinguished by simple observation or physiographic mapping.[17]

Another consequence of lacking a landscape component within soil-related data could be found at the policy makers level. Modern agriculture, especially in Western Europe, is directed toward sustainable development within the physical and social landscapes. However, many aspects of sustainability are investigated and evaluated at the soil profile level, ignoring and eliminating the effective processes operating at the higher landscape level. This implies that international, national, and regional legislation is affected by generalization, which may have undesirable and unrealistic effects. For example, at the regional level the present-day Dutch nitrate leaching legislation does not take into account site-specific soil properties or the location of a farmer in the landscape, therefore neglecting the fact that the landscape is dynamic and that even the smallest gradient will have one farmer leaching his nitrate away to the other farmer. At the international level of the European Community, price controls and subsidies, or even the withdrawal of subsidies, increasingly control land use and land use changes. However, for example, in southern Spain subsidy-driven land use conversions, changed tillage practices, and land abandonment enhance significantly land degradation; these are not considered to enhance sustainable development of the Mediterranean landscape.[18,19]

Fortunately, the landscape context (biophysical, physical, and social) surrounding the soil is starting to revive and receive increasing attention again, abandoning the idea of landscape as only a two-dimensional carrier of soil information.[20] Now, by means of information technology, realistic four-dimensional properties (space plus time) can be addressed, thus reinventing the landscape context within the sciences of soils. Since land use management in cultivated areas affects soil-landscapes through time, they should be given special attention.

Recently, studies in the Netherlands have revealed the importance of the time management factor on the soil, showing that many years of agricultural management can alter significantly the most important soil physical and soil chemical properties and system functions.[21–23]

Nevertheless, in many fields of environmental sciences, including soil science and geology, the landscape context is often underestimated or even lacking at the different levels of investigation (both spatial and temporal). Therefore, this chapter will explicitly address the landscape following the issue raised by American scientists Jacob and Nordt[9] at the end of the last century: "the soil-landscape paradigm is the natural path for pedology to follow."

15.2.1 Soil-Landscape Modeling and Scale Issues

Including the temporal component, landscape can be considered as having four dimensions: length, width, height, and time. Therefore, as with all systems with more than one dimension, scale issues or scale problems are a common point of discussion in environmental sciences. In addition to the multifunctional and sometimes confusing use of the word *scale* (e.g., hierarchical level, temporal and spatial resolution, temporal and spatial extension), these problems refer to the differences in observation, interpretation, and calculation of processes at different levels of organization within the landscape. For example, we can consider here the relationships between the detailed level of individual processes, such as infiltration, sediment and water redistribution, etc., as opposed to processes at global levels, such as climate change and land use change.[24,25]

Both in geomorphology and hydrology the issue of scale has been an important topic over the past years.[26,27] Different causes of scale problems can be identified concerning the behavior or description of processes at different scales:[28]

1. Emerging properties and new processes emerging at different levels
2. Spatial heterogeneity of processes influenced by all sorts of spatial factors, such as topography, soils, and land use
3. Nonlinear behavior of process rates in time
4. Threshold dependency to trigger a process
5. Varying dominant processes at different levels
6. Differing responses to disturbances

Typical scale effects in geomorphology at different spatial resolutions are, for example, the decreasing erosion rates from plot, hillslope, and catchment to basin scale, where spatial heterogeneity of key processes, local resedimentation, sediment transport distances, and the resolution of the measuring techniques play an important role. Another type of scale effect can be found in investigations concerning the impacts of rainfall and flooding events on soil and landscape evolution. Here an aspect of temporal resolution-extension is introduced: the magnitude-frequency distribution. For example, at the scale of a slope, relations can be found between hillslope erosion, parent material, and magnitude-frequency distribution of rainfall.[29] At the catchment scale, there are still uncertainties on the exact role of magnitude and frequency of floods, considering the impact of large catastrophic floods with a very low frequency vs. the cumulative effect of many minor flooding events. The large catastrophic floods seem important in the long-term evolution of the catchment, whereas the smaller floods, depending on the timing, can have a larger cumulative impact.[30]

In the geological sciences, scale effects seem more accepted. For example, geological landscape evolution is driven by three major components: climate, sea level, and tectonics. Long-term dynamics of these components can be traced throughout the Earth's history, although one could state that their temporal and spatial resolutions become coarser going farther back in time. This increasing coarseness of geological observations and the resulting stratigraphical framework are in the first place a result of preservation of sediments and fossils. Second, even modern dating techniques show increasing coarseness in precision and validity. Compared with the global-to-regional character of plate tectonics, climate, and sea level changes, local uplift rates are spatially much more variable, since in active areas uplift or subsidence rates often depend, in addition to the past and present positions on the continents, on local fault systems.

Returning to the present-day landscape-forming processes, for many years, the tendency has been to model the processes with the most detail possible, small-scale short time, and to simply aggregate results to larger areas and longer time spans. However, as stated by Beven,[26] the hydrological or geomorphological modeler will have to accept that it is virtually impossible to model either larger systems including the smallest details or by simple aggregation. Therefore, a model must be assembled with only those effective parameters at the scale of interest (a certain resolution and extension) where the smaller subscale details can safely be ignored.[31]

Going back to the central landscape context, its evolution can be simulated by combining different processes, depending on the chosen spatial and temporal resolutions. In hydrology and geomorphology, different groups of scientists are working at different spatial and temporal scales. In hydrology, there are two major groups focusing on (1) slope and catchment behavior dealing with event-based predictions of runoff and hydrographs[32] and (2) global and regional circulation models including climate change.[33,34] The same division is found in geomorphology, since the hydrologic behavior is one of the inputs for geomorphologic modeling: (1) slope and catchment event-based erosion models[35–37] and (2) landscape evolution models including climate and tectonics.[38–41]

To investigate the soil in the landscape context, a spatially explicit model is needed at the multicatchment level, where the spatial and temporal resolutions will depend on the processes and observations involved. Contrasting modeling approaches exist, ranging from simple empirical models (e.g., Universal Soil Loss Equation (USLE) or its derivatives[42]) to complex physically based process models (e.g., the Water Erosion Prediction Project (WEPP)[32,35]). Since experimental plot data were used to develop the USLE, this model has limited use for modeling at the landscape scale. Furthermore, there is no empirical basis for semiarid to subhumid Mediterranean conditions, and in the previous sections on scale problems, major constraints have been discussed about using plot data at any level other than the plot itself. The decision not to use sophisticated models like TOPMODEL or WEPP,[32,35] in addition to the larger amount of data needed, is the more detailed temporal spatial resolution that, for the simple examples in this chapter, is not really needed. We prefer a simple model, which requires less parameters, variables, and input data, and which makes the modeling exercise in this chapter more transparent and understandable.

Actual field data are needed to calibrate and validate geomorphic landscape process models. Traditionally, the type of field data that is collected shows a mixture of different spatial and temporal resolutions. These types of resolutions (level of observation, grain size) include point measurements of rainfall, profile data of infiltration, runoff and sediment yield of various plot sizes, catchment discharge and sediment load in flumes, and so on. All these types of data have a variable spatial and event-based temporal resolution. Consequently, model calibration and validation becomes a difficult and uncertain procedure. One of the recently developed methods that shows a coarser temporal resolution is the ^{137}Cs technique.[43–45] This technique enables the monitoring of net soil redistribution over the last 30 to 40 years by measuring the soil-related redistribution of the anthropogenic ^{137}Cs radionuclide, deposited in the environment in the 1960s. However, the obtained rates are limited to point data and local soil profile conditions (e.g., clay content, soil texture, absorption rates, etc.), which need interpolation and calibration techniques to provide broad-scale data of limited spatial resolution. Enhancing of the spatial resolution is achieved by increasing the number of sample sites (transects or grids), which increases research costs. Still, these costs are considerably lower than the costs involved in long-term monitoring with traditional techniques to achieve the same spatial and temporal resolutions.

15.2.2 THE FIFTH DIMENSION AND SUSTAINABILITY

As discussed in the previous sections, the landscape in environmental sciences is considered to have three spatial dimensions. Together with the temporal dimension, landscape becomes a four-dimensional entity within its biophysical boundaries. However, at various spatial and temporal levels the human influence has become an important factor of consideration. At first this human factor was only considered to be important at short-term temporal and limited spatial levels of local land use change and management practices. Recently, many research efforts were directed toward the human role in past and future global environmental changes.[25,46] Gradually we become aware that anthropogenic influences may affect even geological development at regional-to-global scales.[47] Therefore, the human influence is introduced here as the fifth dimension of landscape.

Over the past years the consequences of such human impact upon the environment have led to the need for sustainable development and the concept of sustainability. Numerous definitions of sustainability can be found for all the various disciplines involved in environmental sciences. However, the FAO[48] uses one of the most elaborate definitions, defining sustainability as: "The management and conservation of the natural resource base, and the orientation of technological and institutional change in such a manner as to ensure attainment and continued satisfaction of human needs for present and future generations; such sustainable development conserves water, plant and animal genetic resources, is environmentally non-degrading, technically appropriate, economically viable and socially acceptable." This definition combines the ecological aspects of sustainability with the economic and social aspects, emphasizing that sustainability comprises various dimensions. In other words, different perceptions on sustainable development exist, and there is no single meaning or concept for, for example, sustainable agriculture and rural development.

It becomes clear that the concept of sustainability is not without discussion because of the various disciplines involved, from policy maker to Earth scientist. All these disciplines have their own understanding, definition, and implementation of the concept. However, for the Earth scientists there are certain aspects of sustainable development that need some attention since by nature the Earth is constantly changing — both gradual and apparently catastrophic changes. Therefore, sustainability in the form of environmental protection and nature conservation needs a framework of process rates and the interaction with the four-dimensional landscape. As stated by van Loon,[49] it is the task of Earth scientists to raise awareness and to inform the public and policy makers of the spatial and temporal impacts of this geological evolution.

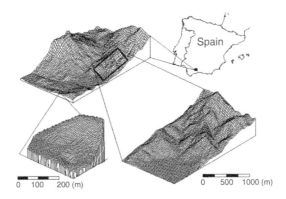

FIGURE 15.1 Location of the case study area, including an overview of the Chopillos slope (lower left graph) and the Clavelinas catchment (lower right graph).

15.2.3 OBJECTIVES

The general objective of this chapter is to illustrate the role of soils and geomorphological processes in the multiscale soil-landscape context. To investigate this multiscale role, a simple model is used, which can be applied at different scales without introducing new parameters. Consequently, multiscale stands for different spatial levels represented by different spatial resolutions of a DEM and temporal levels (i.e., number of time steps). Examples are given, following the list of most important causes of scale problems as introduced in the previous section, by comparing different model runs under controlled parameter conditions.

15.3 MODELING

15.3.1 CASE STUDY AREA FOR THE MODELING EXAMPLE

The case study area is situated in southern Spain near the village of Álora in the river basin of the Guadalhorce, province of Málaga, Andalucía Álora (longitude 04°41′57″W, latitude 36°49′10″N). This research area (see Figure 15.1) shows a dynamic landscape of mountains and hills,[50-52] an interesting complex geological history,[41] and active landscape processes ranging from tectonics and land use changes[53,54] to land degradation.[19,44,45] Soils in the studied areas range from Leptosols and Regosols to Cambisols.[7,55,56]

The area around Álora has a summer-dry Mediterranean climate (Csa) with a mean annual temperature of 17.5°C and a mean yearly rainfall of 534 mm, mainly in the period from October to April. However, in the Mediterranean region, annual rainfall is often variable and historical records show that the mean annual rainfall varied roughly between 250 and 1050 mm over the past 50 years.[44]

15.3.2 Model Introduction and Description

To demonstrate an example of the multiscale effects within soil-landscape modeling, a simple landscape evolution model, LAPSUS,[53,55,57] has been used. This model is a basic surface water erosion model for DEMs and is based on the continuity equation for sediment movement.[58–60] LAPSUS evaluates the sediment transport rate by calculating the downslope transport capacity of water as a function of runoff and slope gradient. If the capacity is higher than the actual transport rate, this capacity is filled by the detachment of soil particles from the soil surface. Detachment is highly dependent on the erodibility K_{es} (m^{-1}) of the surface.[61] This detachment of soil causes lowering of the surface or erosion. When the actual transport rate exceeds the local capacity, for example, because of lower gradients, then the surplus of sediment in transport will be deposited, causing a surface heightening.

Assuming the annual transport capacity C (m^2 a^{-1}) and the detachment capacity D (m a^{-1}) or settlement capacity T (–m a^{-1}) remain constant within one finite element and after integration of the continuity equation, the sediment transport rate is evaluated as follows:[58–60]

$$S = C + (S_0 - C) \cdot e^{-dx/h} \tag{15.1}$$

where the sediment transport rate S (m^2 a^{-1}) along dx length (m^1) of a finite element is calculated as a function of C compared with the amount of sediment already in transport S_0 (m^2 a^{-1}). The term h (m^1) refers to C/D or C/T. Dividing S by the length of the finite element gives an expression for surface lowering (m^1 a^{-1} or m^3 m^{-2} a^{-1}), which when multiplied with the bulk density Bd gives the erosion or sedimentation (10^0 kg m^{-2} a^{-1} or 10^1 t ha^{-1}a^{-1}).

The direction and distribution of the runoff and the resulting model calculations are conducted with a multiple-flow algorithm to allow for a better representation of divergent properties of the convex topography.[62–64] For further details and model application, see References 44, 45, 53, 54, and 57.

15.3.3 Parameters and Assumptions

Modeling experiments for this case study have been executed for two different extensions or subareas (Table 15.1 and Figure 15.1): (1) a segment of a hillslope (9 ha) in the Chopillos area (Chop) and (2) a single catchment (250 ha) of the Clavelinas River (Clav). For both areas, multiple-resolution DEMs have been constructed with spatial resolutions ranging from 1 to 25 m (Table 15.1). The 1-m DEM was the result of a survey with an automated Theodolite or Total Station; the 7.5-m DEM was produced by digital processing of stereo aerial photographs (1:20,000). Both 20- and 25-m DEMs were provided by the Spanish National Geographical Institute and the military, respectively, produced by interpolation of contour lines from their topographic maps at the 1:50,000 scale.

TABLE 15.1
Input DEMs and Model Parameters

DEMs	Resolution (m)	Area (m²)	Number of cells	Sdp A[a] (m)	Sdp B[b] (m)
Chop01	1	85,173	85,173	1.2	0.002
Chop07	7.5	88,818.75	1579	1.2	0.002
Chop20	20	87,600	219	1.2	0.002
Chop25	25	89,375	143	1.2	0.002
Clav07	7.5	2,522,137.5	44,838	1.2	0.002
Clav20	20	2,544,000	6360	1.2	0.002
Clav25	25	2,550,000	4080	1.2	0.002

[a] Sdp A = soil depth scenario A.
[b] Sdp B = soil depth scenario B.

With these different DEMs several LAPSUS runs have been executed, simulating for a period of 10 years. To compare results from different model runs, all input parameters have been kept constant and equal. In other words, rainfall, infiltration, and resulting runoff were the same for all grid cells as well as the K_{es} factor and other model input parameters.[53,54,57] Therefore, the only variable input was the altitude information of the DEMs and the resulting gradients and flow routing. After each time step, a new DEM was calculated, which provided the new gradients and flow routing paths as input for the next run. To investigate the role and influence of the soil, two different scenarios were used (Table 15.1): (1) transport-limited scenario A, giving the whole area a uniform soil depth of 1.2 m, and (2) detachment-limited scenario B, giving the whole area a uniform soil depth of only 0.002 m.

15.4 RESULTS AND DISCUSSION

To discuss the results of the modeling exercise, which are given in the following sections, we will make reference to the generally encountered scale problems occurring in hydrological and geomorphological modeling approaches as mentioned in the introduction: (1) emerging properties, (2) spatial heterogeneity of processes, (3) nonlinear behavior of process rates in time, (4) threshold dependency, (5) Varying dominant processes, and (6) differing responses to disturbances. Results in the following sections are given in meters surface lowering, means for separate grids in Figure 15.2 and Figure 15.3 (considering different grid extents), and means for the whole area in Table 15.2, Figure 15.4, and Figure 15.5 (two different total extents). Expected modeling error will be on the order of 0.1 metric tons per hectare (see also Schoorl and colleagues[53,54]).

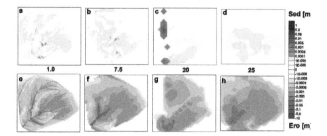

FIGURE 15.2 Sedimentation (upper graphs) and erosion (lower graphs) patterns for the Chopillos area after one time step (1 year) given in meters surface change for each of the four spatial resolutions (from left to right). (See color version on accompanying CD.)

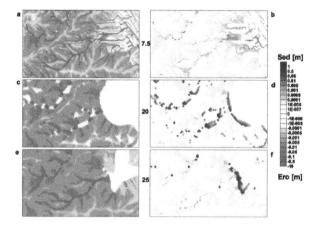

FIGURE 15.3 Erosion (left graphs) and sedimentation (right graphs) patterns for the Clavelinas area after one time step (1 year) given in meters surface change for each of the three different resolutions. (See color version on accompanying CD.)

15.4.1 Spatial Patterns and Scale Effects

Figure 15.2 gives the resulting patterns of 1 year of soil redistribution modeling for the area with the smallest extension (Chopillos). A general overview gives similar patterns of areas with more and less erosion (Figure 15.2e to h). An important scale effect causing these patterns is spatial heterogeneity of processes. The finer resolutions reveal more detail of gully formation and areas of sediment production. However, the patterns of sedimentation reveal more differences between the different resolutions (Figure 15.2a to d). The 20-m DEM especially shows a particular pattern of increased resedimentation at the left side of this graph (Figure 15.2c). A closer look at the 20-m DEM data shows several grids in this area with approximately the same altitude. These grids at similar altitudes result in lower slope gradients providing the drop in capacity to transport sediments, therefore causing resedimentation of soil.

TABLE 15.2
Mean Soil Loss Rates (Lost from the Area), Percentages of Change (Compared with the Coarsest Resolution), Percentages of Redeposition within the Area, and Comparison for Modeled Scenarios A and B

	Scenario A					Scenario B			
	Soil Loss		Bal[a]	Dep[b]	Red[c]	Soil Loss		Bal[a]	Dep[b]
	(m)	(t/ha)	(%)	(%)	(%)	(m)	(t/ha)	(%)	(%)
Chop01	−7.38E-04	−11.4	−56.1	43.6	2.3	−7.21E-04	−11.1	−41.1	1.4
Chop07	−1.59E-03	−24.5	−5.4	0.8	38.7	−9.74E-04	−15.0	−20.5	1.2
Chop20	−9.66E-04	−14.9	−42.5	17.7	32.2	−6.55E-04	−10.1	−46.5	24.1
Chop25	−1.68E-03	−25.9	0	0.99	27.1	−1.22E-03	−18.9	0	1.3
Clav07	−5.09E-03	−78.5	5.2	1.9	71.2	−1.47E-03	−22.7	−4.8	2.7
Clav20	−3.57E-03	−55.0	−26.3	25.2	70.9	−1.04E-03	−16.0	−32.6	36.9
Clav25	−4.84E-03	−74.7	0	7.6	68.2	−1.54E-03	−23.8	0	15.4

[a] Relative difference in soil loss between the coarsest resolutions.
[b] Redeposition of sediments within the area.
[c] Reduction in soil loss from scenario A to B.

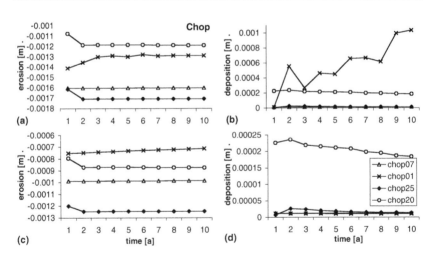

FIGURE 15.4 Modeled soil erosion and deposition estimations for the Chopillos DEMs (four resolutions) over 10 years with graphs (a) and (b) under normal conditions (scenario A) and graphs (c) and (d) under soil depth-limited conditions (scenario B).

Figure 15.3 shows sedimentation and erosion results after 1 year of soil redistribution modeling for the Clavelinas area. Again, erosion patterns are more detailed for the finest resolution, showing more water divides and different local watersheds than for the other resolutions. In the 20-m erosion map (Figure 15.3c), several white grid cells or areas of no erosion are visible throughout the map.

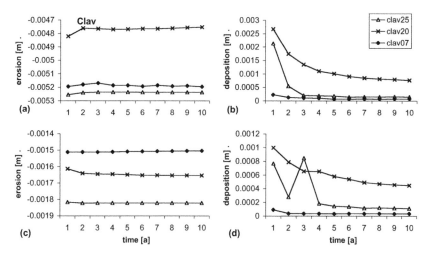

FIGURE 15.5 Modeled soil erosion and deposition estimations for the Clavelinas DEMs (three resolutions) over 10 years with graphs (a) and (b) under normal conditions (scenario A) and graphs (c) and (d) under soil depth-limited conditions (scenario B).

Checking the altitude data of the grid cells in these areas reveals that several grids of equal altitudes can be found, providing the potential for resedimentation, as can be seen from the graph on the right (Figure 15.3d). Another major difference can be found in the upper right corner of the 20- and 25-m erosion maps, where a large area of no erosion exists (Figure 15.3c and e). This large area of no erosion is actually the location of a former river meander belt and, nowadays, a high terrace level.[41] At this resolution (both 20 and 25 m), this river terrace consists of numerous grids giving the same altitude, which therefore have no potentials for erosion. As can be seen from the sedimentation graphs (Figure 15.3d and f), the sediment load is deposited along the edge of this terrace area where the change of slope gradient is most pronounced. Consequently, at the edge of this terraced area the beginning of alluvial fan formation is occurring.

In Table 15.2 the results of the modeling exercise (means of a 10-year simulation) are given for the two scenarios (both transport limited and detachment limited), using the two DEM extensions and varying resolutions. Comparing the results within each of the areas reveals major differences of mean soil loss (e.g., total amount of soil lost from the area) between different resolutions, where the 7.5- and 25-m resolutions show more soil loss than the 1- and 20-m resolutions in both scenarios. These lower amounts of soil loss for the 1- and 20-m DEMs can be explained by the higher amounts of redeposited soil within the area at these resolutions. Comparing the two areas (Chop vs. Clav), more elevated erosion rates were found in the Clavelinas area, where the slopes are longer and steeper than in the Chopillos area.

Comparing the two scenarios, we found a general reduction in soil loss rates of around 70% for the Clavelinas area and around 30% for the Chopillos area. This difference is attributed to the difference in length of slope (amount of running water) and slope angles (the strength of the running water) between the two areas. These types of parameters, however, are critical properties of different landscapes throughout the world. Considering the limiting soil depth in this case as a proxy for surface or detachment characteristics, one can state that the actual response of such a simple parameter as soil depth is not linear, but landscape dependent.

This general decrease of 30 to 70% is not valid for the 1-m resolution, which is only reduced by a few percent. At first this seems remarkable; the reduced soil depth is expected to limit the total amount of soil erosion. However, in this case, for the finer resolution this means that the transport capacity is filled much more slowly and almost no sediment is redeposited within the simulated area, whereas for scenario A more than 43% of the sediment was redeposited within the catchment, so actual erosion rates have been much higher than the sediment load or soil loss given in Table 15.2. For all other DEMs, actual redeposition rates as percentage of total erosion rates are higher for the soil-limited scenario B. This means that although the sediment transport rates are filled much more slowly and with less sediments, they provide sufficient sediment for some of the depositional areas.

Furthermore, Table 15.2 gives an indication of the nonlinear and intrinsic behavior of each of the DEMs. Expected trends, for example, increasing amounts of resedimentation with finer resolutions,[6] are less pronounced due to the differences in DEM precision. In addition, soil loss rates vary considerably, indicating that general slope characteristics (angle and length) differ between the used DEM resolutions. Comparing both areas, the example of the soil-limited scenario indicates a landscape-driven scale resolution- and extension- dependent response, and therefore a certain reduction in soil depth, as the source of sediment in transport will lead to different nonlinear responses for different types of landscapes and resulting DEMs.

15.4.2 SCALE EFFECTS AND TEMPORAL EXTENSION

To investigate the temporal effects of the scenarios for the different DEM resolutions and extensions simulations of 10 years were executed. In Figure 15.4 and Figure 15.5 the annual mean erosion and deposition rates are given for the 10-year modeling example. Both figures indicate a variety of trends. Each of the DEMs develops a particular response, thus indicating nonlinear behavior of process rates in time. Contrary to what would be expected (less erosion, thus less materials to deposit), the DEM with the lowest erosion rate does not necessarily show the lowest redeposition rate. In addition, the DEM with the highest erosion rate under scenario A is not the same as that under scenario B, as would be expected since the erosion potentials (gradients) remain the same. The location and spatial relation with other grids within the DEM determine how much the transport capacities are filled (see Figure 15.4a and c, Figure 15.5a and c). In this case, the original number of grids for each DEM that already is showing transport-limited erosion, even with enough supply (due to interplay of gradient and discharge), is variable. Consequently, this modeling example, with a simple erosion model and taking into account that many input factors and parameters remained the same, indicates that the interplay of transport capacity and filling the transport rates as a function of the slope gradient and local runoff is revealing a complex nonlinear behavior. Filling the sediment transport rates more slowly in one area can provoke increased erosion or additional sedimentation in another area, depending on the local discharges and gradients. All grid cells are linked within the landscape system and can influence each other's behavior.

For the Clavelinas area, general trends of scenario A indicate slightly decreasing erosion rates for the first years, after which they tend to vary less, but never become constant (Figure 15.5a). This indicates that the soil depth of 1.2 m throughout the whole area provides a constant supply of sediment that is not depleted during the simulation. In contrast, the sedimentation rates in general show an exponential decrease (Figure 15.5b and d). This is probably due to the limited accommodation space for sediment within the present catchment. Since all model parameters are constant, no dryer or wetter year was incorporated to flush the system and move the deposited sediments to another location. The system seems to be heading toward a steady state. Adding an extreme event could provoke the sixth scale effect, different responses of the DEMs to disturbances. However, it is still possible that after a certain period of time some of the former erosion areas are flattened out and provide a new location to store local sediments, the so-called threshold scale effect.[26] This seems, for example, the case for the 25-m DEM (Figure 15.5d), where suddenly in the third simulation year an extra amount of sediment is stored within the area.

Considering the Chopillos area (more limited extent), the general trends (Figure 15.4) are comparable to those discussed in the previous sections for the Clavelinas area (Figure 15.5), with the exception of the first simulation years, where both the 20- and 25-m resolutions show first a small increase in erosion rates. This temporal increase for the first time steps is probably due to some artifacts in these DEMs (originating from the interpolation techniques), which are rapidly removed (Figure 15.4a and c). However, concerning the temporal behavior of deposition in the Chopillos area, the most striking difference can be found in Figure 15.4b. In this case, the 1-m resolution DEM reveals a complex response to the changing sediment transport capacities, showing a general increase over 10 years, with some yearly temporal decreases. These temporal decreases and the decreasing erosion rates in the first few simulation years (Figure 15.4a) seem to contradict the general trends found so far. In addition, the assumption that if less soil is eroded, less soil would also be deposited does not hold true. Again, this demonstrates the complex behavior of the system, threshold effects, emerging properties, and varying dominant processes. Gradients, runoff production, and flow routing provide each time step with different patterns and areas within the DEM suitable for resedimentation of soil lost from upslope areas.

15.5 CONCLUSIONS

Besides the specific geomorphic properties of the landscape, both resolution and extension of the DEM (different scales) are important properties in modeling the soil-landscape continuum. Furthermore, differences in DEM precision or the accuracy at which the landscape is digitally represented directly influence model behavior, outcomes, and consequently calibration and validation. The example presented in this chapter of the soil-limited scenario, although with only a limited set of variables, suggests a spatial and temporal scale resolution- and extension-dependent response for different DEMs to the same set of input parameters, thus illustrating the multiscale character of the landscape and the existence of many of the well-known scale problems within the soil-landscape context.

The modeling example given in this chapter is merely a simple and limited exercise, with many potentials for further investigation. For example, what was not dealt with in this chapter, and which could be of equal interest, is the effect of changing temporal resolution by means of variable time steps. Another comparable exercise could be investigating the effect of upscaling of the DEMs from one single source, contrary to the DEMs from different sources used in this chapter.

Depending on the goal of modeling, e.g., land use, planning, soil redistribution, or the complete landscape system, the example described in this chapter involves only one of many aspects, parameters, or variables that are important when modeling landscapes and can cause scale effects. Therefore, to deal with all the different feedbacks and multiscale effects in soil-landscape modeling, we advocate a modeling approach that can deal with these kinds of effects. Such a multiscale model should be able to handle the need for dynamic feedbacks and incorporate dynamic boundary conditions, modeling processes at their appropriate scales.

REFERENCES

1. C King, Engineer Department U.S.A. *Geological and Topographical Atlas Accompanying the Report of the Geological Exploration of the Fortieth Parallel Made by the Authority of the Honorable Secretary of War under the Direction of Brig. and Brvt. Major General A.A. Humphreys, Chief of Engineers U.S.A. by Clarence King, U.S. Geologist in Charge, 1876*. Julius Bien Lith, 1876.
2. N Sibirtzev. Etude des sols de la Russie. In *Memoires presentes a 7me session congres geologique international, 1897, St. Petersbourg*, A Bittner, Ed. Petersburg, USSR, 1897, pp. 100–110.
3. H Margulis. *Aux sources de la pedologie (Dokoutchaiev-Sibirtzev)*. FR: Ecole Nationale Superieure Agronomique de Toulouse, Toulouse, France, 1954, 85 pp.
4. R Felix. Bodemkartering voor 1943 het geologisch perspectief. In *Van bodemkaart tot informatiesysteem*, P Buurman, J Sevink, Eds. Wageningen Pers, Wageningen, Netherlands, 1995, pp. 1–17.
5. H De Bakker. Veertig jaar bodemkartering 1943–1983: introductie van bodemkundige concepten. In *Van bodemkaart tot informatiesysteem*, P Buurman, J Sevink, Eds. Wageningen Pers, Wageningen, Netherlands, 1995, pp. 18–27.
6. G Milne. Normal erosion as a factor in soil profile development. *Nature* 138: 548–549, 1936.

7. FAO. *FAO/Unesco Soil Map of the World, Revised Legend*, World Resources Report 60. FAO, Rome, 1989. (Reprinted as Technical Paper 20, 1988, 138 pp.)

8. USDA. *Soil Taxonomy: A Basic System of Soil Classification for Making and Interpreting Soil Surveys*, 2nd ed. U.S. Goverment Printing Office, Wasinghton, DC, 1999, 869 pp.

9. JS Jacob, LC Nordt. Soil and landscape evolution: a paradigm for pedology. *Soil Sci Soc Am J* 55: 1194–1195, 1991.

10. J Bouma. The new role of soil science in a network society. *Soil Sci* 166: 874–879, 2001.

11. J Bouma, AE Hartemink. Soil science and society in the Dutch context. *Neth J Agric Sci* 50: 133–140, 2002.

12. Soil Survey Staff, USDA. *Soil Classification, 7th Approximation*. U.S. Government Printing Office, Washington, DC, 1969, 265 pp.

13. J Bouma, HWG Booltink, A Stein. Reliability of soil data and risk assessment of data applications. *Soil Sci Soc Am* 47: 63–79, 1996.

14. S de Bruin, A Stein. Soil-landscape modelling using fuzzy c-means clustering attribute data derived from a digital elevation model (DEM). *Geoderma* 83: 17–33, 1998.

15. A Saldana, A Stein, JA Zinck. Spatial variability of soil properties at different scales within three terraces of the Henares River (Spain). *Catena* 33: 139–153, 1998.

16. RM Lark, PHT Beckett. A geostatistical descriptor of the spatial distribution of soil classes, and its use in predicting the purity of possible soil map units. *Geoderma* 83: 243–267, 1998.

17. SJ Park, K McSweeney, B Lowery. Identification of the spatial distribution of soils using a process-based terrain characterization. *Geoderma* 103: 249–272, 2001.

18. JL Rubio, E Bochet. Desertification indicators as diagnosis criteria for desertification risk assessment in Europe. *J Arid Environ* 39: 113–120, 1998.

19. J de Graaff, LAAJ Eppink. Olive oil production and soil conservation in southern Spain in relation to EU subsidy policies. *Land Use Policy* 16: 259–267, 1999.

20. A Veldkamp, K Kok, GHJ De Koning, JM Schoorl, MPW Sonneveld, PH Verburg. Multi-scale system approaches in agronomic research at the landscape level. *Soil Till Res* 58: 129–140, 2001.

21. P Droogers, J Bouma. Soil survey input in exploratory modeling of sustainable soil management practices. *Soil Sci Soc Am J* 61: 1704–1710, 1997.

22. MM Pulleman, J Bouma, EA van Essen, EW Meijles. Soil organic matter content as a function of different land use history. *Soil Sci Soc Am J* 64: 689–693, 2000.

23. MPW Sonneveld, J Bouma, A Veldkamp. Refining soil survey information for a Dutch soil series using land use history. *Soil Use Manage* 18: 157–163, 2002.

24. DL Peterson, T Parker. *Ecological Scale: Theory and Application*. Columbia University Press, New York, 1998, 615 pp.

25. EF Moran. News on the Land project. *Global Change Newsl* 54: 18–20, 2003.

26. KJ Beven. Linking parameters across scales: subgrid parameterizations and scale dependent hydrological models. *Hydrol Process* 9: 507–525, 1995.

27. JD Kalma, M Sivapalan. *Scale Issues in Hydrological Modelling*. Wiley, Chichester, U.K., 1995, 589 pp.

28. R Schulze. Transcending scales of space and time in impact studies of climate change on agrohydrological responses. *Agric Ecosyst Environ* 82: 185–212, 2000.

29. J De Ploey, MJ Kirkby, F Ahnert. Hillslope erosion by rainstorms: a magnitude-frequency analysis. *Earth Surf Process Landforms* 16: 399–409, 1991.

30. TJ Coulthard, MG Macklin, MJ Kirkby. Modelling the impacts of different flood magnitudes and frequencies on catchment evolution. In *River Basin Sediment Systems: Archives of Environmental Change*, D Maddy, MG Macklin, JC Woodward, Eds. A.A. Balkema Publishers, Tokyo, 2001, pp. 285–503.

31. MJ Kirkby, AC Imeson, G Bergkamp, LH Cammeraat. Scaling up processes and models from the field plot to the watershed and regional areas. *J Soil Water Conserv* 51: 391–396, 1996.

32. KJ Beven, MJ Kirkby, N Schoffield, A Tagg. Testing a physically based flood forecasting model (TOPMODEL) for three UK catchments. *J Hydrol* 69: 119–143, 1984.

33. J Alcamo. *IMAGE 2.0 integrated modeling of global climate change.* Kluwer Academic Publishers, Dordrecht, Netherlands, 1994, 75 pp.

34. P Doll, F Kaspar, B Lehner. A global hydrological model for deriving water availability indicators: model tuning and validation. *J Hydrol* 270: 105–134, 2003.

35. MA Nearing, GR Foster, LJ Lane, SC Finckner. A process-based soil erosion model for USDA Water Erosion Prediction Project technology. *Trans Am Soc Agric Eng* 32: 1587–1593, 1989.

36. MJ Kirkby. An erosion-limited hillslope evolution model. *Catena Suppl* 23: 157–187, 1992.

37. RPC Morgan, JN Quinton, RJ Rickson. Modelling methodology for soil erosion assessment and soil conservation design: the EUROSEM approach. *Outlook Agric* 23: 5–9, 1994.

38. AD Howard. Equilibrium and time scales in geomorphology: application to sand-bed alluvial streams. *Earth Surf Process Landforms* 7: 305–325, 1982.

39. LA Tebbens, A Veldkamp, JJ Van Dijke, JM Schoorl. Modeling longitudinal profile development in response to late Quaternary tectonics, climate and sea-level changes: the River Meuse. *Global Planetary Change* 27: 165–186, 2000.

40. GR Willgoose, GS Hancock, GA Kuczera. A framework for the quantitative testing of landform evolution models. In *Predictions in Geomorphology*, PR Wilcock, RM Iverson, Eds. American Geophysical Union, Washington DC, 2003, pp. 195–216.

41. JM Schoorl, A Veldkamp. Late Cenozoic landscape development and its tectonic implications for the Guadalhorce Valley near Alora (South Spain). *Geomorphology* 50: 43–57, 2003.

42. WH Wischmeier, DD Smith. Rainfall energy and its relationship to soil loss. *Trans Am Geophys Union* 39: 285–291, 1958.

43. DE Walling, TA Quine. The use of caesium-137 measurements in soil erosion surveys. In *Erosion and Sediment Transport Monitoring Programmes in River Basins* (Proceedings of the Oslo Symposium, August 1992), IAHS Publication 210, J Bogen, DE Walling, TJ Day, Eds. IAHS, 1992, pp. 143–153.

44. JM Schoorl, C Boix Fayos, RJ de Meijer, ER van der Graaf, A Veldkamp. The 137Cs technique on steep Mediterranean slopes. Part I. Analysing effects of lithology and slope position. *Catena*, 57: 15–34, 2004.

45. JM Schoorl, C Boix Fayos, RJ de Meijer, ER van der Graaf, A Veldkamp. The 137Cs technique on steep Mediterranean slopes. Part II. Landscape evolution and model calibration. *Catena*, 57: 35–54, 2004.

46. BL Turner II, D Skole, S Sanderson, G Fischer, LO Fresco, R Leemans. Land-Use and Land-Cover Change, Science/Research Plan, IGBP Report 35, HDP Report 7. IGBP, HDP, Stockholm, Geneva, 1995, 132 pp.

47. AJ van Loon. Changing the face of the earth. *Earth Sci Rev* 52: 371–379, 2001.

48. FAO. *The Den Bosch Declaration and Agenda for Action on Sustainable Agriculture and Rural Development*, report of the FAO/Netherlands conference on agriculture and the environment, 's-Hertogenbosch, Netherlands, April 15–19, 1991. FAO, Rome, 1992, 55 pp.

49. AJ van Loon. Towards the geological past. *Earth Sci Rev* 51: 203–209, 2000.

50. S de Bruin, WG Wielemaker, M Molenaar. Formalisation of soil-landscape knowledge through interactive hierarchical disaggregation. *Geoderma* 91: 151–172, 1999.

51. S de Bruin, BGH Gorte. Probabilistic image classification using geological map units applied to land-cover change detection. *Int J Remote Sensing* 21: 2389–2402, 2000.

52. S de Bruin. Predicting the areal extent of land-cover types using classified imagery and geostatistics. *Remote Sensing Environ* 74: 387–396, 2000.

53. JM Schoorl, A Veldkamp. Linking land use and landscape process modelling: a case study for the Alora region (South Spain). *Agric Ecosyst Environ* 85: 281–292, 2001.

54. JM Schoorl, A Veldkamp, J Bouma. Modelling water and soil redistribution in a dynamic landscape context. *Soil Sci Soc Am J* 66: 1610–1619, 2002.

55. JA Ruiz, E Ortega, C Sierra, I Saura, C Asensio, A Roca, A Iriarte. *Proyecto LUCDEME: mapa de suelos escala 1:100.000, Alora-1052*. ICONA, Granada, Spain, 1993, 76 pp.

56. WG Wielemaker, S de Bruin, GF Epema, A Veldkamp. Significance and application of the multi-hierarchical land system in soil mapping. *Catena* 43: 15–34, 2001.

57. JM Schoorl, MPW Sonneveld, A Veldkamp. Three-dimensional landscape process modelling: the effect of DEM resolution. *Earth Surf Process Landforms* 25: 1025–1034, 2000.

58. MJ Kirkby. Hillslope process-response models based on the continuity equation. In *Slopes, Forms and Processes*, 3rd ed., D Brunsden, Ed. Institute of British Geographers, London, 1971, pp. 15–30.

59. GR Foster, LD Meyer. A closed-form soil erosion equation for upland areas. In *Sedimentation: Symposium to Honour Professor H.A. Einstein*, HW Shen, Ed. Colorado State University, Fort Collins, 1972, pp. 12.1–12.19.

60. GR Foster, LD Meyer. Mathematical simulation of upland erosion by fundamental erosion mechanics. In *Present and Perspective Technology for Predicting Sediment Yields and Sources* (Proceedings of the Sediment Yield Workshop, Oxford, 1972), Anonymous, Ed. U.S. Department of Agriculture, Washington, DC, 1975, pp. 190–207.

61. D Torri, JWA Poesen, L Borselli. Predictability and uncertainty of the soil erodibility factor using a global dataset. *Catena* 31: 1–22, 1997.

62. TG Freeman. Calculating catchment area with divergent flow based on a regular grid. *Comput Geosci* 17: 413–422, 1991.

63. P Holmgren. Multiple flow direction algorithms for runoff modelling in grid based elevation models: an empirical evaluation. *Hydrol Process* 8: 327–334, 1994.

64. P Quinn, KJ Beven, P Chevallier, O Planchon. The prediction of hillslope flow paths for distributed hydrological modelling using digital terrain models. *Hydrol Process* 5: 59–79, 1991.

16 Space–Time Geostatistics

Gerard B.M. Heuvelink
and Judith J.J.C. Snepvangers

CONTENTS

ABSTRACT

Conventional geostatistics deals with variables that vary exclusively in space. However, many branches within the earth and environmental sciences, including soil science, frequently have to deal with variables that vary not only in space, but also in time. Recently, additional effort has been made to develop spatiotemporal statistical models and to apply spatiotemporal kriging. This chapter reviews the main approaches to extending conventional geostatistical methods to the space–time domain. Whenever possible, one should try to explain part of the temporal variation by including drift functions that represent dynamic process knowledge. The residual may then be modeled as a realization of a stationary random function, which will usually have geometric as well as zonal anisotropies.

Space–time interpolation is performed using standard kriging algorithms. The theory is illustrated with an example of space–time kriging of soil water content.

16.1 INTRODUCTION

While its origins are in mining, the use of geostatistics in the earth and environmental sciences is by now widespread. Over the last decades geostatistical techniques have been successfully applied to various fields, such as hydrology, meteorology, geology, forestry, and geomorphology.[1-3] Geostatistics is also extensively used in soil science[4,5] (also see other Chapters 1, 9, and 10 in this book). Traditionally, geostatistics deals with static spatial attributes, which is due to its roots in mining (geologic deposits do not change very rapidly in time). Yet, most of the new research fields do deal with attributes that change in space as well as in time. For instance, within soil science there are numerous dynamic spatial attributes, such as soil water content, infiltration rate, heat flow, water pressure, and solute concentration.

The extension of kriging (and other geostatistical techniques) to the space–time domain is not straightforward. Incorporating time is more than just adding a third (or fourth) dimension, because the behavior of a variable over time often is entirely different from its behavior over space. Consequently, if a geostatistical model characterizing the spatiotemporal behavior of an attribute is to be of any value, then it has to take these fundamental differences into account. In recent years, progress has been made in building and applying spatiotemporal geostatistical models.[6-16]

The purpose of this chapter is to review the main approaches in space–time geostatistics, with a focus on application to soil science. Readers that wish to know more about the details of and specific topics within space–time geostatistics are referred to the literature above.

16.2 CHARACTERIZING SPACE–TIME VARIABILITY

Consider an attribute $z = \{z(\mathbf{s}, t) \mid \mathbf{s} \in S, t \in T\}$ that is defined on a geographical domain $S \subset \Re^2$ and a time interval $T \subset \Re$. Let z be observed at n space–time points $(\mathbf{s}_i, t_i)$, $i = 1, \ldots, n$. These space–time observations may be a series of observations at multiple time points at a fixed spatial location, spatial samples collected at fixed points in time, any arbitrary set of space–time points, or combinations of all these. Although the number of observations n may occasionally be very large (for example, for the application discussed later on in this chapter, $n = 96,236$), it is impossible to observe z at each and every combination of time and space. To obtain complete space–time coverage of z requires some form of interpolation. The objective then is to obtain a prediction of $z(\mathbf{s}_0, t_0)$ at a point $(\mathbf{s}_0, t_0)$ where z was not observed. To do this, we assume that z is a realization of a random function Z, and we capitalize on the space–time variability of Z to predict $Z(\mathbf{s}_0, t_0)$. Sometimes knowledge about the space–time variability of Z will be a purpose in itself. For instance, it is instructive for the exploration, comparison, interpretation, and explanation of the magnitude of temporal, spatial, and spatiotemporal variation, and it is an essential prerequisite for optimizing spatiotemporal monitoring designs.

16.2.1 Decomposition of the Space–Time Signal into a Drift and Stochastic Residual

To characterize the space–time variability of Z, we first decompose it into a deterministic drift m and a zero-mean stochastic residual ε:

$$Z(\mathbf{s},t) = m(\mathbf{s},t) + \varepsilon(\mathbf{s},t) \qquad (16.1)$$

The drift m is a deterministic, structural component representing large-scale variation. The residual is a stochastic component representing small-scale, noisy variation. It is useful to think of the drift as that part of z that can be explained physically or empirically, using auxiliary information. The residual then is the leftover part. It is important to realize that the decomposition of Z into a drift and residual is a subjective choice made by the modeler. There is no unique way to decompose Z, and different modelers will come to different choices. Also, different decompositions are made at different scales and at different levels of auxiliary information.

In the introduction to this chapter we observed that the behavior of a variable over time often is entirely different from its behavior over space. Dynamic processes govern the temporal behavior, and these processes may be very hard to capture in a statistical description of a stochastic residual. For instance, crop growth is causally dependent on solar radiation and water and nutrient availability. It would be unwise to ignore this information when interpolating crop growth observations in space and time. This is not only because exploiting the information is likely to reduce the space–time interpolation errors, but also because it is very difficult to define a multivariate probability distribution for the space–time distribution of crop growth that acknowledges its complex, nonstationary dynamic behavior.

It is thus recommended to incorporate a drift in the space–time statistical model. Ideally, the drift would be a process-oriented, physical-deterministic model, such as a soil water infiltration model, a soil acidification model, a dynamic soil erosion or sedimentation model, or a soil weathering model. However, when deterministic modeling is not feasible due to a lack of understanding of the governing processes or because external forces and boundary conditions that govern the behavior of the target variable are unknown, one may also rely on a regression-type model relating the dependent to the explanatory variables in an empirical way.

After the drift has been selected and obtained, it may be subtracted from z so that attention may be directed to the space–time residual ε. However, in this way uncertainties in the detrending procedure would not be taken into account in the subsequent analysis. This causes the interpolation uncertainty to appear lower than it is. This problem can be circumvented by accounting for uncertainties in the drift coefficients, such as is done in universal kriging and in kriging with external drift.[16,17] In this chapter we will assume that the drift is perfectly known and directly subtracted from z.

16.2.2 STATISTICAL CHARACTERIZATION OF THE STOCHASTIC RESIDUAL

The second-order properties of the zero-mean stochastic residual ε are completely characterized by its space–time variogram:

$$\gamma(\mathbf{s}_i, t_i, \mathbf{s}_j, t_j) = \tfrac{1}{2} E[(\varepsilon(\mathbf{s}_i, t_i) - \varepsilon(\mathbf{s}_j, t_j))^2] \qquad (16.2)$$

To facilitate the estimation of $\gamma(\mathbf{s}_i, t_i, \mathbf{s}_j, t_j)$ from observations $(\mathbf{s}_i, t_i)$, $i = 1, \ldots, n$, it is necessary to make some assumptions about it. One very common assumption is second-order stationarity, which states that the variogram depends only on the distance between the space–time points:

$$\gamma(\mathbf{s}_i, t_i, \mathbf{s}_j, t_j) = \gamma(\mathbf{s}_i - \mathbf{s}_j, t_i - t_j) = \gamma(\mathbf{h}_s, h_t) \qquad (16.3)$$

In Equation 16.3, $\mathbf{h}_s$ is distance in space and h_t is distance in time. Note that $\mathbf{h}_s$ is a vector because space is multidimensional. However, if we also assume isotropy in space, which we will do in this chapter, then the vector $\mathbf{h}_s$ may be replaced by the scalar h_s, defined as the Euclidean distance between the two points in space.

With these assumptions we have much simplified the space–time correlation structure of ε, but in practice, a further simplification is still needed to be able to estimate a variogram model from the observations. Two main modes of operation can be distinguished here, which we will now discuss.

16.2.2.1 Product-Sum Model

The product-sum model[13,14] assumes that the variogram of ε may be written as

$$\gamma(h_s, h_t) = \gamma_S(h_s) + \gamma_T(h_t) + a \cdot \gamma_S(h_s) \cdot \gamma_T(h_t) \qquad (16.4)$$

It can be demonstrated that this is a statistically valid model, provided γ_S and γ_T are valid variogram models and provided that the parameter a satisfies certain conditions.[13] When a is zero, the product-sum model reduces to the sum model, which is a highly unrealistic model, because it assumes complete temporal persistence of spatial patterns and complete spatial persistence of temporal patterns.[18]

Fitting the product-sum model to the space–time experimental variogram is fairly easy. First, γ_S is fitted to the experimental marginal spatial variogram (i.e., the variogram computed from pairing observations that have no time difference). Next, γ_T is fitted to the experimental marginal temporal variogram (i.e., the variogram computed from pairing observations that have zero separation in space). Finally, parameter a is fitted by adjusting the sill of the space–time model to the sill of the experimental variogram, in the joint space–time direction. The product-sum model has been shown to perform well in practice, in spite of its lack of physical support and somewhat artificial structure.

16.2.2.2 Bilonick Model

The Bilonick model[16,19] reduces the general space–time variogram to

$$\gamma(h_s, h_t) = \gamma_1(h_s) + \gamma_2(h_t) + \gamma_3(h_s + \alpha \cdot h_t) \tag{16.5}$$

The first two terms on the right-hand side of Equation 16.5 allow for the presence of zonal anisotropies (i.e., variogram sills are not the same in all directions). Zonal anisotropy occurs when the amount of variation in time is smaller or larger than that in space, or that in the joint space–time. The third term on the right-hand side of Equation 16.5 represents a joint space–time structure. It contains a geometric anisotropy ratio α to convert distances in time into distances in space. Geometric anisotropy is needed because a unit distance in space is not the same as a unit distance in time. For instance, if $\alpha = 20$ m/day, then two points that are 100 m separated in space and zero days in time have the same correlation as two points that are 5 days apart in time and zero meters in space, or as two points that are 60 m separated in space and 2 days apart in time, and so on. The Bilonick model is also termed the metric model because the third component γ_3 assumes a certain metric in the joint space–time domain.

The Bilonick model makes sense physically, but its structural analysis (i.e., the estimation of variogram parameters) is far from easy. This is because the variograms are exchangeable to some extent. Temporal variability is represented by both γ_2 and γ_3, and spatial variability by γ_1 and γ_3. For the product-sum model, variogram fitting is much easier because γ_S is the marginal spatial variogram and γ_T is the marginal temporal variogram. This is because the product term in Equation 16.4 is always zero when either h_s or h_t equals zero. The same does not hold for the Bilonick model, because γ_3 need not be zero when either h_s or h_t is zero.

Fitting the parameters of the Bilonick model has so far been tried with *ad hoc* trial-and-error methods[8] or by relying on numerical optimization techniques.[16]

16.2.3 SPACE–TIME KRIGING

Once the drift has been determined and the variogram of the residual has been specified, space–time interpolation can be done in the usual way. The residual is interpolated using simple kriging (because its mean is known to be identical to zero):

$$\hat{\varepsilon}(\mathbf{s}_0, t_0) = \sum_{i=1}^{n} \lambda_i \cdot \varepsilon(\mathbf{s}_i, t_i) = \sum_{i=1}^{n} \lambda_i \cdot (z(\mathbf{s}_i, t_i) - m(\mathbf{s}_i, t_i)) \qquad (16.6)$$

where the λ_i are simple kriging weights obtained from solving the linear system

$$\sum_{j=1}^{n} \lambda_j \cdot C(|\mathbf{s}_i - \mathbf{s}_j|, t_i - t_j) = C(|\mathbf{s}_i - \mathbf{s}_0|, t_i - t_0) \qquad \textit{for all } i = 1, \ldots, n \quad (16.7)$$

where $C(h_s, h_t)$ is the autocovariance function of ε, which can be easily derived from its variogram provided the variogram is bounded. Note that these are the standard simple kriging equations.[4] Thus, from a statistical prediction theory perspective, space–time interpolation is not different from spatial interpolation.

After the residual has been interpolated, it is added to the drift to obtain predictions of the target variable z. As mentioned before, uncertainty in the drift coefficients may be incorporated by using universal kriging or kriging with external drift. This is also done in the usual way. Although kriging in space–time is not very different from kriging in space, one possible difference worth mentioning is that in the space–time situation, predictions will often be made ahead of time. When predictions for the future are based on observations in the past and present, space–time kriging becomes an extrapolator alongside the time axis, instead of an interpolator. This has no effect on the kriging equations, but it will cause the predictions to become less accurate (i.e., a larger kriging variance), because observations come from one direction only.

16.3 SPACE–TIME INTERPOLATION OF SOIL WATER CONTENT IN A GRASSLAND PLOT IN MOLENSCHOT, THE NETHERLANDS

The example of space–time geostatistics given here is a condensed version of the case study presented in Snepvangers et al.[16]

Knowledge of the soil water content at any given point in time and space is useful for many purposes. It is used for monitoring dynamic processes that take place in the soil, such as the movement of water in the soil and the uptake of water by plants. It is also used to identify important soil physical properties, such as the unsaturated hydraulic conductivity function and the moisture retention characteristic.[20,21]

Soil water content is traditionally determined by measuring the weight loss after oven drying a soil sample collected in the field.[20] This measurement technique is laborious and destructive. In recent years, several alternative methods have been introduced that allow cheap and nondestructive measurements of soil water content. One method that is particularly useful makes use of time domain reflectometry (TDR).[22,23] A TDR probe is inserted into the soil, after which the soil water content is derived from the measured velocity of electromagnetic waves in the soil. In combination with a data logger, high-resolution time series of soil water content at one or more locations within a given area can be obtained. Alternatively, within a few hours one can manually collect measurements of soil water content at hundreds of locations within the area of interest.

Although measuring soil water content using TDR technology is cheap, clearly it is impossible to measure the soil water content at each and every combination of time and space points. To obtain a complete space–time coverage of soil water content therefore requires some form of interpolation.

In the summer of 2000, an irrigation experiment was carried out on a 60 × 60 m grassland located in Molenschot, the Netherlands. The soil was classified as a Plaggept on sandy loam.[24] Sprinklers with different ranges and intensities created a spatial pattern of soil water content on two occasions in a 30-day monitoring period (August 16 to September 14, 2000). Drying and rewetting by natural precipitation caused the spatial pattern to change over time. The soil water content was monitored with vertically installed 10-cm TDR probes, both manually and automatically. For the manual measurements, a probe was manually installed at each of 229 locations at every measurement time (Figure 16.1). In total, there

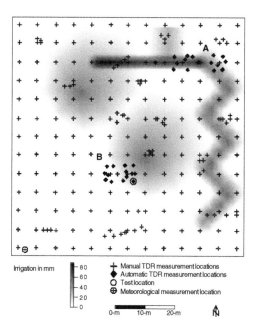

FIGURE 16.1 Locations of the TDR and meteorological measurements. Shades of gray represent the irrigation pattern.

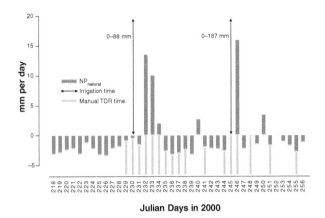

Julian Days in 2000

FIGURE 16.2 Course of the daily average natural net precipitation during the experiment, displayed together with the irrigation and manual TDR measurement times.

were 19 manual measurement rounds (Figure 16.2). The probes for automatic measurements remained installed throughout the whole experimental period at 34 locations (Figure 16.1). These probes were connected to two computer-controlled measurement systems. The two systems, A and B, caused clustering of the automatic probes because the quality of the TDR measurements decreases with cable length.[23] Automatic measurements were carried out every 15 minutes.

Meteorological measurements provided information on the net precipitation. Net precipitation was defined as the input to the topsoil from precipitation and irrigation minus the output from the topsoil through actual evapotranspiration. The latter was calculated using the Penman equation (for details, see Snepvangers et al.[16]).

Because of the irrigation, the net precipitation varied not only in time, but also in space. Irrigation took place on August 17 (day 230, 5:42 A.M.) and September 1 (day 245, 4:00 A.M.). Both irrigations lasted 4 hours. We assumed that all sprinklers of one type had the same irrigation characteristics. This allowed us to measure the distribution of irrigated water around one sprinkler per sprinkler type and translate this to a total irrigation pattern. The irrigation patterns are shown in Figure 16.1. In Figure 16.2, the course of the daily natural net precipitation is shown together with the irrigation dates and the manual TDR rounds.

16.3.1 SPACE–TIME VARIOGRAPHY

The structural analysis and subsequent kriging considered three different cases. In the first case, no drift was included and space–time ordinary kriging (ST-OK) was employed. The second and third cases did include a drift and hence used space–time kriging with external drift (ST-KED). The second case (ST-KED (linear)) incorporated a drift that was a linear combination of net precipitation and time-delayed net precipitation. In the third case (ST-KED (logarithmic)), the drift was a linear combination of log-transformed net precipitation and its time delays.

Figure 16.3 (top graph) shows the experimental variogram for the ST-OK case. There are clear differences in the behavior of the variogram in the space and time directions. There is a strong increase in variogram value in the space direction up to a distance of 5 m and a less pronounced increase up to 10 m. In the time direction, a periodicity with a period of approximately 15 days is clearly visible. This is explained from the fact that both the irrigation days and the heavy rainstorms occurred with intervals of approximately 15 days (Figure 16.2). The sill of the marginal temporal variogram is found at a distance of about 9 days. In the marginal variograms for ST-OK (Figure 16.4), the differences in the variogram behavior in the space and time directions are more clearly visible. A substantial nugget effect is present in the space direction, whereas it is absent in the time direction. The spatial nugget is caused by small-scale spatial variation due to texture and vegetation differences, as well as by animal activity (among others, molehills).

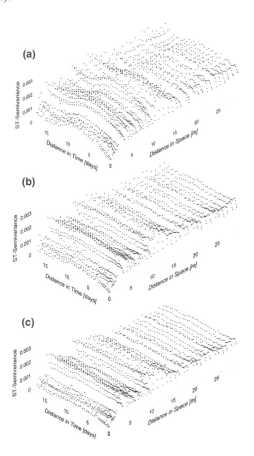

FIGURE 16.3 Experimental variograms for ST-OK (a), ST-KED (linear) (b), and ST-KED (logarithmic) (c).

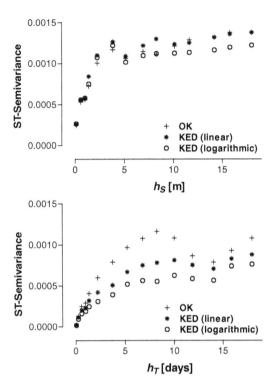

FIGURE 16.4 Marginal experimental variograms for ST-OK, ST-KED (linear), and ST-KED (logarithmic): space direction (a) and time direction (b).

The space–time experimental variogram was fitted using the Bilonick model. The spatial, temporal, and spatiotemporal components were all included and represented by exponential structures. No temporal nugget was incorporated. This resulted in a joint variogram model for the ST-OK case that had eight parameters (ranges and sills for the spatial, temporal, and spatiotemporal parts, a spatial nugget, and a space–time anisotropy ratio). To estimate these parameters, a standard Matlab subspace trust region algorithm based on the interior reflective Newton method was used.[25]

For the ST-KED cases, the structural analysis was applied to the residuals from multiple regression using net precipitation and time-delayed net precipitation as explanatory variables. First the experimental space–time variograms were calculated (Figure 16.3). Similar to the ST-OK case, exponential models for the spatial, temporal, and spatiotemporal parts of the variogram model were chosen. Comparison with the ST-OK case shows that for the ST-KED cases there is a strong decrease in the temporal and spatiotemporal sills (Figure 16.3 and Figure 16.4). Also, the temporal periodicity had largely disappeared due to the incorporation of the net precipitation-driven drift. For the space direction, the effect of detrending is hardly visible. For the spatial nugget, this can easily be explained

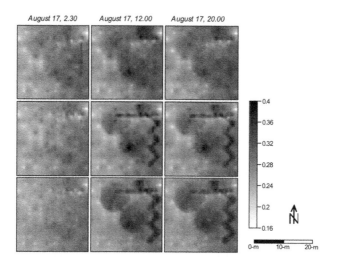

August 17, 2.30   August 17, 12.00   August 17, 20.00

FIGURE 16.5 Interpolated soil water content for the whole study area using ST-OK, ST-KED (linear), and ST-KED (logarithmic) for three points in time on day 230 (August 17, 2000): 2:30 A.M., 12:00 A.M., and 8:00 P.M.

because no information about small-scale spatial variability was used in the spatiotemporal drift. With regard to the spatial sill, it must be concluded that the contribution of the irrigation pattern to spatial variation in soil water content was not very strong. Apparently, there are more important sources of spatial variation in soil water content, such as soil texture, soil physical properties, and vegetation.

16.3.2 SPACE–TIME KRIGING

Space–time kriging was performed using the Gstat software.[26] The top row of Figure 16.5 gives ST-OK interpolations at three time points on August 17. Recall that August 17 is the first irrigation day. Due to local dryer and wetter areas, caused by small-scale spatial variation, all three maps show some spotting. The overall pattern is clear, though. Most striking is that the pattern of irrigation (see Figure 16.1) is already visible (as a darker area in the northeastern part of the area) at 2:30 A.M. in the morning (left), 3 hours before the irrigation started. This physically unrealistic result is caused by the fact that in kriging, future measurements influence predictions as much as past measurements. At 12:00 A.M. (middle) and 8:00 P.M. (right), it can be seen that the irrigation caused a strong increase in soil water content in a large part of the field. The dry spot in the northeastern corner of the field is caused by a large tree, located 5 m outside the study area. The middle and bottom rows of Figure 16.5 give the results for ST-KED (linear) and ST-KED (logarithmic) interpolation, respectively. Most striking is the clear imprint of the irrigation pattern in the time points following irrigation. Before irrigation, at 2:30 A.M., the interpolated map of soil water content shows a fairly even picture; this is in contrast to the ST-OK map, but in accordance with measurements.

To examine the temporal behavior of interpolated soil water content, a test location (x = 31.55, y = 19.00; Figure 16.1) was selected to examine how well measured time series of soil water content could be reproduced by space–time kriging. At the test location, soil water content was observed using automated TDR measurements. The automated TDR measurements of system B were not included in the interpolation because the coverage in the temporal domain is very high around automated TDR locations, leaving little challenge for interpolation. In Figure 16.6, measured and predicted soil water content at the test location are displayed. At times without sudden precipitation, measurements and predictions are within the uncertainty limits of plus or minus 1 standard deviation, but close to irrigation times and heavy rainstorms, the measurements are much greater than the predictions. This is due to the smoothing effect of kriging.

Because all three interpolation cases used the same measurements of soil water content, the average behavior of the ST-KED predictions closely resembles that of the ST-OK predictions, with a slight underestimation of the soil water content at the test location. It is remarkable, though, that ST-KED predictions show more variation in soil water content in between manual measurement rounds than the ST-OK predictions. At some time points, even the daily cycle of soil water content becomes clear. This may be explained from the daily fluctuation in the net precipitation-driven drift. Figure 16.6 also shows that the ST-KED predictions follow the sudden increases in soil water content much better than the ST-OK predictions.

16.4   DISCUSSION AND CONCLUSIONS

This chapter has shown that much of the existing geostatistical theory for spatial interpolation may also be used for space–time interpolation. The kriging itself does not change. This is because in space–time interpolation the objective is to optimally predict the value of a variable at an unobserved point from nearby observations, just as is the case in spatial kriging. The predictions are weighed linear combinations of the observations, whereby the weights are chosen such that the expected squared prediction error is minimized, under the condition of unbiasedness. Again, this is the same as in spatial kriging. The best linear unbiased predictor (BLUP) from mathematical statistics applies to both cases. However, to apply BLUP, one must specify the correlations between the variables at unobserved and observed points, and to do this realistically turns out to be a difficult task in the space–time setting. Approaches that ignore the fundamental differences in spatial and temporal variation and that fit isotropic stationary variogram models to space–time variables are clearly inappropriate. It is crucially important to take the differences between spatial and temporal variations into account. One important means of doing just that is by incorporating a (temporal) drift in the model, ideally representing dynamic process knowledge in one way or another. In the example of soil water content presented in this chapter, this was done by including

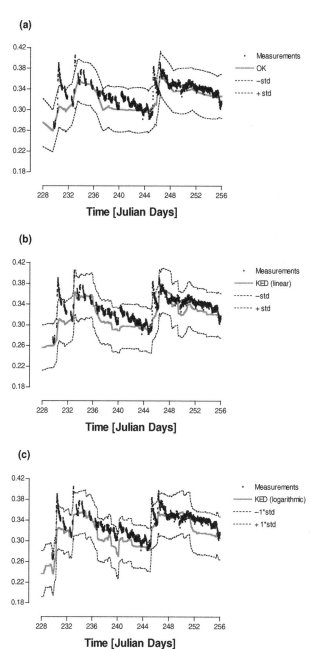

FIGURE 16.6 Measured vs. predicted soil water content for the whole time at location (x, y) = (31.55, 19.00), using ST-OK (a), ST-KED (linear) (b), and ST-KED (logarithmic) (c). The dashed lines show the ±1 standard deviation bands.

a drift driven by net precipitation, which is a driving factor behind the dynamic behavior in soil water content.

Although incorporating a drift is an appropriate and efficient way to capture the specific (causal) structure of temporal variation, this does not guarantee that the residual resulting from the detrending will behave as isotropic. Geometric and zonal anisotropies will often occur, and it is a true challenge to define valid space–time variogram models that can realistically describe the space–time variability at hand and whose parameters can be estimated reliably. Both models presented in this chapter have their weaknesses. The product-sum model is fairly easy to fit, but it is doubtful that it represents a physically plausible and sufficiently flexible structure. The Bilonick model is more realistic because it has a clear physical interpretation, but its parameters cannot easily be estimated because the three components are, to a large degree, dependent. Therefore, in the near future much research should be devoted to improving these models or the way in which their parameters may be estimated, or to developing new and better models.

Completely relying on automatic fitting procedures, such as was done in the example in this chapter, is not without risk. It would be advisable to have an experienced geostatistician exercise control on the outcome of the fitting procedure, because the choice of the number and type of model structures is difficult to automate and because automatic fits sometimes need manual adjustment. However, this necessitates the availability of tools that allow an easy visualization of anisotropic three-dimensional experimental variograms and the models fitted to them. Currently, some commercial packages provide considerable functionality for this operation (e.g., GS+, SAGE 2001), yet further development and a more widespread availability of these tools would be most welcome.

ACKNOWLEDGMENTS

The authors thank three anonymous reviewers for valuable and constructive comments.

REFERENCES

1. N Cressie. *Statistics for Spatial Data*, revised edition. Wiley, New York, 1993.
2. EY Baafi, NA Schofield, Eds. *Geostatistics Wollongong '96*. Kluwer, Dordrecht, Netherlands, 1997
3. PA Burrough, RA McDonnell. *Principles of Geographical Information Systems.* Oxford University Press, Oxford, 1998.
4. P Goovaerts. Geostatistics in soil science: state-of-the-art and perspectives. *Geoderma* 89: 1–45, 1999.
5. R Webster, MA Oliver. *Geostatistics for Environmental Scientists.* Wiley, Chicester, U.K., 2001.
6. V Comegna, C Vitale. Space-time analysis of water status in a volcanic Vesuvian soil. *Geoderma* 60: 135–158, 1993.

7. P Bogaert, G Christakos. Stochastic analysis of solute content measurements using a regressive space/time model. *Stochastic Hydrol Hydraul* 11: 267–295, 1997.
8. GBM Heuvelink, P Musters, EJ Pebesma. Spatio-temporal kriging of soil water content. In *Geostatistics Wollongong '96*, EY Baafi, NA Schofield, Eds. Kluwer, Dordrecht, Netherlands, 1997, pp. 1020–1030.
9. N Cressie, HC Huang. Classes of nonseparable, spatio-temporal stationary covariance functions. *J Am Stat Assoc* 94: 1–53, 1999.
10. PC Kyriakidis, AG Journel. Geostatistical space-time models: a review. *Math Geol* 31: 651–684, 1999.
11. G Christakos. *Modern Spatiotemporal Geostatistics*. Oxford University Press, New York, 2000.
12. PC Kyriakidis, AG Journel. Stochastic modeling of atmospheric pollution: a spatial time-series frame-work. Part I. Methodology. *Atmos Environ* 35: 2331–2337, 2001.
13. L De Cesare, DE Myers, D Posa. Estimating and modeling space-time correlation structures. *Stat Probab Lett* 51: 9–14, 2001.
14. L De Cesare, DE Myers, D Posa. Product-sum covariance for space-time modeling: an environmental application. *Environmetrics* 12: 11–23, 2001.
15. S De Iaco, DE Myers, D Posa. Nonseparable space-time covariance models: some parametric families. *Math Geol* 34: 23–42, 2002.
16. JJJC Snepvangers, GBM Heuvelink, JA Huisman. Soil water content interpolation using spatio-temporal kriging with external drift. *Geoderma* 112: 253–271, 2003.
17. P Goovaerts. *Geostatistics for Natural Resources Evaluation*, Applied Geostatistics Series. Oxford University Press, Oxford, 1997.
18. GBM Heuvelink, R Webster. Modelling soil variation: past, present, and future. *Geoderma* 100: 269–301, 2001.
19. RA Bilonick. Monthly hydrogen ion deposition maps for the northeastern U.S. from July 1982 to September 1984. *Atmos Environ* 22: 1909–1924, 1988.
20. D Hillel. *Fundamentals of Soil Physics*. Academic Press, New York, 1980.
21. W Bouten, TJ Heimovaara, A Tiktak. Spatial patterns of throughfall and soil water dynamics in a Douglas fir stand. *Water Resour Res* 28: 3227–3234, 1993.
22. GC Topp, JL Davis, AP Annan. Electromagnetic determination of soil water content: measurements in coaxial transmission lines. *Water Resour Res* 16: 574–582, 1980.
23. TJ Heimovaara, JI Freijer, W Bouten. The application of TDR in laboratory column experiments. *Soil Technol* 6: 337–377, 1993.
24. USDA. *Soil Taxonomy: A Basic System of Soil Classification for Making and Interpreting Soil Surveys*, Agricultural Handbook, Vol. 436. USDA Soil Conservation Service, Washington, DC, 1975.
25. TF Coleman, Y Li. An interior, trust region approach for nonlinear minimization subject to bounds. *SIAM J Optimization* 6: 418–445, 1996.
26. EJ Pebesma, CG Wesseling. Gstat, a program for geostatistical modelling, prediction and simulation. *Comput Geosci* 24: 17–31, 1998.

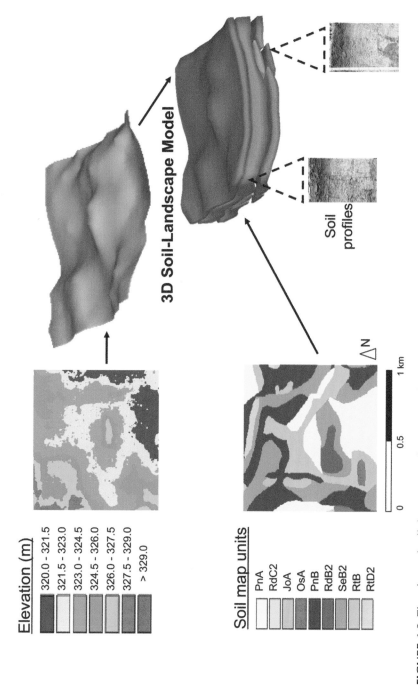

Elevation (m)

- 320.0 – 321.5
- 321.5 – 323.0
- 323.0 – 324.5
- 324.5 – 326.0
- 326.0 – 327.5
- 327.5 – 329.0
- > 329.0

3D Soil-Landscape Model

Soil profiles

Soil map units

- PnA
- RdC2
- JoA
- OsA
- PnB
- RdB2
- SeB2
- RtB
- RtD2

△N

0   0.5   1 km

FIGURE 1.2 Elevation and soil data were used to create a three-dimensional soil-landscape model for a site in southern Wisconsin.

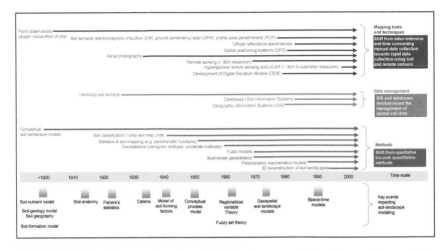

FIGURE 1.4 Pivotal events that shaped soil-landscape modeling history. Time periods and placement of events are approximate.

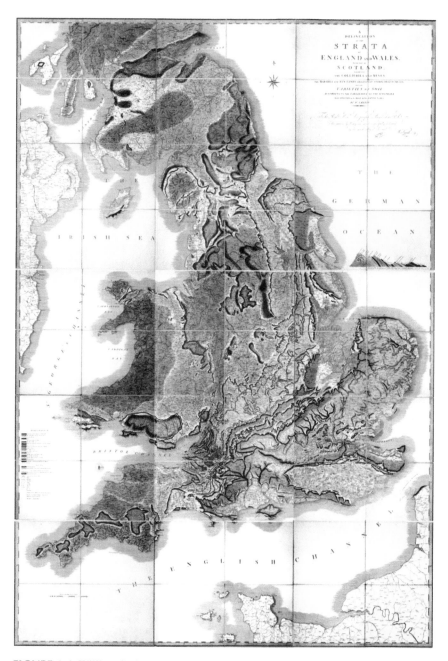

FIGURE 3.1 William Smith's 1815 geological "map that changed the world."[12] The legend indicates that the map expresses "varieties of soil according to the variations in the substrata," a common approach in the early 19th century. (Reprinted from Schneer, C.J., William "Strata" Smith on the Web, 2004, available at http://www.unh.edu/esci/wmsmith. html.[13])

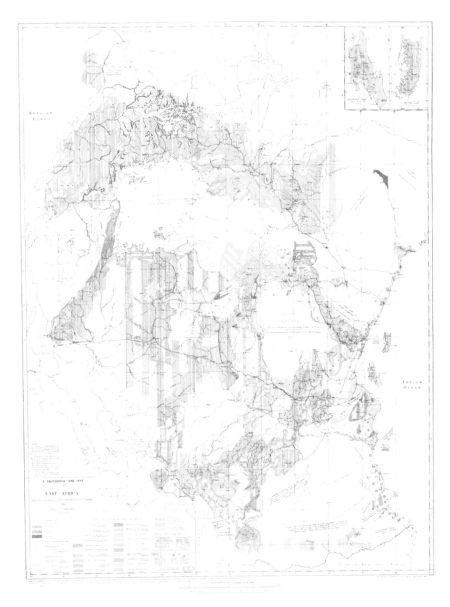

FIGURE 3.8 Geoffrey Milne's 1935 *Soil Map of East Africa*. Note the "pajama striping" for the catenas and large land areas left blank or partially blank where soil investigations were absent or incomplete. (Reprinted from Milne, G. et al., *A Provisional Soil Map of East Africa (Kenya, Uganda, Tanganyika and Zanzibar) with Explanatory Memoir*, Amani Memoir 31, East African Agricultural Research Station, Amani, Tangayika, 1936.)

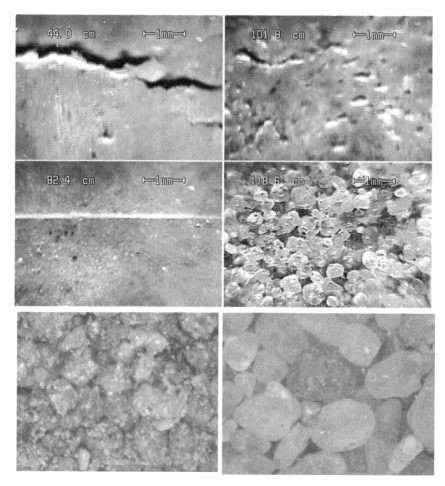

FIGURE 5.5 Images collected with a soil imaging penetrometer. (Courtesy of Soil and Topography Information, LLC, Madison, WI.)

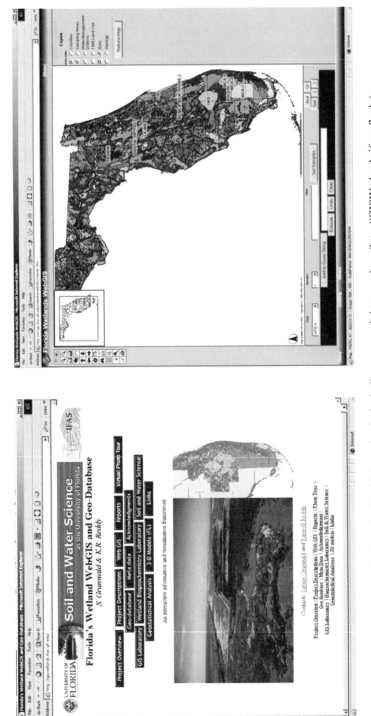

FIGURE 5.6 WebGIS and geodatabase for Florida's wetlands, including map and data services (http://GISWetlands.ifas.ufl.edu).

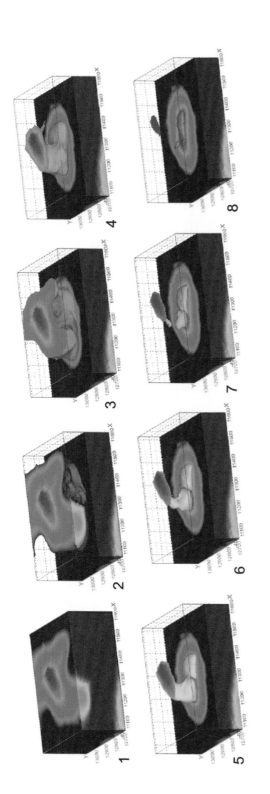

FIGURE 5.8 Nitrate–nitrogen plume at different periods (n = 1 to 8) implemented within a GIS (EVS-PRO). The space–time model shows change in attribute values and geometry of the plume.

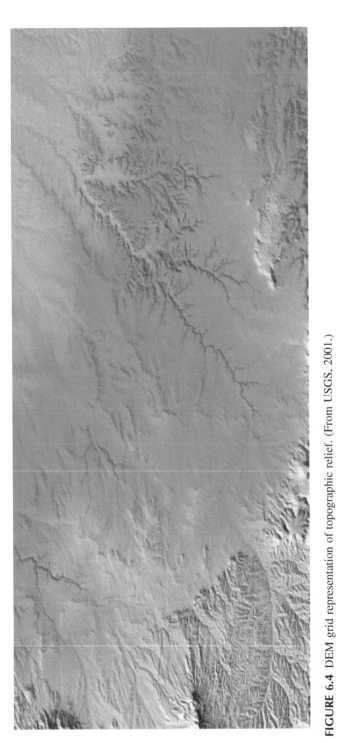

FIGURE 6.4 DEM grid representation of topographic relief. (From USGS, 2001.)

FIGURE 6.8 Orthophotograph/lidar drape of Devil's Millhopper sinkhole in Alachua County, Florida: View 1. (Courtesy of 3001 Spatial Data Corporation.)

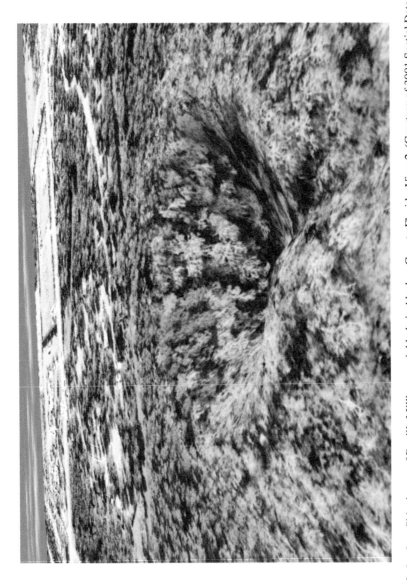

FIGURE 6.9 Orthophoto/lidar image of Devil's Millhopper sinkhole in Alachua County Florida: View 2. (Courtesy of 3001 Spatial Data Corporation.)

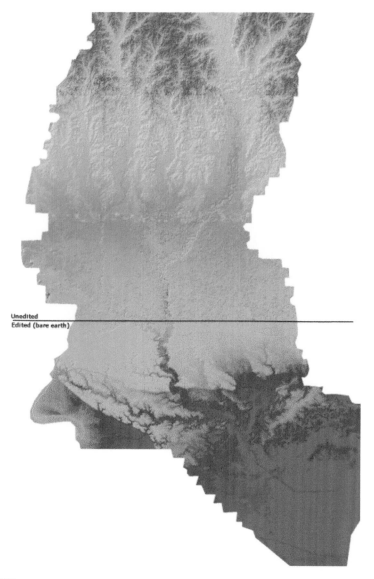

FIGURE 6.10 Shaded relief map with 30-m drape of Amite River Basin in Louisiana. (Courtesy of 3001 Spatial Data Corporation.)

FIGURE 6.11 Five-meter DEM derived from lidar data of Profit Island in the Mississippi River Basin. (Courtesy of 3001 Spatial Data Corporation.)

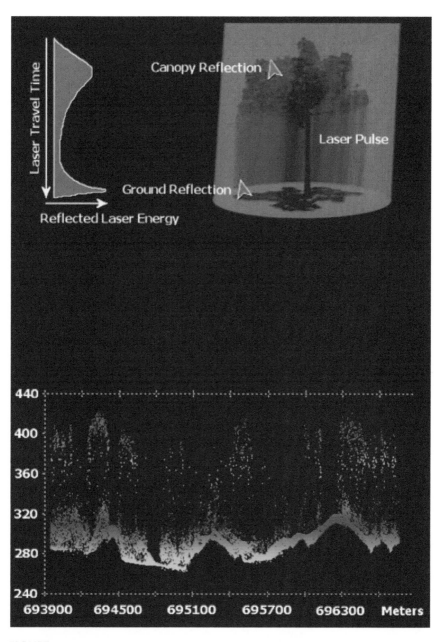

FIGURE 6.12 Lidar profile of forested area. (Courtesy of 3001 Spatial Data Corporation.)

FIGURE 6.13 Combined lidar and aerial photograph. (Courtesy of 3001 Spatial Data Corporation.)

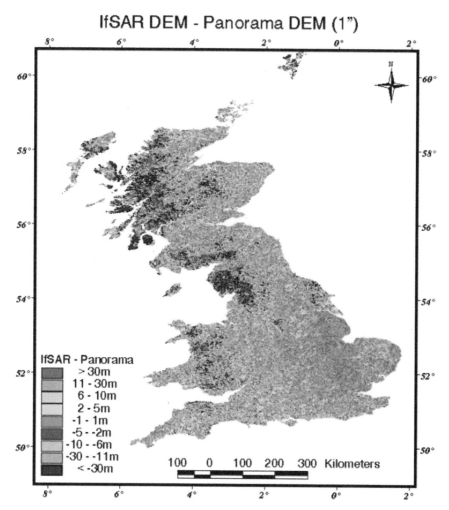

FIGURE 6.14 SRTM topographic image. (From the Landmark Project, University of Manchester, U.K.)

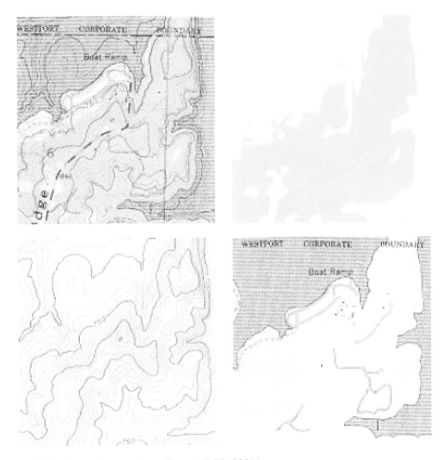

FIGURE 6.15 Map separates. (From USGS, 2001.)

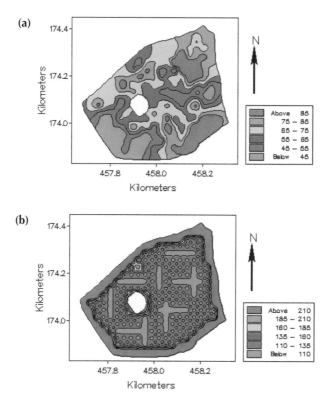

FIGURE 9.13 Block kriged map of (a) topsoil Mg and (b) its associated map of the kriging variances.

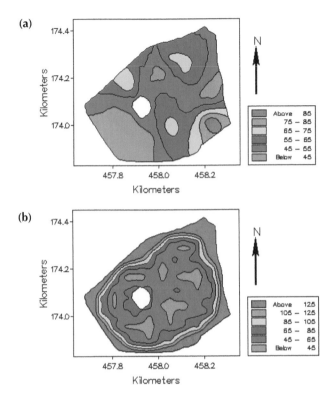

FIGURE 9.14 Maps of (a) kriged estimates of subsoil magnesium, Mg_{30}, and (b) the associated kriging variances.

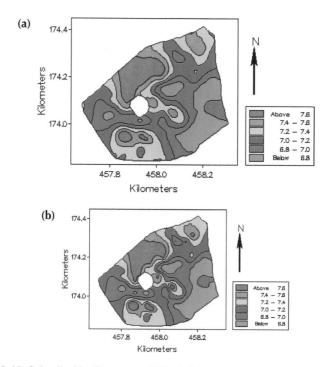

FIGURE 9.15 Subsoil pH, pH_{30}, map of block kriged predictions using (a) the penta-spherical model and (b) the exponential function.

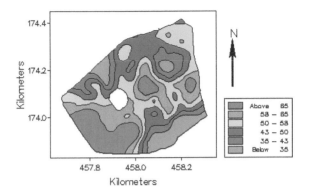

FIGURE 9.16 Block kriged map of topsoil sand.

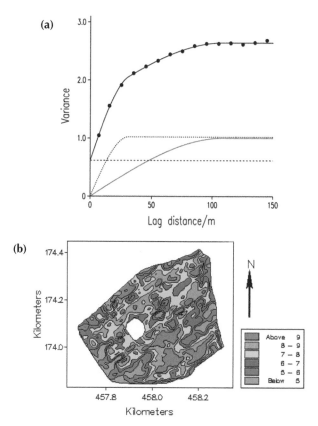

FIGURE 9.17 (a) Variogram of yield in 1995 fitted, with nugget and short- and long-range components shown separately. (b) Map of ordinary kriged predictions made with this model.

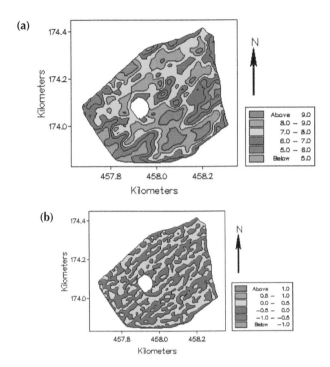

FIGURE 9.18 Maps of yield made by factorial kriging. (a) Predictions based on the long-range component of the variogram, a_2. (b) Predictions based on the short-range component of the variogram, a_1.

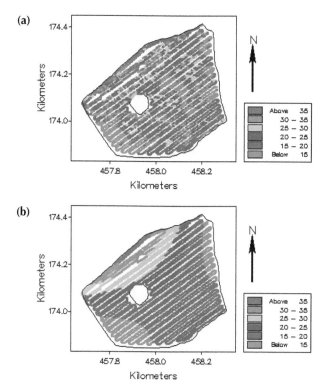

FIGURE 9.19 Pixel maps of apparent electrical conductivity of the soil, EC_a, of (a) raw values and (b) the quadratic trend.

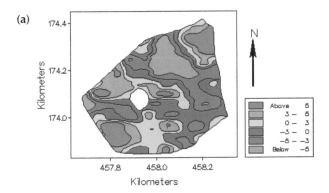

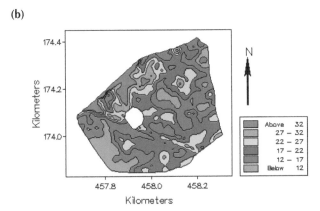

FIGURE 9.21 (a) Map of block kriged residuals of EC$_a$ from the quadratic trend surface. (b) Map of block kriged predictions with the trend added back.

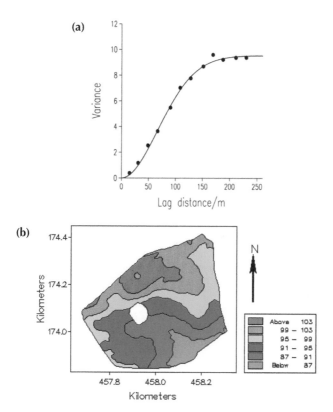

FIGURE 9.22 Elevation. (a) Experimental variogram and fitted stable model with exponent 1.965 for the residuals from a linear trend. (b) Map of kriged predictions made with this model with the trend added back.

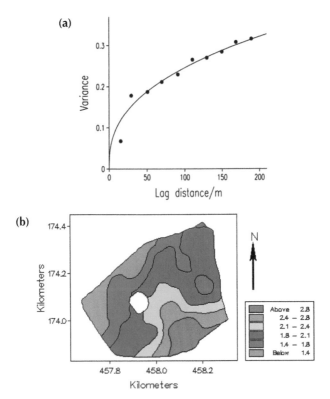

FIGURE 9.23 Leaf area index. (a) Experimental variogram and fitted power function for the residuals from a linear trend. (b) Map of kriged predictions made with this model with the trend added back.

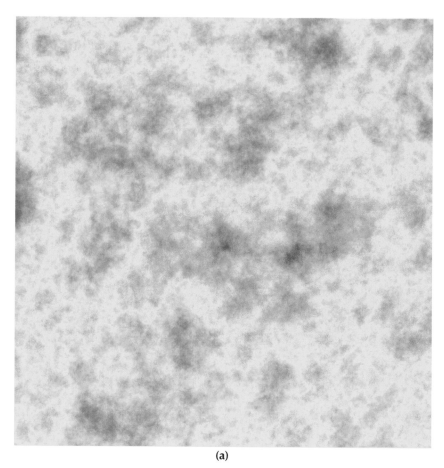

(a)

FIGURE 10.1 Simulations showing the same histogram and the same exponential covariance: (a) Gaussian RF, (b) mosaic RF with Gaussian marginal (tessellation by Poisson polygons).

(b)

FIGURE 10.1 *Continued.*

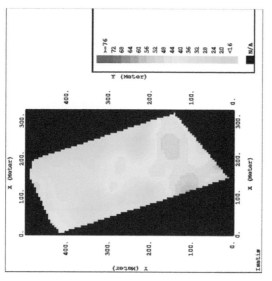

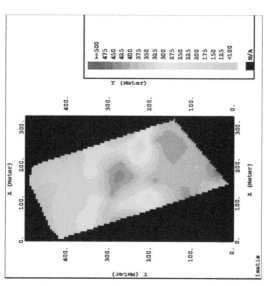

FIGURE 10.7 Kriging maps of (a) WC and (b) N.

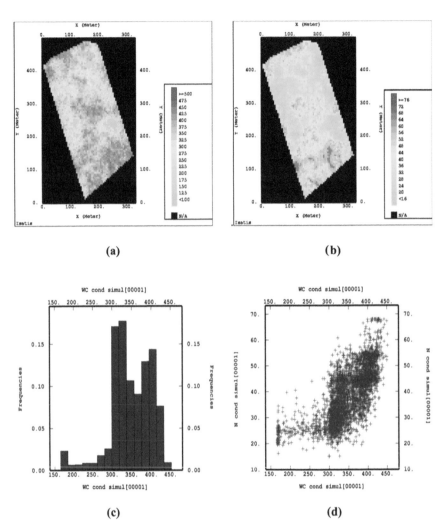

FIGURE 10.10 Conditional simulations of (a) WC and (b) N. (c) Histogram of WC. (d) Cross-plot of simulated N vs. WC.

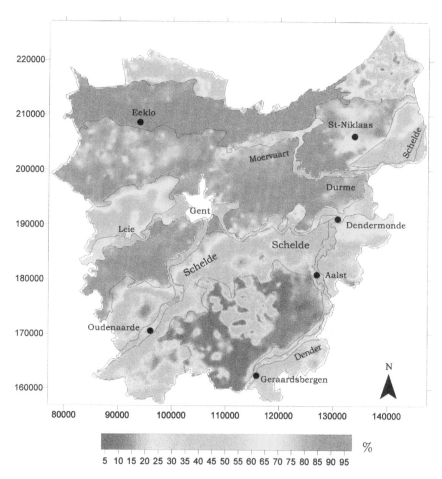

FIGURE 11.9 Topsoil sand content obtained by stratified compositional ordinary block kriging (block dimensions: 250 × 250 m) taking into account the nature of the boundaries used for stratification (lines).

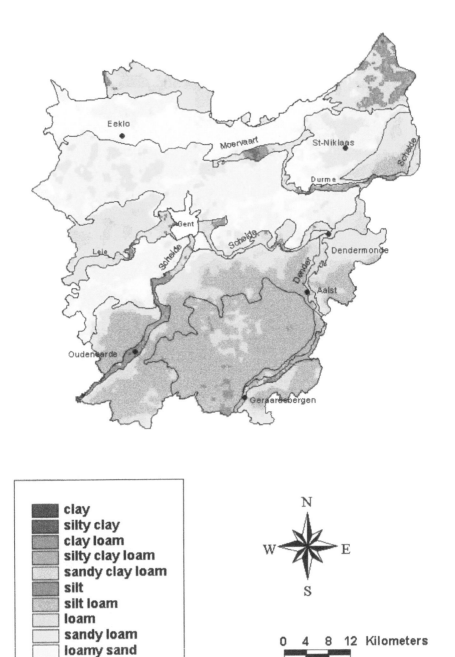

■	clay
■	silty clay
■	clay loam
■	silty clay loam
□	sandy clay loam
■	silt
■	silt loam
□	loam
□	sandy loam
□	loamy sand
□	sand

FIGURE 11.10 Soil texture classes according to the USDA texture triangle of East Flanders, Belgium.

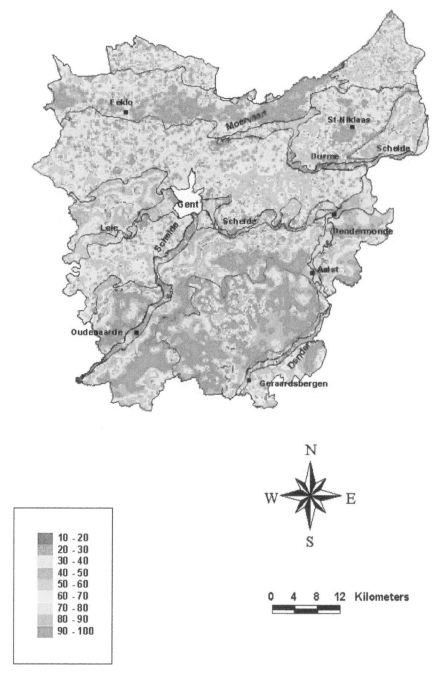

FIGURE 11.11 Frequency of obtaining the same texture class as predicted by COK (Figure 11.10) with the Latin hypercube sampling of the three textural distributions (using 7500 samples). Large frequencies indicate a stable classification.

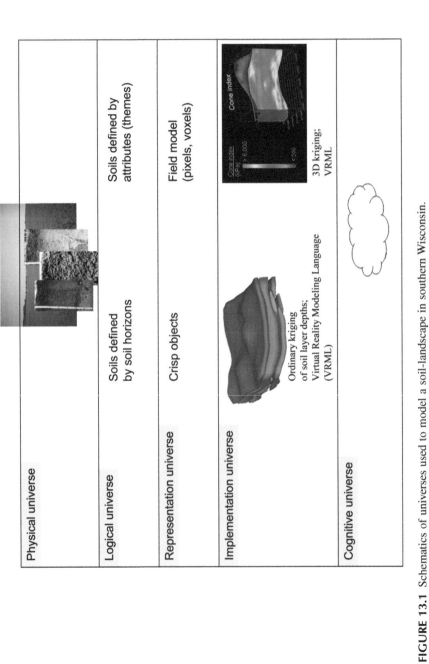

Physical universe

Logical universe

Soils defined
by soil horizons

Soils defined by
attributes (themes)

Representation universe

Crisp objects

Field model
(pixels, voxels)

Implementation universe

Ordinary kriging
of soil layer depths;
Virtual Reality Modeling Language
(VRML)

3D kriging;
VRML

Cone index

Cone index
(kPa)   > 6.000

< 500

Cognitive universe

FIGURE 13.1 Schematics of universes used to model a soil-landscape in southern Wisconsin.

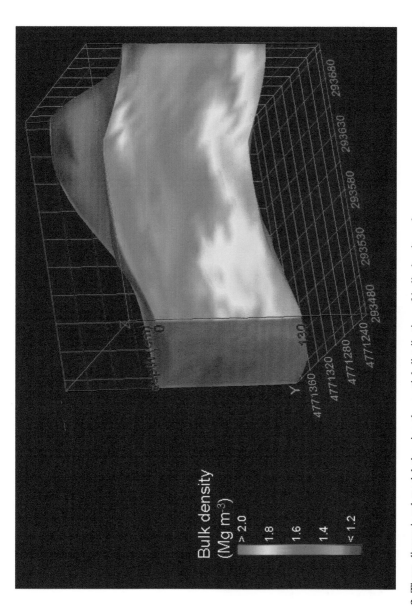

FIGURE 13.2 Three-dimensional model showing the spatial distribution of bulk density values.

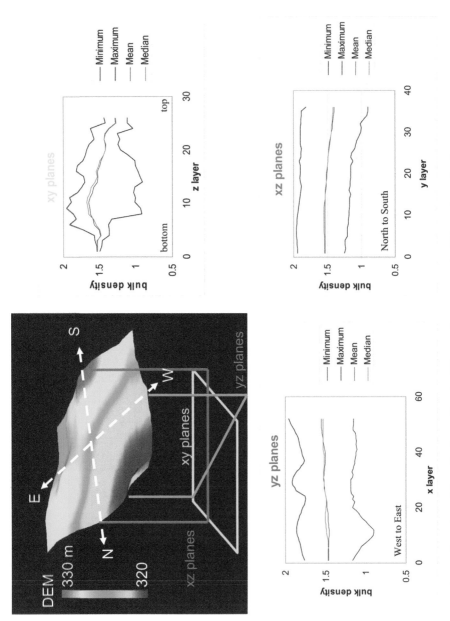

FIGURE 13.3 Bulk density values (Mg m⁻³) along slices in different planes across the Wisconsin study site.

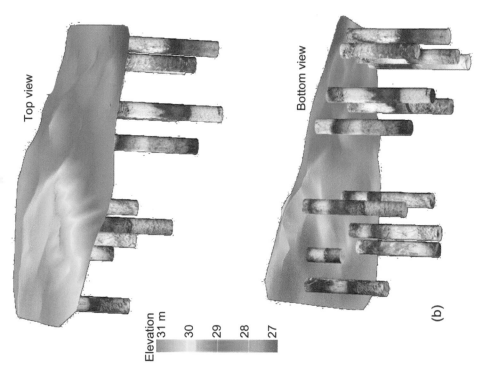

Top view

Bottom view

Elevation
31 m
30
29
28
27

(b)

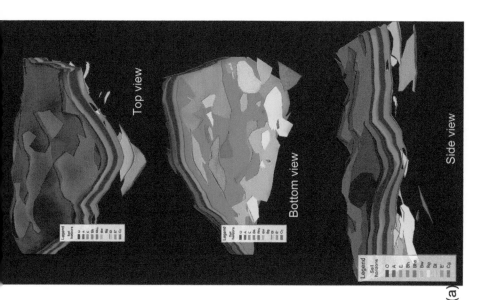

Top view

Bottom view

Side view

(a)

FIGURE 13.4 (a) Three-dimensional soil horizon model. (b) Soil profile model.

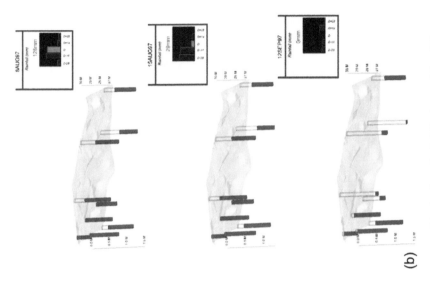

(a)

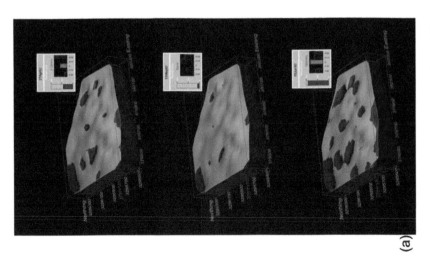

(b)

FIGURE 13.5 (a) Space–time models of inundation. (b) Space–time models show water table depth and rainfall observed for different periods.

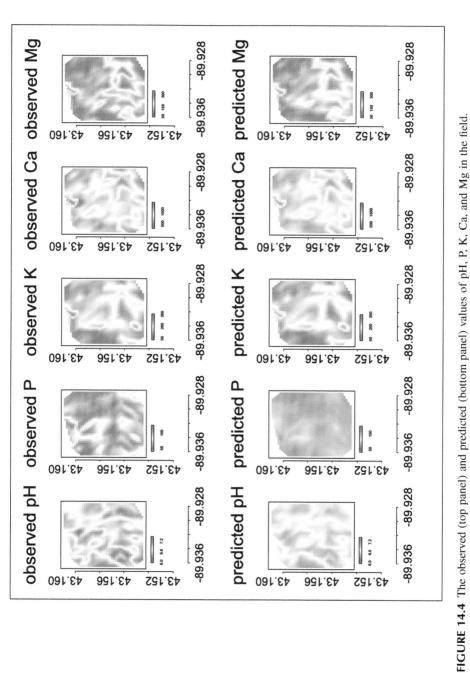

FIGURE 14.4 The observed (top panel) and predicted (bottom panel) values of pH, P, K, Ca, and Mg in the field.

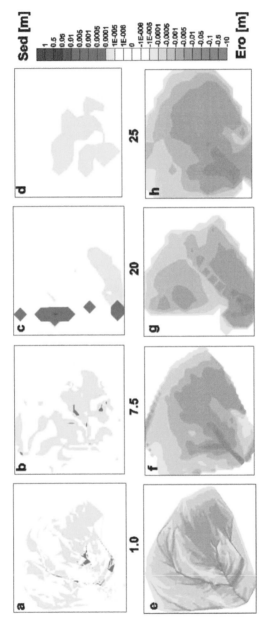

FIGURE 15.2 Sedimentation (upper graphs) and erosion (lower graphs) patterns for the Chopillos area after one time step (1 year) given in meters surface change for each of the four spatial resolutions (from left to right).

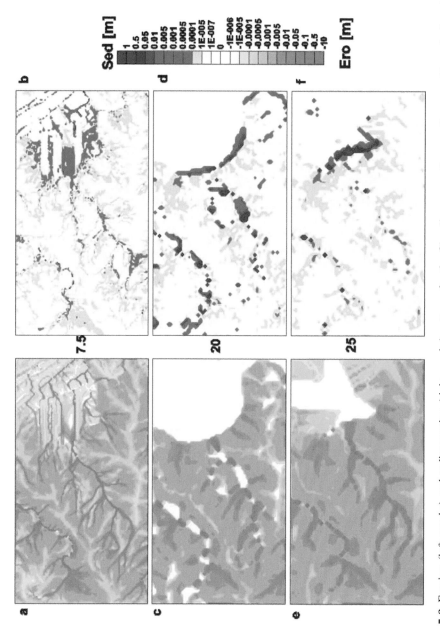

FIGURE 15.3 Erosion (left graphs) and sedimentation (right graphs) patterns for the Clavelinas area after one time step (1 year) given in meters surface change for each of the three different resolutions.

Index

A

Aboveground attributes, 4
Abrupt
  boundary, 330, 333, 334, 340
  variation, 7, 9, 221
Absolute
  orientation, 166, 167, 168
  position, 246
Absorption, 138, 422
Abstraction, 5, 7, 48
  data, 381
  skills, 383
Accumulated flow index, 189
Accuracy, 22, 25, 31, 32, 81, 84
  of digital elevation model, 186, 197, 207
  of global positioning systems, 134
  horizontal, 134, 172
  map, standards for, 179–180
  spatial, 44, 93
  submeter, 134
  target, 50
  vertical, 134, 172, 181
Acidity, 7, 93
  of catalysts, 130, 133, 139, 227, 229
  indication of, 192
  processes, 40
Aerial digital cameras, 171–172
Aerial photography, 92–94, 160–161
  stereoscopic vision and, 165
Aerial triangulation, 167–168
Aforestation, 5
Aggregation, 133, 134, 395
Agriculture, 419, 423
  precision, 4, 129, 290, 311
Air photo interpretation, 53
Airborne Visible/Infrared Imaging
          Spectrometer, 13, 140, 141
Akaike's information criterion, 401
Algorithm(s), 21, 188, 203, 237
  change-of-resolution, 393–394, 396, 398,
          400, 401, 403, 413, 415
  conditional simulation, 311
  expectation-maximization, 396, 400, 403
  flow direction, 187
  fuzzy c-means, 226, 229, 237

Gaussian, 300
  grouping, 21
  iterative, 310
  Kalman filter, 394, 396, 398, 401, 403, 415
  kriging, 438
  Matlab subspace trust region, 446
  multiple-flow, 425
  neighborhood search, 298
  one-pass change-of-resolution, 400, 401,
          413
  prediction, 396
  pyramid, 350, 352
  sequential indicator, 300, 305
  simulation, 294, 295, 296, 299, 300, 303,
          311
  spatial, 378
  turning band, 300, 318
Alluvial, 43, 52, 329, 330, 331, 332, 334, 336,
          338, 367, 429
Altitude, 167, 171, 195, 426, 427, 429
Aluminum, eluviation of, 26
Amplitude, 135, 346, 347
Analog photogrammetry, 166, 168
Analytical photogrammetry, 166, 168–169
Anatomy, 62, 68, 87–92
Anisotropy, 7, 253–254, 255, 300, 446
Area(s)
  agricultural, 45, 94
  alluvial, 338
  complex, 53
  cultivated, 420
  definition, 91
  delineation, 54
  dispersal, 188, 195
  distribution of soil-landscapes in, 42
  erosions in, 43
  forested, 73, 156, 173, 175, 178
  glaciated, 55
  heterogeneous, 134
  hilly, 73
  homogeneous, 135
  humid, 44
  large, 325
  large land, 49, 62, 77, 82, 83, 86, 128, 140,
          156, 169, 171, 230
  leaf, 270, 274, 275, 283, 284

Printed and bound by CPI Group (UK) Ltd, Croydon, CR0 4YY

23/10/2024

01778263-0009

)